Landscapes of Human Evolution

Landscapes of Human Evolution

Contributions in honour of John Gowlett

edited by

James Cole, John McNabb, Matt Grove
and Rob Hosfield

ARCHAEOPRESS ARCHAEOLOGY

ARCHAEOPRESS PUBLISHING LTD
Summertown Pavilion
18-24 Middle Way
Summertown
Oxford OX2 7LG

www.archaeopress.com

ISBN 978-1-78969-379-9
ISBN 978-1-78969-380-5 (e-Pdf)

Cover: Handaxe from Amanzi Springs, © Caruana and Herries, this volume;
photograph © James Cole - John Gowlett at Chesowanje (Kenya) 2014;
background photograph © James Cole – Kilombe (Kenya) 2014.
Back Cover: photograph © Clive Gamble – John Gowlett at Klithi (Greece) 1983.

Printed in England by Holywell Press, Oxford

This book is available direct from Archaeopress or from our website www.archaeopress.com

Contents

Contents

Foreword

James Cole, John McNabb, Matt Grove and Rob Hosfield

In 1984's *Ascent to Civilisation* John Gowlett observed: 'As the present is but an instant, we [must] depend upon past experience for strategies [to] cope with the future' (Gowlett 1984: 196). John's reflection on the value of our collective past is especially relevant in regards to the challenges that humans as a global society face today. The impacts of rapid, and in some senses unpredictable, climate change are global issues that impact us all. In this regard, the study of human evolution remains an exciting, and important, field of study to help humanity navigate the challenges of the future.

As the seemingly deeply interconnected origins of our species are being realised at a geographic, genetic and cultural level, understanding these relationships with regard to other human ancestral species (e.g. the Neanderthals), and appreciating how they navigated periods of climatic and social instability in the past, must underpin our strategies for the global future and our notions of 'self' and 'other'. Furthermore, gaining insights into the origins of the complex social and cultural behaviours that underline our entire way of life in the present (e.g. language, symbolism, landscape manipulation, social networks, and the creation of distinct cultural identities) are fundamental to charting a more sustainable and inclusive future for ourselves and our planet. This *Landscapes of Human Evolution* volume is therefore highly timely in its drawing together of some of the world's leading scholars in human evolution and related disciplines in order to present cutting-edge research papers, and is in honour of Prof. John Gowlett, who has played a pivotal, although often typically understated, role in developing the fields of human origins research, spanning lithic material culture, Plio-Pleistocene landscapes and environments, pyrotechnologies, and the archaeology of the social brain, through projects across Europe and Africa. The landscapes of human evolution covered in this volume are therefore broad; incorporating physical topography, socio-cultural and cognitive structures that stretch back into the past and before us into the future.

Landscapes of Human Evolution has therefore invited contributions that fit within four main themes that John has pioneered throughout his career. (1) The biological development of early hominins, especially members of the genus *Homo*, and their characteristic features of large brains, bipedal locomotion and behavioural adaptation. Within this theme the volume addresses how encephalisation is related to increasing behavioural and cultural sophistication, and how this trajectory can be mapped throughout the course of human evolution (Du and Wood; Bilsborough and Wood; Crompton). (2) The strategies employed for dealing with, and ultimately manipulating, heterogeneous landscapes and environments, with a particular focus on how the changing environments of the Plio-Pleistocene have influenced hominin adaptation, and how this might have impinged on early hominins' biological development (Kübler et al; Hoare et al; Gamble). (3) The origins of controlled use of fire as a mechanism for survival and an incubator of increasing social complexity within hominin groups, clarifying how the use and control of fire accords with other cultural developments, and how it changed the dynamics of hominin society (Dunbar; Shankland). (4) The role of lithic technologies in developing hominin behavioural complexity, and the ways in which the contemporary classification of these technologies frames the understanding of the past in the present. This final theme seeks to address how data on lithic technology – the most durable record of ancestral human behaviour – can be mined to elucidate changes in hominin cognition, behaviour, and their interactions with ancestral environments (McNabb; de la Torre and Mora; Caruana and Herries; Wynn; Foley and Lahr). Finally, Sinclair presents a citation network analysis from two major areas of human evolution research (*Palaeolithic Archaeology*; *Evolutionary Anthropology*) that serves to explicitly demonstrate the tremendous impact that John has had on the discipline of human origins research.

As well as having a profound impact in framing the way current researchers seek to engage with and interpret the complex behaviours of our human ancestors with his research, John has always been a champion of early career researchers and has left a valuable legacy of friendship and training networks in the many corners of the world in which he has worked. John is always ready to give his time and knowledge to any who ask for it and it is this generosity of spirit that has been recognised time and again by all who come into contact with him, and which has inspired this collection of papers. In return we, the editors, hope that this volume goes some way in demonstrating the high professional and personal standing in which Prof. John Gowlett is held.

Thank you John.

A Good Man in Africa: John Gowlett's Writings on Africa and its Hominin Archaeology from the Late 1970s to the Early 2000s

John McNabb

Introduction

This overview of selected aspects of John Gowlett's work from the first two decades of his research career is not intended to be definitive or comprehensive. It is a personal perspective on Gowlett's thinking and writing focusing on just a few of the many aspects of archaeology that he is interested in. Topics like fire and chronology I will barely touch on, nor will I discuss the papers arising from his Beeches Pit excavations in the 1990s.

I have elected to review aspects of Gowlett's thinking prior to the formal beginning of the British Academy's Centenary Research Project, 'Lucy to Language: the Archaeology of the Social Brain', in 2003. He was one of the principle investigators, and along with Clive Gamble, he was invited to participate by Robin Dunbar. My choice of selected themes and time frame is a result of the strong synergy between the aspects of cognitive evolution that the Social Brain project was focused upon, and the fact that Gowlett was writing about these themes in the quarter of a century prior to the formal beginnings of the project. In a sense the Social Brain project brought together what were an already coalescing group of related research interests under a single umbrella, uniting them via the idea of sociality as the driver for evolutionary change. This latter had been implicit in Gowlett's own theory building long before the Millennium, but it was only in the middle-late 1990s that sociality became an explicit element in the development of his ideas.

I should point out that my interpretations of John Gowlett's work are not necessarily ones he would agree with; nor does he necessarily hold the same views today as he once did.

Theory and mind in the African years

1978 to the late 1980s saw Gowlett's entry into the research culture of the East African Pleistocene with excavations at Kilombe (Gowlett 1978), Kariandusi (Gowlett and Crompton 1994) and Chesowanja (Gowlett, et al. 1981; Harris, et al. 1981), all in Kenya. The first two were Acheulean sites and represented major contributions to his Ph.D (Gowlett 1979b), whereas the last was primarily an Oldowan site, with a later intrusive Acheulean component. Chesowanja

became a significant element in the debates on early fire that Gowlett was heavily involved with in the 1980s. Kariandusi was considered a disturbed factory site and did not really contribute much to discourse during this first decade of Gowlett's research .

Across this decade Gowlett's writings broadly fall into two camps. Firstly, those that represent a basic *reportage* of sites, descriptions of assemblage character and composition, and interim site reports. Secondly, those which tackle broader theoretical concerns. Right from the outset it is clear how data from his excavations fed directly into his theory building. The influence of other thinkers is evident too; Glyn Isaac, in particular his work on the reasons for variation in assemblage composition (Isaac 1977), and artefact character (Isaac 1972); Mary Leakey at Oldupai Gorge (Leakey 1971), but also of Julian Huxley (Huxley 1955) and Ralph Holloway (Holloway 1969), two names that repeatedly crop up in the more theoretical papers and whose theory building has persistently informed Gowlett's own ideas across these two decades . It is worth pointing out that younger researchers today may assume that theory building is synonymous with 'theoretical archaeology'. However in the Palaeolithic archaeology of the 1970s and early 1980s it was usually focused on practical epistemology; what should we call so-and-so, why and how do we know so-and-so is real; what does the reality of so-and-so mean for everything else?

From the decade 1978 - 1988, these are the major themes that to my mind thread their way through Gowlett's work. I have parsed them out here, but they are clearly interlinked.

- The human way of thinking has deep roots and this can be seen in stone tools and through the analysis of their manufacture.
- The nature of the relationship between culture, the mind and material culture. This engaged Huxley's concept of the psychosocial (Huxley 1955), which in today's terminology might equate with cognitive evolution and its relationship with sociality.
- Arising from the previous point, Gowlett argued passionately that stone tools provided key insights into the 'psychosocial sector', stemming from both the tools themselves but also their patterned distribution in time and space.

- Developing from the last, epistemological concerns about the recognition, description and quantification of temporal and spatial variability seen in stone tools. This was exemplified by the Acheulean vs Developed Oldowan debate.

From the late 1980s to the beginning of the Social Brain project in 2003, these basic themes continued to underwrite much of Gowlett's research output, although in some cases the themes changed as the research questions of the decade changed. Whether topics were in or out of the frame, the primary data and theoretical foundations behind Gowlett's work remained the same, although many were progressively developed through the 1990s.

The main themes, as I see them for the late 80's to the early 2000s, are as follows.

- Elucidating the procedural templates / rule sets behind handaxe production and handaxe assemblages. This continues the focus on variability and cognitive evolution, as well as addressing questions about the way the human mind worked. However, concerns with demonstrating the deep roots of human thought processes were less visible.
- A stronger interest in the relationship between culture and cognition – with cognitive development providing the bridge between biological and cultural evolution.
- As the Developed Oldowan / Acheulean debate faded, Gowlett championed the continued study of stone tools and their spatio-temporal context. This was in the face of strong epistemological challenges to the information potential inherent in Palaeolithic artefacts. Isaac's own students were arguing that the tool types of the older typologies were not genuine design norms, while others advocated the position that shaped tools were actually just cores; functional arguments had challenged cultural interpretations of variability, and the concept of a 'finished artefact' was under the microscope of critique.
- Although allometry (adjustments to shape with changes in size) was a technique for engaging with some of the above themes, it was also a reflection of the increasing sophistication (maturity?) with which these research questions were being interrogated as the Millennium approached. Gowlett was an early enthusiast of the use of complex multivariate analysis.

Deep roots to the human way of thinking

As noted, the research questions of the post-Millennium Social Brain project were already being addressed by Gowlett from the late 1970s onwards. He was responding directly to the negative views of a number of colleagues who were downgrading the importance of stone tools (a reaction to excessive typological studies), in addition to denying any real evolutionary significance to pre-modern humans (Gowlett 1984; Gowlett 1986). The view advocated was that important changes in human evolution all happened with modern humans in the last one hundred thousand years. Gowlett championed the importance of Early Pleistocene stone tool analysis by demonstrating that the fundamental basis of how humans think – the foundations of our thought processes - had a long evolutionary history and they were present in earlier *Homo*.

He argued that modern humans conceptualise the external world through the creation of internal mental visualisations, images in the mind's eye. In modern terminology - we construct an internalised mental model of external reality (Gowlett 1982; Gowlett 1984; Gowlett 1986). Hominins possessed the same capacity and it was demonstrable through material culture. The process of making Oldowan core tools and Acheulean handaxes showed that hominins possessed different *templates* (mental models) for both procedure (process of making) and form (final product). These models were unitary – one internal visualisation for one aspect of the outside world. These unitary representations were then chained together to form the procedural templates themselves, and were embedded in a visualisation of the final tool – the form template. So the stone tools of early *Homo* were proof positive of the ancient roots of modern human thought processes.

The psychosocial link to culture was that these unitary internal visualisations were ideas, but they were generated and learnt in a social context, as was the construction and maintenance of the various procedural and form templates.

Without these unitary internal models the ability to make tools would not exist. These templates or routines guided and informed the knapper's actions (Gowlett 1982). However, Gowlett was also clear that the stages of tool production had to be embedded within (evolving) concepts of space and time (Gowlett 1984), as these too were key features in the way modern people thought. So here, procedural templates were embedded in forward planning, anticipation of need, resource procurement and the distribution of activity across landscapes.

A key early insight of Gowlett's into cognitive evolution was the recognition of the degree to which the different stages within the procedural template were integrated with each other (Gowlett 1986). One of the defining traits of the psychosocial sector was management of complexity. It was the level at which this occurred that really characterised hominins and humans. This was clearly emphasised by the differences between ourselves and our nearest cousins in nature, the apes (another recurrent theme of Gowlett's (Gowlett 1986; Gowlett 1993)). Our extant relatives show occasional glimpses of these traits. They make tools, they curate anvils, and modify sticks to make termite fishing tools, all of which suggests that they have procedural and form templates of their own. However, they do not possess the complexity in manipulating and integrating the elaborate procedural templates that we and our Acheulean and Oldowan making ancestors share. Knapping an Oldowan tool shares basic procedural and form templates with a chimp's termite fishing stick, but the utility of the comparison stops there. Integration of the templates is what makes hominins different. This is evident in the following quotation.

> 'We have seen that early human beings, over a million years ago, had minds that could handle extents of time and space, much as we can, and construct long chains of activity through them, using set routines, but able to rewrite these flexibly in detail.' (Gowlett 1984, 214)

Complexity in one area of hominin behaviour bespoke the potential for complexity in other areas too. Fire making (Gowlett, et al. 1981; Harris, et al. 1981) and the skilful butchery of animal tissue (Gowlett 1984), even the very fact of the imposition of arbitrary form (Holloway 1969) on the world (Acheulean handaxes and cleavers, discoids from Oldupai's site DK; i.e. form templates): all of these provide evidence of the complex integration of elaborate internalised mental visualisations of the external world.

Across the second decade of Gowlett's research career this emphasis on the deep roots of cognition receded. This is curious as the context of Gowlett's polemic, the belief that advanced cognition was restricted only to modern humans, had crystalized into a formal and popular theory - the Human / Upper Palaeolithic Revolution (Mellars and Stringer 1989) in the late 1980s. Nevertheless, the psychosocial element and the notion of the procedural templates continued to inform Gowlett's ideas (Gowlett 1984; Gowlett 1995a; Gowlett 1995b). Responding to the challenges of Nick Toth's work (Toth 1985) which argued that the shapes of Oldowan cores were fortuitous, Gowlett asserted that even if this was the case, they were still knapped to a complex procedural template (termed instruction sets by the mid-90s) which structured them from acquisition of the cobble to use of the flake as a tool (Gowlett 1995a).

Explaining variability – taxonomies and spatio-temporal patterning – the Developed Oldowan/Acheulean debate

From the initial publication of Kilombe (Gowlett 1978) questions concerning assemblage composition, artefact taxonomy and what the variability sampled across time and space actually meant, were major elements in Gowlett's thinking for the simple reason that they informed so many other aspects of our understanding of the deep past. An upper horizon at Kilombe, post-dating the main Acheulean floor at locality EH, revealed a flake assemblage associated with a palaeosol, prompting the possibility of a non-handaxe Acheulean facies, accompanied by all the definitional chaos that that concept entailed (Gowlett 1978; Harris, et al. 1981). On two occasions Gowlett predicted the Lomekwian (Gowlett 1986; Gowlett 1996b) as an earlier facies of the Oldowan.

Mary Leakey (Leakey 1971) had interpreted Oldupai Gorge in terms of the monolithic conception of culture that she had grown up with in the 1930s (de la Torre and Mora 2014). Artefacts were realisations of specific design forms, and consequently culture had a somewhat fixed and invariant character. Almost by definition significant variability in assemblage composition would imply different cultures or industrial traditions. Glyn Isaac had challenged this (Crompton and Gowlett 1993; Isaac 1972), arguing that there was considerable handaxe variability in the supposedly broadly contemporary localities revealed in his Olorgesailie excavations (Isaac 1977). Earlier, Isimila had also raised the spectre of spatial variability across contemporary localities (Howell 1961; Howell, et al. 1962). The Developed Oldowan (DO) vs Acheulean debate encapsulated the problems that emerged when an overlap in tool types was present in a rigidly imposed cultural framework. Following Maxine Kleindienst's scheme (Kleindienst 1962), Mary Leakey had allowed for a small number of handaxes (albeit smaller and cruder than Acheulean ones) in the DO.

From the outset, Gowlett eschewed a typological and even a technological definition of a handaxe, preferring instead a psychosocial narrative. It is the long axis of the tool (point to base) that is the key to differentiating the Acheulean handaxe, with both bifacial flaking and bilateral symmetry arranged laterally in respect of the long axis. Whether the concept of the axis came first, and then big flakes were made to accommodate this, or it was the other way around was not clear. The reason for distancing the understanding of these large cutting tools from a typo-technological one was the very evident presence of bifacial flaking on cores (and handaxes) in the DO, and Leakey's acceptance of a small handaxe component to the DO. Gowlett's 1979 paper on the DO/Acheulean debate, *Complexities of Cultural Evidence* (Gowlett 1979a), offered no solution to the question but noted how variability in terminology, definition and artefact taxonomy, and even in the theory and practice of sampling at inter- and intra-site levels made answering the question of whether the DO and Acheulean were culturally distinct phenomena, or activity variants within the same tradition, impossible.

(The philosophy and practice of sampling threads its way through many of Gowlett's papers in the 1980s although I will not delve any deeper into them here.)

The topic was back on the menu in 1986 and 1988 but with more sophisticated analytical procedures; multivariate statistical techniques in the form of principle components analysis (PCA) and cluster analysis (CA). This was also the time that acronyms became more noticeable in archaeology. Using the frequency of occurrence of different tool types (assemblage composition) across a range of Acheulean and DO sites, Gowlett demonstrated that there was a real difference between the Oldupai Gorge DO sites when compared to Acheulean sites from elsewhere in Eastern Africa (Gowlett 1986). As well as frequency differences, the PCA & CA showed that there was a decrease in the use of core tools at DO sites over time. However, a particular type of CA conducted on measured data demonstrated the presence of two sub-groups of LCTs at Kilombe (Gowlett 1982; Gowlett 1986), one with large handaxes more classically Acheulean, the other with smaller handaxes which on examination were often cruder in finish – more akin to those of the DO at Oldupai. Gowlett's interpretation was typical, arguing in favour of a more nuanced approach suggested by both sets of results (Gowlett 1988). At Oldupai the DO/Acheulean distinction was real (whatever its explanation), but at Kilombe the distinction was in all likelihood functional because the two handaxe variants occurred on a single contemporary land surface. He speculated that It might be possible to trace a development from the shorter Oldowan discoid, through the short-ish DO biface, to the elongated axis and bilateral symmetry of the handaxe proper.

The DO/Acheulean debate did not continue into the 1990s as other non-cultural research questions took centre stage. In some respects the DO vs Acheulean debate was a ripple from earlier debates on culture vs function from other areas of Prehistory, debates that by the 1980s had already been played out, or had just ground to a halt. Nevertheless, it was a valid exercise in the epistemology of how archaeologists quantify, describe and analyse variability in stone tool assemblages.

Explaining variability – taxonomies and spatio-temporal patterning in handaxe manufacture – allometry

On the other hand the question of artefact variability as revealed by the size distinctions in the Kilombe handaxes did persist into the 90s. The implications of a series of allometric studies (size-related variation in shape) on handaxes from Kariandusi, Kilombe and elsewhere unexpectedly exposed some of the procedural templates (Gowlett 1982; Gowlett 1984;

Gowlett 1986) already described. Gowlett had always been a little uncomfortable with the label 'procedural template' believing it to be too rigid, and open to misinterpretation. From the start of the new decade he began to reformulate the terminology and its implications, preferring to see them as sets of shared instructions or 'reference routines' held in the brain (Gowlett 1990). This had coalesced into 'instruction sets' by the middle of the decade (Gowlett 1996b), with the old form template now subsumed within the concept.

The psychosocial element of ideas being culturally learnt and passed on remained implicit in this.

The presence of standardization in handaxe making - as an indicator of the existence of instruction sets - had been a theme of Gowlett's work on the Kilombe handaxes from early in the 80s. He identified a consistently reoccurring handaxe width/length ratio of c. 0.6 from all the different contemporary areas on the Kilombe land surface. He argued this recurrent pattern was a deliberately imposed design feature – part of an instruction set, and akin to the 'golden ratio' of artists and architects (Gowlett 1982). It indicated the dawning of a sense of proportion in hominin psychology, another instruction set and a further insight into the origins of the modern human thought process.

Gowlett was joined by his Liverpool colleague Robin Crompton in the early 1990s to conduct an allometric analysis of the Kilombe bifaces (Crompton and Gowlett 1993). Allometry added to the list of instruction sets that could be identified in handaxe making and provided independent proof of their validity. The concept originates in biology (Crompton is a specialist in bipedalism). When the physical size of an organism changes, and all aspects of the organism change in direct proportion, this is akin to geometric scaling and the organism and its various component parts are said to be in isometry. However when size changes but some elements do not scale appropriately (i.e. they are bigger or smaller than they should be) then this is allometric scaling. The size of the human brain may be thought of as allometrically scaled as it is far larger than it should be for a mammal of our average body size.

The presence of allometric scaling at Kilombe (Crompton and Gowlett 1993) demonstrated that the handaxe knappers did not share a single common handaxe template or design norm which they imposed on every handaxe. This in itself this was a significant observation for cultural interpretations. Rather, allometric differences in various measured features of handaxes (width of the tip, thickness of the base etc.) were imposed by the knappers as size changed. The two basic groups of handaxes, large and small, identified in the earlier analysis, were confirmed in this new research. Even if no site-wide handaxe template existed, what was clear was that allometric adjustments to handaxe shape and thickness were applied in a similar way across the site. So the knappers in different parts of the site were responding in the same ways to changes in handaxe size, making some aspects of the axes thinner and others narrower as size altered. In particular, the larger handaxes tended to be thinner and the tips were always thinner than isometry would require; the larger specimens were narrower in plan at the tip; and the most isometrically stable part of the handaxe was always the base probably because it was the handle for use. So even if there was no one culturally generated signature handaxe type at the site, there were commonly shared understandings (instruction sets) of how to adjust shape when size changed. But there were important differences too. One locality at Kilombe, Z, had LCTs that were significantly thicker than isometry predicted.

The explanation of these various allometric changes was cautiously accepted as functional.

The following year allometry was applied to Kariandusi (Gowlett and Crompton 1994). Its assemblages were in secondary context, but unlikely to have been moved far. Both levels of the site were dated to a similar time range to Kilombe, c. 0.7 – 1.0 mya. A series of obsidian bifaces from Louis Leakey's upper site were compared with lava examples from the excavated lower site recovered by Gowlett. These data sets were then compared with Kilombe and with lava LCTs from the Kapthurin Formation, also in Kenya. This latter site dated to about 0.5-0.4 kya and its LCTs were made on Levallois blanks. There were some strong similarities between the LCTs from all three sites, despite the fact that Kapthurin post-dated the other two by many hundreds of thousands of years. Intriguingly the obsidian artefacts from Kariandusi upper site were allometrically similar in many respects to the Kapthurin lava axes. The lava axes from Kariandusi lower also showed some similarities but were markedly more asymmetric with thinner tips, these being interpreted as two allometric changes in the procedural templates for handaxes in that assemblage. The Kariandusi lava axes were noted for their butt size - another example of allometric scaling, in this case specific to the knappers of that assemblage. Once more function was seen as potentially a driver for these allometric differences.

In summing up Gowlett and Crompton noted that across the East African Acheulean (at least for their data) allometry was a significant factor in LCT production, and across a significant time depth. This meant that *Homo erectus* at Kariandusi and Kilombe, and early *Homo sapiens* (now more likely African *Homo heidelbergensis*) at Kapthurin, were applying similar rule sets to their material culture in similar ways. The pervasiveness of the pattern was proved when allometric adjustments

similar to those in East Africa were found in the handaxes of the Acheulean Casablanca sequence in Morocco (Crompton and Gowlett 1997). The Casablanca sites showed *Homo erectus* (STIC quarry possibly similar in age to Kilombe and Kariandusi) adapting to size changes as they did in East Africa; a later Casablanca handaxe site (Cunette) continued this pattern. Heading southwards, allometry was clearly at work in the handaxes of the Zambian site Kalambo Falls (Gowlett, et al. 2001). Here allometric analysis was able to confirm the old typological distinction between handaxes and the bigger picks of the Sangoan, showing that these remained deliberately thicker and heavier. This may have been related to function and the way they were held – two handed with the extra weight for increased power.

Why was allometry important to John Gowlett? In a decade when form templates were out of fashion and tools as finished forms were a 'fallacy', it was hard to convince people that stone tools were a worthwhile pursuit. Allometry provided an objective answer, one that was independent of typology, and squarely rooted in the psychosocial. Allometric changes were knappers adjusting their instruction sets (or perhaps accessing sub-sets) to ensure that what they made was still a viable tool. Allometry proved the existence of instruction sets, and showed that lithic analysis could contribute to the new research agendas emerging in the 1990s. I suspect there was also a pleasing 'human' element here too. We can empathise with an Erectine knapper more than a million years ago that has to make allowances to keep edges sharp and tips thinner, at the same time as trying to keep the handaxe's butt big enough to hold on to.

Individuals, their societies and their psychosocial worlds

I will finish this rather personal overview of selected aspects of Gowlett's earlier work by looking a little at the glue that held it all together – the relationship between culture and the psychosocial. During the 1990s Palaeolithic archaeology saw the acceleration of two sub-disciplines within the field of human origins research, cognitive evolution and hominin social archaeology. At the risk of generalising, culture had been out of fashion across the 1980s (Gowlett 1990) and hominin 'behavioural' interpretations had taken its place. The broad umbrella of behaviour could be broken down into specific sub-sets of behaviours and empirically tested. The concept of behaviours had a more scientific and contemporary feel to it. Behavioural studies offered the chance to promote single causes for later social patterns (Gowlett 1984) – food sharing for Glyn Isaac, or the hunting hypothesis for other researchers. Single behavioural solutions had a simplicity to them, and they were more amenable to empirical testing than 'culture'. While acknowledging the importance of these questions Gowlett kicked back against the trend for downgrading the importance of culture as a concept, if for no other reason than that stone tools were cultural artefacts, and allometry was proving their worth in the emerging studies of cognitive evolution.

From almost the beginning of his research output Gowlett had combined Julian Huxley's notion of the psychosocial (Huxley 1955) with Ralph Holloway's insight that culture reflected an 'imposition of arbitrary form' on the natural world (Holloway 1969). These are frequent references in Gowlett's papers of the 1980s and 1990s. 'Form' did not just mean modifying elements of the natural world to make material culture, it also meant imposing ideas on lives lived in the outside world to structure the actions carried out by those lives - culture or sociality in our terms. 'Arbitrary', in this sense, meant imposing something that was not 'known in nature' (Gowlett 1995a) i.e. not present in the outside world – so it originated within the mind.

A quotation will suffice to make this point.

> 'Further insights come from Julian Huxley's view that cultural development represents a fundamental change of evolutionary level, from ordinary biological evolution to psychosocial evolution, in which change can happen much more rapidly and in which, ideally, it can be guided by the species concerned. In this sense, culture, as a concept, embodies not just material objects but all the abstracted rule systems ...[Holloway's imposed arbitrary form]...by which human beings operate, and which are handed down from individual to individual. It has become widely accepted that in such a system, biological evolution and cultural evolution affect one another in a positive feedback relationship, thus providing both change and its cause. This view has never been effectively challenged...' (Gowlett 1984, 202; my square brackets).

I sense a strong gene-culture co-evolution element to Gowlett's theory building in these years (Gowlett 1984; Gowlett 1986; Gowlett 1990; Gowlett 1996b), and it is interesting that in his *Mental Abilities of Early Homo* paper in 1996 he explicitly rejects such a link (p193), but the rejection is more about that stripe of co-evolution promoted by E.O Wilson in his now renowned (or infamous) *Sociobiology* book (1975). My gut feeling is that Gowlett's early writings are more in line with the modern gene-culture co-evolution of Joseph Henrich (Henrich 2016), or perhaps more specifically with Cecilia Heyes (Heyes 2018) since she and Gowlett both place social learning at the very heart of the socio-cognitive relationship (Gowlett 1984, 1986; see below).

Going out on a limb, I suggest that Gowlett eschewed behavioural archaeologies in favour of more psychosocially orientated ones because for him these were synonymous with culture. This was implicit in his 1984 *Mental Abilities of Early Man* paper; culture was an integral part of biology and the mind mediated between the two (Gowlett 1984). Evolution in one meant evolution in the other, and so evolution of the mind that linked them. It was through the internalised mental models that this was achieved. They were adaptive. Increase the effectiveness of internal representations and the effectiveness of instruction sets in the outside world increased. This enhanced the inclusive fitness of an organism as it made it better equipped to be successful in demanding environments. This was clearly reiterated in his 1995 paper *Psychological Worlds Within and Without* (Gowlett 1995b). Gowlett also added an evaluative element to this, mentalising future possibilities:

> 'To experiment in the head is cheaper than to experiment in actuality.' (Gowlett 1995b, 37)

And more specifically,

> 'Efficiency can only be ensured through mental simulation – that is a planning of activities in which alternatives can be evaluated, and discarded if found wanting.' (Gowlett 1995b, 38)

In this sense Gowlett effectively tied culture and biology together through selection pressures on cognition.

Paraphrasing Gowlett's (1990) interpretation, Julian Huxley (1955) characterised evolutionary biology by an 'interlocking trinity of subject matter':

- The mechanisms of maintaining existence
- The basis of reproduction and variation
- The modes of evolutionary transformation

Gowlett (*ibid*, 90) argued that modern culture need not reflect each of these individually (Julian Huxley's paper was on the relationship between culture and biology, and was the T.H. Huxley memorial lecture for 1955), but it nevertheless achieves the same result as they do, although I suspect the modern gene-culture co-evolutionists would have little trouble in tying these biological principles directly to cultural activities.

> '...my conclusion is that at its most basic the cultural system is an adjunct system of living, set up through process that are genetically controlled, and dependent for its operation on brain store, brain process, and coded electrical signals. This reduction does little to reach towards the higher

levels of mind, but it does suggest the possibility of analysing objectively the information content of ancient technology.' (Gowlett 1990, 90)

This 'objective analysis' is made possible because the reductionist view of culture sees it as a way of storing, processing and transferring information. The information is located in the brain, which as it evolves (i.e. more effective mental models and improved instruction sets which are culturally learned), is able to process and store ever more information and to transfer it to subsequent generations through learning (imposed arbitrary form is the channel through which processing is made relevant, and transfer is facilitated). Hence the evolving mind is the mediator of culture and biology (*ibid* 1990).

It is interesting to note, in concluding, that sociality and the individual did not really come to prominence in Gowlett's work before the mid-1990s. Both I think were implicit in his thinking but they had not been formally expressed as such. Both of course were key elements in the Social Brain project. They were inherent to the concept of the psychosocial and cultural learning; allometry allowed the individual to be recognised as a thinking actor. Of the papers I have had access to, the individual is only acknowledged as such from 1996 onwards (Gowlett 1996b); and sociality as a structuring factor also occurs first in the same year (Gowlett 1996a). It is therefore perhaps not so surprising to discover that the year before, Gowlett actually anticipated sociality as the key driver for brain expansion, what would become the core theory of the Social Brain,

'How then can we justify the very expensive human brain, that uses so much energy? I can think of two solutions:

1. That it is necessitated by an environment which is largely the human social environment – no other animal has this...' (Gowlett 1995b, 37)

The second reason was that the brain payed for itself with ever more effective mental models.

In his 1996 *Mental Abilities of Early Homo* paper Gowlett presented the concept of the personal pointer (1996a), a zone of free-play (see also Isaac 1972), within which an individual knapper could express themselves in the handaxes they made without stepping outside of their society's understanding of what material culture was. He expressed it as the relationship between the individual Acheulean knapper and the group. Twenty three years after this was articulated, and nine years after the Social Brain project ended, this remains one of the highest research priorities in Lower and Middle Pleistocene archaeology.

References

Crompton, R.H., Gowlett, J.A.J. 1993. Allometry and multidimensional form in Acheulean bifaces from Kilombe, Kenya. *Journal of Human Evolution* 25:175 - 199.

Crompton, R.H., Gowlett, J.A.J. 1997. The Acheulean and the Sahara: Comparisons between North and East African Sites. In: Sinclair, A., Slater, E., Gowlett, J.A.J., editors. *Archaeological Sciences 1995. Proceedings of a Conference on the Application of Scientific Techniques to the Study of Archaeology.* Oxford: Oxbow Monograph 64, Oxbow Books. p 400-405.

de la Torre, I., Mora, R. 2014. The Transition to the Acheulean in East Africa: an Assessment of Paradigms and Evidence from Olduvai Gorge (Tanzania). *Journal of Archaeological Method and Theory* 214:781-823.

Gowlett, J.A.J. 1978. Kilombe - An Acheulean Site Complex in Kenya. In: Bishop, W.W., editor. *Geological Background to Fossil Man.* Edinburgh: Scottish Academic Press. p 337-360.

Gowlett, J.A.J. 1979a. Complexities of cultural evidence in the Lower and Middle Pleistocene. *Nature* 278:14 - 17.

Gowlett, J.A.J. 1979b. A Contribution to Studies of the Acheulean in East Africa with Especial Reference to Kilombe and Kariandusi. University of Cambridge: Unpublished Ph.D.

Gowlett, J.A.J. 1982. Procedure and Form in a Lower Palaeolithic Industry: Stoneworking at Kilombe, Kenya. *Studia Praehistorica Belgica* 2:101-109.

Gowlett, J.A.J. 1984. Mental Abilities of Early Man. *Culture Education and Society* 38:199-220.

Gowlett, J.A.J. 1986. Culture and Conceptualisation: the Oldowan-Acheulean Gradient. In: Bailey, G.N., Callow, P., editors. *Stone Age Prehistory. Studies in Memory of Charles McBurney.* Cambridge: Cambridge University Press. p 243-260.

Gowlett, J.A.J. 1988. A case of Developed Oldowan in the Acheulean? *World Archaeology* 20:13 - 26.

Gowlett, J.A.J. 1990. Technology, Skill, and the Psychosocial Sector in the Long Term of Human Evolution. *Archaeological Review from Cambridge* 9:82-103.

Gowlett, J.A.J. 1993. Chimpanzees Deserve More than Crumbs of the Palaeoanthropological Cake. *Cambridge Archaeological Journal* 3:297-300.

Gowlett, J.A.J. 1995a. A Matter of Form: Instruction Sets and the Shaping of Early Technology. *Lithics* 16:2-16.

Gowlett, J.A.J. 1995b. Psychological Worlds Within and Without: Human-Environment Relations in Early Parts of the Palaeolithic. In: Ullrich, H., editor. Man and Environment in the Palaeolithic. Liège: *Etudes et Recherches de l'Université de Liège* 62, Université de Liège. p 29-42.

Gowlett, J.A.J. 1996a. Chapter 5. The Frameworks of early Hominid Social Systems. How Many

Useful Parameters of Archaeological Evidence can we Isolate. In: Shennan S., Steele J., editors. *The Archaeology of Human Ancestry; Power, Sex and Tradition*. London: Routledge. p 135-183.

Gowlett J.A.J. 1996b. Mental Abilities of Early *Homo*: Elements of Constrain and Choice in Rule Systems. In: Mellars, P., Gibson, K., editors. *Modelling the Early Human Mind*. Cambridge: McDonald Institute for Archaeological Research. p 191-215.

Gowlett, J.A.J., Crompton, R.H. 1994. Kariandusi: Acheulean morphology and the question of allometry. *The African Archaeological Review* 12(3 - 42).

Gowlett, J.A.J., Crompton, R.H., Yu, L. 2001. Allometric Comparisons between Acheulean and Sangoan Large Cutting Tools at Kalambo Falls. In: Clark, J.D., editor. *Kalambo Falls Prehistoric Site III*. Cambridge: Cambridge University Press.

Gowlett, J.A.J., Harris, J.W.K., Walton, D.A., Wood, B.A. 1981. Early archaeological sites, hominid remains and traces of fire from Chesowanja, Kenya. *Nature* 294:125 - 129.

Harris, J.W.K., Gowlett, J.A.J., Walton, D.A., Wood, B.A. 1981. *Palaeoanthropological studies at Chesowanja*. p 64 - 100.

Henrich, J. 2016. *The Secret of Our Success*. Woodstock Oxfordshire: Princeton University Press.

Heyes, C. 2018. *Cognitive Gadgets*. Cambridge Massachusetts: Bellknap Press of Harvard University Press.

Holloway, R.L. 1969. Culture: a Human Domain. *Current Anthropology* 10:395-412.

Howell, F.C. 1961. Isimila: A Palaeolithic site in Africa. *Scientific American* 205:119-129.

Howell, F.C., Cole, G.H., Kleindienst, M.R. 1962. Isimila, an Acheulean occupation site in the Iringa highlands. In: Mortelmans, J., Nenquin, J., editors. *Actes du IVᵉ Congres Panafricaine de Prehistoire et de l'Etude du Quaternaire*. Tervuren: Musee Royale de l'Afrique Centale. p 43-80.

Huxley, J. 1955. *Evolution. Cultural and Biological*. Guest Editorial: Wenner Gren.

Isaac, G.L. 1972. Early Phases of Human Behaviour: Models in Lower Palaeolithic Archaeology. In: Clarke, D.L., editor. *Models in Archaeology*. London: Methuen. p 167-199.

Isaac, G.L. 1977. *Olorgesailie: Archaeological Studies of a Middle Pleistocene Lake Basin in Kenya*. London: University of Chicago Press.

Kleindienst, M.R. 1962. Components of the East African Acheulean Assemblage. *Actes du IVe Congres Panafricain de Prehistoire et de l'Etude du Quaternaire*: 81 - 103.

Leakey, M.D. 1971. *Olduvai Gorge: Excavations in Beds I and II 1960 - 1963*. Cambridge: Cambridge University Press.

Mellars, P., Stringer, C., editors. 1989. *The Human Revolution: Behavioural and Biological Perspectives on the Origin of Modern Humans*. Edinburgh: Edinburgh University Press.

Toth, N. 1985. The Oldowan Reassessed: A Close Look at Early Stone Artefacts. *Journal of Archaeological Science* 12:101-120.

John McNabb
Department of Archaeology, University of Southampton, Avenue Campus, Southampton, SO17 1BF, UK.
J.McNabb@soton.ac.uk

Brain Size Evolution in the Hominin Clade

Andrew Du and Bernard Wood

Introduction

Paleolithic artefacts are one of the few ways we have of accessing the minds of early hominins, and John Gowlett has been a pioneer of the use of tools to infer the cognitive level of the hominins who made them. After briefly reviewing what artefacts might be able to tell us about the cognitive abilities of their authors, this contribution will explore another proxy for cognition, brain size, and within the context of Gowlett's research program, we discuss what existing data about endocranial volume can tell us about how hominin brain size has evolved within the hominin clade.

More than three decades ago, John Gowlett addressed how the archaeological record might be used to make inferences about the mental abilities of early hominins. He suggested that artefacts provide unique insights into the mind of extinct hominins and thus have the potential to 'cast much light on its evolution' (Gowlett, 1984, p. 169). Although he conceded that the earliest stone industries 'are simple by our standards' (*ibid*, p. 180), he proposed that their manufacture implied substantial knowledge about the physical properties of the raw material from which they were fashioned, as well as evidence of the ability to conceptualize form in three dimensions, the mastery of multi-step routines, and the ability to 'transfer that knowledge to other individuals' (*ibid*, p. 180). When he returned to this topic just over two decades later, Gowlett stressed how the form of Acheulian artefacts, specifically hand-axes, implies a sophisticated knowledge of design principles, and he suggested their manufacture implies the ability to handle complex interactions among multiple variables (Gowlett, 2006).

Brains and tool manufacture

Interest in the role played by brain size in demarcating modern humans from our close living relatives has a long and distinguished history. In *The Descent of Man*, Charles Darwin remarked on 'the large proportion which the size of man's brain bears to his body, compared to the same proportion in the gorilla or orang' (Darwin, 1871, p. 55), and suggested that our relatively, and absolutely, large brains resulted in the 'intellectual faculties' that mark modern humans out from our closest living relatives— which we now know are chimpanzees and bonobos.

One of those intellectual faculties is the propensity to conceive of and manufacture tools. In the preface to the first edition of *Man the Tool-Maker*, Kenneth Oakley explains that booklet was the result of being commissioned to provide a palaeontologist's view of the culture of early humans to accompany an eponymous wartime exhibit of early implements (Oakley, 1949). Oakley proposed that while living apes are capable of 'improvising a tool to meet a given situation [...] the idea of shaping a stone or stick for use in an imagined future eventuality is beyond the capacity of any known apes' (*ibid*, p. 3). He also suggested 'if man is defined as the tool-making animal, then the problem of the antiquity of man resolves itself into the question of the geological age of the earliest known artifacts' (*ibid*, p. 3). In a later publication Oakley refers to a 'close correlation between culture and cerebral development' (Oakley, 1957, p. 208), but he provided little in the way of detail about what he meant by 'cerebral development', nor did he specify the nature of the alleged close correlation. But he did express the belief that while *Australopithecus* may have been a tool-user, it was not capable of manufacturing tools because 'systematic tool-making requires a larger brain than the ape-size brain of *Australopithecus*' (*ibid*, p. 207), so we must assume that Oakley's 'cerebral development' includes, or is even synonymous with, brain size. Oakley did have the good sense to suggest that his proposition 'may prove to be ill-founded' (*ibid*, p. 207), which was especially prescient given claims there are cut-marked bones in 3.4 Ma sediments at Dikika in Ethiopia (McPherron *et al.*, 2010) and the suggestion that some of the stones recovered from 3.3 Ma sediments at West Turkana in Kenya were deliberately shaped (Harmand *et al.*, 2015). Thus, presently at least, the manufacture of stone tools seems to have preceded the appearance in the fossil record of larger-brained taxa at *c*.2.5 Ma that at least some researchers are comfortable assigning to the genus *Homo* (Wood and Collard, 1999).

The manufacture of tools, stone or otherwise, requires the manufacturer to have a mental image of the end product, a sense of the sequence the maker needs to follow in order to fashion the object, an understanding of the types of judgments and decisions that must be made as they follow that sequence, a knowledge of the whereabouts of suitable raw materials and familiarity with their physical properties, plus a hard and soft-tissue phenotype that enables the maker to locate, transport, manipulate and modify that raw material. The maker also needs to be able to make decisions about when the tool should be used. Although hominins are currently the only taxa known to be capable of successfully and regularly manufacturing and using *stone* or *bone* tools, the proficiencies described above are not confined to hominins. For example, New Caledonian crows (Hunt, 1996) and chimpanzees (Sanz *et al.*, 2009) fashion plant stems to extract insect prey from otherwise inaccessible locations. The requirements set out above can be broken down into two categories, cognitive and non-cognitive. Having a plan is no good if your phenotype does not allow you to execute it, nor is it any good to have the required dexterity in the absence of the cognitive ability to exploit that facility (i.e., both dexterity and cognitive ability are necessary for tool manufacture and use, and neither is sufficient in isolation).

It is perhaps not surprising that the organs at opposite ends of the complex sequence of events involved in the manufacture and use of tools—the brain and the hand—have received most attention with respect to reconstructing whether a particular type of early hominin could make tools, and if so, what sorts of tools were they capable of making and using (Panger *et al.*, 2002). Two years after Oakley's initial contribution, the roles played by the brain and the hand in the evolution of tool manufacture were considered by Washburn (1959). With respect to the hand, he proposed that the distinctively shorter fingers and the longer opposable thumb of the hand of modern humans are the result of culture, not part of the explanation for it. However, a more recent review of the evolution of hand morphology in apes by Almécija and Sherwood (2017), which had access to a wider comparative sample than the one available to Washburn, suggests that modern human hand morphology has a much deeper evolutionary history than Washburn suggested. As for the brain, Washburn focused on both brain size and on differences in the relative size (i.e., surface area) of the separate areas of the motor cortex of the cerebral hemispheres that contain the nerve cells, called neurons, which control the movements of the hand and foot. As well as emphasizing the three-fold size difference between the brain size of chimpanzees and australopiths, on the one hand, and the brain size of modern humans (*ibid*, Table 1, p. 27) on the other, Washburn also stressed the potential for differences in the numbers of neurons devoted to the motor control of the small muscles of the hand. He cited the work of Woolsey and Settlage (1950) who had demonstrated that in the monkey motor cortex the areas devoted to the hand and the foot were approximately equal in size (Washburn, 1959, Fig. 3, p. 29), but when Penfield and Rasmussen (1950) investigated this in modern humans they showed that the surface area of the part of the motor cortex devoted to the hand is an order of magnitude larger than that devoted to the foot (Washburn, 1959, Fig.

2, p. 28). Washburn linked these differences with the proportionally larger frontal cortex of modern humans to suggest that 'our brains are not just enlarged, but the increase in size is directly related to tool use, speech, and to increased memory and planning' (*ibid*, p. 29). However, in 1959 the closeness of the relationship between modern humans and chimpanzees/bonobos was not known, and although Washburn implicitly assumed the larger cortical area devoted to the hand seen in modern humans was something that had evolved within the hominin clade since it diverged from its closest living relative, he had no direct evidence to support his assumption. But if one follows Washburn's logic, and assumes the common ancestor of modern humans and chimpanzees/bonobos had a monkey-like relationship between the cortical areas devoted to the hand and the foot—and therefore the modern human cortical relationship evolved within the hominin clade—then this has potential consequences for dexterity. This is because modern humans have fewer intrinsic hand muscles than both chimpanzees (Diogo and Wood, 2011) and bonobos (Diogo *et al.*, 2017), so unless the fewer muscles in the hand of modern humans are bulkier than those of chimpanzees/bonobos, any additional motor neurons in modern humans compared with chimpanzees/bonobos would result in modern humans having smaller motor units (i.e., each motor neuron would control fewer muscle fibers) with the potential for more precise motor control. Yet, even if that was the case, unless differences in the amount of motor cortex devoted to the hands and feet are reflected in quantifiable differences in endocranial shape, the tempo and mode of any changes in these proportions between the common ancestor of chimpanzees/bonobos and modern humans is presently, and is always likely to be, unknowable.

Brain size within the hominin clade

While Washburn warned that brain 'capacity is a very crude measure of brain size evolution' (*ibid*, p. 27), researchers have traditionally used a reliable proxy for brain size—endocranial volume (ECV)—to investigate the tempo and mode of changes in brain size within the hominin clade. During the course of human evolution, from the earliest unambiguous hominin species to *Homo sapiens*, endocranial volume has more than tripled. But when and how did hominins acquire their unusually large brains?

Perhaps the most influential image related to hominin brain size evolution is a scatter plot in Aiello and Wheeler's (1995) much cited paper proposing the 'expensive tissue hypothesis.' This figure depicting the ECVs of individual hominins from the last 3.5 million years plotted against their ages is the basis for the conventional wisdom—now incorporated into textbooks (e.g., Striedter, 2005)—that there was an episode of particularly rapid increase in ECV at *c*.1.8 Ma. Researchers have sought explanations of one sort or another (e.g., Potts, 1996; deMenocal, 2004) for this apparent sudden increase in brain size, and they have attempted to link this episode of saltation with evidence for climate change as inferred from mammal community turnover (Vrba, 1996), changes in diet (Aiello & Wheeler, 1995), and/or behavioral innovation (Isaac, 1972; Ambrose, 2001). All of these explanations may be plausible, but to our knowledge, no one has questioned whether the *c*.1.8 Ma 'pulse' of increased brain size is a genuine evolutionary signal or the result of sampling error/biases in the data. For example, there is a paucity of ECV data from 2.3-2.0 Ma, right before the dramatic increase in brain size. Thus, one cannot reject the possibility that intermediate ECVs linking older/smaller and younger/larger crania may have existed, thus more sophisticated analytical techniques that can handle sampling error are needed to properly interpret these data (e.g., Hunt, 2006).

The figure in Aiello and Wheeler (1995) does not include error bars, but there are at least two potential sources of error for each data point—error in the endocranial volume estimate and in the ascribed age (Figure 1).

Most of the error in the endocranial volume estimates is the result of the incompleteness of the crania used to generate those estimates—Spoor *et al.* (2015) provide a good example of how awry estimates of a crucial data point, in this case the early *Homo* calotte belonging to OH 7, can be. As for the ages of the specimens, absolute dates have a troubling way of being cited and re-cited over time, often from secondary sources, such that error ranges disappear and real and important controversies about the ages of specimens are glossed over (Millard, 2008).

We can interpret the *c*.1.8 Ma ECV saltation in the context of Gowlett's body of research. As mentioned previously, Gowlett (2006) remarked on the complexity of Acheulian artefacts and hypothesized that their

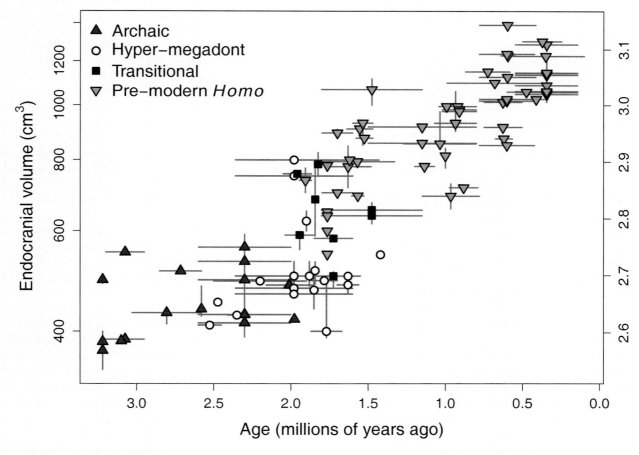

Figure 1. Plot of analyzed hominin ECVs through time. Points represent ECV and age midpoints, and bars represent ranges of error on both estimates. Points are coded by hominin grade. Archaic species include *Australopithecus afarensis*, *Australopithecus africanus*, and *Australopithecus sediba*; hyper-megadont and megadont species include *Australopithecus garhi*, *Paranthropus aethiopicus*, *Paranthropus boisei*, and *Paranthropus robustus*; transitional species include *Homo habilis sensu stricto* and *Homo rudolfensis*; and pre-modern *Homo* species include *Homo erectus sensu stricto*, *Homo ergaster*, *Homo georgicus*, and *Homo heidelbergensis*. The left y-axis is on a logarithmic scale, while the right y-axis' tick labels are log$_{10}$-transformed values. This figure is reproduced from Du et al. (2018).

manufacture requires sophisticated knowledge on behalf of the maker. Assuming that hypothesis to be correct, and given the historical emphasis on the connection between brain size and cultural complexity (e.g., Darwin, 1871; Oakley 1957), one may expect to see a sharp increase in ECV when the Acheulian industry appears in the fossil record (*c.*1.8 Ma; Lepre et al. 2011). This, of course, is not a new observation, for other researchers have noted the correspondence between an ECV increase and the advent of the Acheulian at *c.*1.8 Ma (e.g., Antón et al. 2014). We should note, however, that the current first appearance of the Acheulian is almost certainly an underestimate, and this date will be pushed back as more fieldwork explores older sediments. The same is likely true for the appearance of cranial specimens with large estimated ECVs, so the synchronous appearances of the Acheulian and large ECVs may persist, or disappear, as our knowledge of the hominin record increases. While acknowledging the many pitfalls involved in studying phenotypic patterns through time, we propose converting Gowlett's conjecture into a hypothesis that can be tested using the hominin ECV fossil record. After briefly reviewing what has been written about patterns of hominin ECV evolution, we summarize the inferential issues related to such studies. We then draw on the results of a recent analysis (Du et al., 2018) that can be used to address the evidence for a *c.*1.8 Ma saltation in brain size.

Modes of change in hominin brain size

Over the years, many researchers have studied temporal trends in ECV to investigate the pattern and rate of brain size evolution within the hominin clade. Some interpreted the data as supporting a model of gradual change (Lestrel and Read, 1973; Cronin et al., 1981; De Miguel and Henneberg, 2001; Pagel, 2002; Lee S-H and Wolpoff, 2003; Diniz-Filno et al., 2019; Miller et al., 2019), whereas others supported a punctuated equilibrium model in which hominin ECV experienced short episodes, or bursts, of rapid increase between periods of relative stasis (Tobias, 1971; Gould and Eldredge, 1977; Rightmire, 1981; Hofman, 1983). Other researchers either supported a model that combines elements of both, or they claimed the two models cannot be distinguished (Blumenberg, 1978; Godfrey and Jacobs, 1981; Shultz et al. 2012). The lack of agreement about which evolutionary model best characterizes temporal changes in hominin ECV is unfortunate because in order to generate hypotheses about the evolutionary factors that drove this ECV increase (e.g., establishing a causal relationship between an ECV increase at *c.*1.8 Ma and the onset of the Acheulian industry), we need to accurately characterize the rate and pattern of phenotypic change in brain size. This problem is especially pertinent given recent findings that selection on hominin brain size may have driven an increase in body size, and not the other way around (Grabowski, 2016).

The cause of disagreements about the tempo and mode of the ultimate three-fold increase in brain size in the hominin clade is almost certainly multifactorial. The first factor is the general reliance on qualitative verbal models to describe the pattern of brain size increase. Eldredge and Gould (1972, p. 85), in their landmark punctuated equilibrium paper, noted that 'All observation is colored by theory and expectation,' referring to the tenacity of phyletic gradualism as a theory explaining phenotypic evolutionary patterns. An extreme interpretation of this problem is that qualitative descriptions of hominin ECV evolution through time may be more a function of researchers' preconceived beliefs rather than any underlying reality. This issue led to the development of a number of quantitative methods in paleobiology, all of which were centered on the concept of the random walk as a null model (e.g., Raup, 1977; Bookstein, 1987; Gingerich, 1993; Roopnarine, 2001). The use of random walks to infer evolutionary modes was, strangely enough, not incorporated into the human evolution research toolkit (but see Pagel, 2002). Instead, paleoanthropologists commonly used traditional null hypothesis tests to examine ECV change through time, where non-significant change implied stasis and significant change implied gradualism or some saltation event. However, there are problems with using the null hypothesis inferential framework in this case (*cf.* Smith, 2018), since (1) the usually small ECV sample sizes result in under-powered tests, thereby increasing the risk of Type II error (Levinton, 1982), (2) on a related note, a non-significant p-value (i.e., $p > 0.05$) does not demonstrate the null hypothesis is true (i.e., it is not an adequate test for the presence of stasis [e.g., Rightmire, 1981]), and (3) data points in time series are not independent (i.e., the data are temporally autocorrelated) thus violating an assumption of these tests and increasing the risk of Type I error. For these, and other, reasons we suggest that traditional null hypothesis tests are inadequate and unsuitable for discriminating among evolutionary models.

One might have more success by adopting the strong inference philosophy of Platt (1964; see also Chamberlin, 1980), where instead of testing one null hypothesis, multiple hypotheses are presented and competed against each other on equal footing. In fact, Chamberlin (1890, p. 754) has said 'the dangers of parental affection for a favorite theory can be circumvented' using such a framework, thereby alleviating some of the preconceived biases that Eldredge and Gould (1972) referred to. Recently, Gene Hunt (2006, 2008a, 2008b, Hunt et al., 2015) elegantly operationalized the strong inference philosophy by creating statistical models for each of the commonly studied evolutionary modes (i.e., random walk, stasis, gradualism, punctuated equilibrium, or any combination thereof), which were then discriminated using commonly used model selection techniques (i.e., the Akaike Information Criterion; Burnham and Anderson, 2002).

We recently used Hunt's framework to examine hominin ECV patterns through time (Du et al., 2018) and here we use it to test the idea that ECV evolution experienced a saltation event, or evolutionary mode change, at *c.*1.8 Ma (operationalized at the clade level). Our sample comprised $\log_{10}$-transformed absolute ECVs for 94 specimens ranging from *Australopithecus* to *Homo heidelbergensis* (i.e., beginning with the earliest uncontroversial hominins and ending when ECV values begin to overlap with the range of modern humans). We analyzed absolute and not relative brain size because the former better predicts cognitive ability among primates (Deaner et al., 2007; Shultz and Dunbar, 2010), and because the calculation of relative brain size introduces additional unknown error into any analysis. We binned the ECV data into 0.2 Ma bins and used the R package *paleoTS* (Hunt, 2015) to fit via maximum likelihood a range of evolutionary modes (i.e., random walk, gradualism, stasis, stasis-stasis [clade-level 'punctuated equilibrium'], stasis-random walk, and stasis-gradualism) to the data while explicitly incorporating inter-observer error from ECV estimates for individual fossils (Du et al., 2018). Random walk and gradualism were modeled as having bin-to-bin transitions drawn from a normal distribution; both models have an estimated variance parameter, but the mean is set to zero in the random walk model and is an estimated parameter in the gradualism model (Hunt,

2006; 2008b). Stasis was modeled as normally distributed variation around a mean, both of which stay constant through time (Hunt, 2006; 2008b). The more complex models (i.e., the last three) are permutations of the simpler models, which are separated by an estimated transition age point (Hunt, 2008a; Hunt *et al.,* 2015), something that is relevant here and which we also discuss below. We examined the effect of dating error by resampling dates for each specimen according to their age error distribution and repeating the model fitting and selection procedure 1000 times (Du et al., 2018). Evolutionary models were evaluated against each other using the bias-corrected Akaike Information Criterion (AICc) transformed into relative weights (which sum to one across all models, with higher weights representing more model support) (Burnham and Anderson, 2002).

If the appearance of the Acheulian industry was associated with an increase in hominin brain size or an overall shift in how brain size evolves through time, one would expect that either the stasis-stasis, stasis-random walk, or stasis-gradualism model would be best supported by the data, and if there was a transition between stasis and the next evolutionary mode, it should be at *c.*1.8 Ma (Lepre et al., 2011). Instead, the gradualism model is the best fit for describing hominin ECV change over time (median AICc weight = 0.9) (Figure 2a). All of the mean ECV estimates of the observed time series fall within the 95% probability envelope predicted by the gradualism model (Figure 2b), and model R^2 is 0.676. Multiple sensitivity analyses

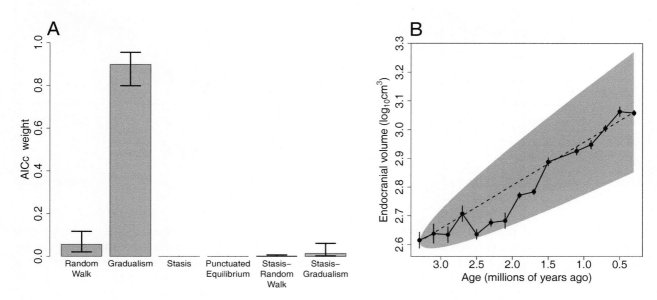

Figure 2. (A) Model selection results testing six evolutionary modes. Bias-corrected Akaike information criterion (AICc; Burnham and Anderson, 2002) weights sum to one across all models, with higher weights representing more model support. Bars represent AICc weight medians, and error bars represent 1st and 3rd quartiles from resampling age estimates. (B) Gradualism model fit for the clade-level ECV time series using 0.2 Ma bins. Binning here was done using observed (not resampled) age midpoints for plotting purposes only. Points are mean ECV estimates, and error bars are ± 1 SE. Dotted line represents the expected evolutionary trajectory of the fitted gradualism model surrounded by the 95% probability envelope in gray. Y-axes as in Figure 1. This figure is reproduced from Du et al. (2018).

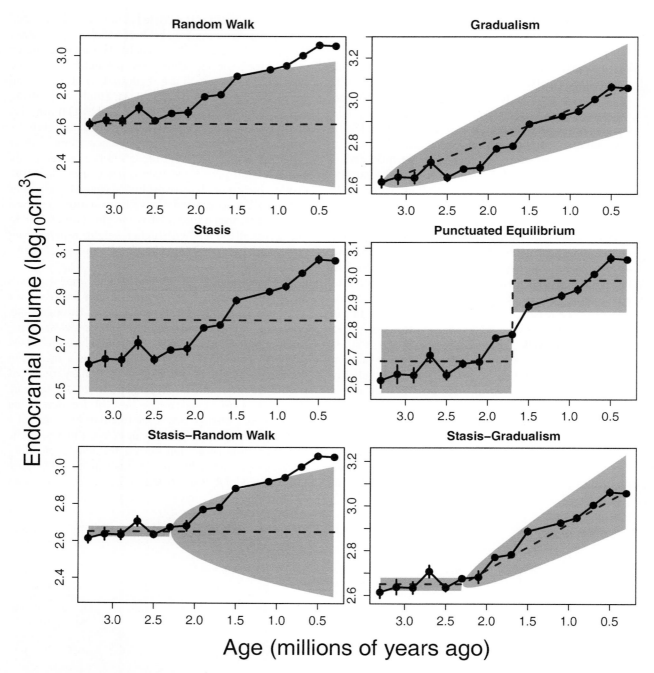

Figure 3. Hominin ECV model fits for all six evolutionary modes. Plotting details for each model are the same as in Figure 2b, and the gradualism model fit is the same. This figure is reproduced from Du et al. (2018).

suggest support for the gradualism model is robust (Du et al., 2018). As for the other models, while the estimated transition point for the stasis-stasis model is noisy (due to dating error) and either estimated to be between 2.0 and 1.8 Ma or 1.8 and 1.6 Ma, for the other two models it is between 2.4 and 2.2 Ma (Table 1; Figure 3). Although future discoveries may push the onset of the Acheulean back to between 2.4 and 2.2 Ma, gradualism's convincing model support indicates that there is no evolutionary mode shift describing clade-level ECV increase through time that coincides with the first appearance of the Acheulean.

It is worth discussing here the potential pitfalls of testing the presence of a c.1.8 Ma ECV saltation event (or any significant rate of ECV increase) at the clade level, if the c.1.8 Ma saltation event is only hypothesized to be in one lineage (i.e., *Homo habilis-Homo erectus sensu lato*). Phenotypic trends are scale-dependent in the sense that species-level saltations and trends can be concealed when looking at the larger clade level, especially if these are randomly distributed through time (Gould, 2002; McShea, 2004; Jablonski, 2007; Novack-Gottshall and Lanier, 2008; Du et al., 2018). Therefore, a clade-level c.1.8 Ma saltation is most clearly observed if, across all

Table 1. Shift point results for the stasis-stasis, stasis-random walk, and stasis-gradualism models. For each model, counts denote the number of times (out of 1,000 iterations) the estimated shift point falls in a given time bin. Iterations involve resampling age estimates for each hominin cranial specimen according to their age error distribution to account for dating error (see Du et al., 2018 for more details).

Models	2.4-2.2 Ma	2.2-2.0 Ma	2.0-1.8 Ma	1.8-1.6 Ma	1.6-1.4 Ma	1.4-1.2 Ma
Stasis-stasis	0	33	371	505	58	33
Stasis-random walk	956	44	0	0	0	0
Stasis-gradualism	943	49	8	0	0	0

lineages, notable upturns in the rate of ECV increase and/or saltation events were all concentrated at *c*.1.8 Ma. However, an ECV increase in one hominin lineage (e.g., as a saltation event) should still be observable at the clade level if it is not swamped by sampling and dating error or by ECV variation arising from changes in other lineages. The most straightforward test for a saltation event would be to analyze the one lineage where ECV is expected to increase, but, at least for the present, the reduction in sample size and temporal coverage that accompanies analyses of single lineages precludes this.

The lack of support for a *c*.1.8 Ma saltation event in brain size weakens the case for linking behavioral innovations, such as the onset of Acheulian tool making, with increases in hominin brain size during this time period. It does not negate Gowlett's suggestion that stone technology provides a window into the minds of the early hominins, and the imaging studies of Stout and colleagues (reviewed in Stout and Khreisheh, 2015) are elegant evidence of how brain activity can be harnessed to explore these connections. But it does suggest that at the level of the hominin clade, the size of the organ that generates what we refer to as cognition was evolving gradually and not in fits and starts. If there were cognitive innovations, then, on the basis of present evidence, at least one of these innovations was not driven by a synchronic spurt in the size of the brain. It seems that any increase in cognitive complexity associated with the manufacture of the Acheulian more likely involved establishing novel neural connections among the existing complement of neurons (e.g., Stout et al., 2015), rather than an increase in the absolute numbers of neurons.

Conclusion

The singular cognitive and technical abilities of modern humans are an obvious and tempting target for investigation. So it is understandable that researchers have sought, and continue to seek, to explain why modern humans have the wherewithal to contribute to and publish volumes about distinguished archeologists, whereas our closest living relatives do not. But 'why' questions are rarely, if ever, scientifically tractable (Smith and Wood, 2017), and it is to John Gowlett's

credit that his career has sensibly focused on 'what' and 'how' questions. Gowlett has shown time and time again that he makes wise choices about his research agenda. Good judgment is the essence of good science.

Acknowledgements

We thank Dietrich Stout for advice, and BW thanks the GW Provost for support for the University Professor of Human Origins.

References

Aiello, L.C. and P. Wheeler. 1995. The Expensive-Tissue Hypothesis: The brain and the digestive system in human and primate evolution. *Current Anthropology* 36: 199–221.

Almécija, S. and C.C. Sherwood. 2017. Hands, Brains, and Precision Grips: Origins of Tool Use Behaviors, in J.H. Kaas (ed.) *Evolution of Nervous Systems*: 299–315. 2nd ed. Elsevier.

Ambrose, S.H. 2001. Paleolithic technology and human evolution. *Science* 291: 1748–1753.

Antón, S.C., R. Potts and L.C. Aiello. 2014. Evolution of early *Homo*: An integrated biological perspective. *Science* 345: 1236828–1236828.

Blumenberg, B. 1978. Hominid ECV versus time: Available data does not permit a choice of model. *Journal of Human Evolution* 7: 425–436.

Bookstein, F.L. 1987. Random Walk and the Existence of Evolutionary Rates. *Paleobiology* 13: 446–464.

Burnham, K.P. and D.R. Anderson. 2002. *Model Selection and Multimodel Inference: A Practical Information-Theoretic Approach*. 2nd ed. New York: Springer.

Chamberlin, T.C. 1890. The method of multiple working hypotheses. *Science* 15: 92–96.

Cronin, J.E., N.T. Boaz, C.B. Stringer and Y. Rak. 1981. Tempo and mode in hominid evolution. *Nature* 292: 113–122.

Darwin, C. 1871. *The Descent of Man, and Selection in Relation to Sex*. London: J. Murray.

De Miguel, C. and M. Henneberg. 2001. Variation in hominid brain size: How much is due to method? *HOMO - Journal of Comparative Human Biology* 52: 3–58.

Deaner, R.O., K. Isler, J. Burkart and C. van Schaik. 2007. Overall brain size, and not encephalization quotient, best predicts cognitive ability across non-human primates. *Brain, Behavior and Evolution* 70: 115–124.

deMenocal, P.B. 2004. African climate change and faunal evolution during the Pliocene-Pleistocene. *Earth and Planetary Science Letters* 220: 3–24.

Diniz-Filho, J.A.F., Jardin, L., Mondanaro, A. and Raia, P. 2019. Multiple Components of Phylogenetic Non-stationarity in the Evolution of Brain Size in Fossil Hominins. *Evolutionary Biology* 46: 47-59

Diogo, R. J. L. Molnar and B. Wood. 2017. First detailed anatomy study of bonobos reveals higher rates of human evolution and bonobos as best model for our common ancestor. *Scientific Reports*. 7: 608.

Diogo, R. and B. Wood. 2011. Soft-tissue anatomy of the primates: phylogenetic analyses based on the evolution of the head, neck, pectoral region and upper limbs, with notes on the evolution of these muscles. *Journal of Anatomy*. 219: 273-359.

Du, A., A.M. Zipkin, K.G. Hatala, E. Renner, J.L. Baker, S. Bianchi, K.H. Bernal and B.A. Wood. 2018. Pattern and process in hominin brain size evolution are scale-dependent. *Proceedings of the Royal Society of London B: Biological Sciences* 285: 20172738.

Eldredge, N. and S.J. Gould. 1972. Punctuated equilibria: an alternative to phyletic gradualism, in T.J.M. Schopf (ed.) *Models in Paleobiology*: 82–115. San Francisco: Freeman, Cooper and Company.

Gingerich, P.D. 1993. Quantification and comparison of evolutionary rates. *American Journal of Science* 293: 453–478.

Godfrey, L. and K.H. Jacobs. 1981. Gradual, autocatalytic and punctuational models of hominid brain evolution: A cautionary tale. *Journal of Human Evolution* 10: 255–272.

Gould S.J. 2002. *The Structure of Evolutionary Theory*. Cambridge, MA: Belknap Press of Harvard University Press.

Gould, S.J. and N. Eldredge. 1977. Punctuated equilibria: The tempo and mode of evolution reconsidered. *Paleobiology* 3: 115–151.

Gowlett, J.A.J., 2006. The early settlement of northern Europe: fire history in the context of climate change and the social brain. *Comptes Rendus Paleovol* 5 (1-2): 299-310.

Gowlett, J.A.J. 1984. Mental abilities of early man: a look at some hard evidence. In Foley, R.A. (ed.) Hominid Evolution and Community Ecology. Academic Press, New York, pp 167-192.

Grabowski, M. 2016. Bigger Brains Led to Bigger Bodies?: The Correlated Evolution of Human Brain and Body Size. *Current Anthropology* 57: 174–196.

Harmand, S., J.E. Lewis, C.S. Feibel, C.J. Lepre, S. Prat, A. Lenoble, X. Boës, R.L. Quinn, M. Brenet, A. Arroyo, N. Taylor, S. Clément, G. Daver, J.-P. Brugal, L. Leakey, R.A. Mortlock, J.D. Wright, S. Lokorodi, C. Kirwa, D.V. Kent and H. Roche. 2015. 3.3-million-year-old stone tools from Lomekwi 3, West Turkana, Kenya. *Nature* 521: 310–315.

Hofman, M.A. 1983. Encephalization in hominids: Evidence for the model of punctuationalism. *Brain, Behavior and Evolution* 22: 102–117.

Hunt, G. 2006. Fitting and comparing models of phyletic evolution: random walks and beyond. *Paleobiology* 32: 578–601.

Hunt, G. 2008a. Gradual or pulsed evolution: When should punctuational explanations be preferred? *Paleobiology* 34: 360–377.

Hunt, G. 2008b. Evolutionary patterns within fossil lineages: Model-based assessment of modes, rates, punctuations and process, in P.H. Kelley and R.K. Bambach (eds) *From Evolution to Geobiology: Research Questions Driving Paleontology at the Start of a New Century*: 117–131. Pittsburgh: Paleontological Society Papers. Paleontological Society.

Hunt, G. 2015. paleoTS: analyze paleontological time-series. R package version 0.5-1. See cran.rproject.org/web/packages/paleoTS/.

Hunt, G., M.J. Hopkins and S. Lidgard. 2015. Simple versus complex models of trait evolution and stasis as a response to environmental change. *Proceedings of the National Academy of Sciences* 112: 4885–4890.

Hunt, G.R. 1996. Manufacture and use of hook-tools by New Caledonian crows. *Nature* 379: 249-251.

Isaac, G.L.L. 1972. Early phases of human behavior: models in Lower Palaeolithic archaeology, in D.L. Clarke (ed.) *Models in Archaeology*: 167-199. Routledge.

Jablonski, D. 2007. Scale and hierarchy in macroevolution. *Palaeontology* 50: 87–109.

Lee, S.-H. and M.H. Wolpoff. 2003. The pattern of evolution in Pleistocene human brain size. *Paleobiology* 29: 186–196.

Lepre, C.J., H. Roche, D.V. Kent, S. Harmand, R.L. Quinn, J.-P. Brugal, P.-J. Texier, A. Lenoble and C.S. Feibel. 2011. An earlier origin for the Acheulian. *Nature* 477: 82.

Lestrel, P.E. and D.W. Read. 1973. Hominid cranial capacity versus time: A regression approach. *Journal of Human Evolution* 2: 405–411.

Levinton, J.S. 1982. Estimating Stasis: Can a Null Hypothesis be too Null? *Paleobiology* 8: 307–307.

McPherron, S.P., Z. Alemseged, C.W. Marean, J.G. Wynn, D. Reed, D. Geraads, R. Bobe and H.A. Béarat. 2010. Evidence for stone-tool-assisted consumption of animal tissues before 3.39 million years ago at Dikika, Ethiopia. *Nature* 466: 857–860.

McShea, D.W. 2004. A revised Darwinism. *Biology and Philosophy* 19: 45–53.

Millard, A.R. 2008. A critique of the chronometric evidence for hominid fossils: I. Africa and the Near East 500–50ka. *Journal of Human Evolution* 54: 848–874.

Miller, I.F., Barton, R-A. and Nunn, C.L. 2019. Quantitative uniqueness of human brain evolution revealed through phylogenetic comparative analysis. *eLife* 8:e41250

Novack-Gottshall P.M. and M.A. Lanier. 2008. Scale-dependence of Cope's rule in body size evolution of Paleozoic brachiopods. *Proceedings of the National Academy of Sciences* 105: 5430–5434.

Oakley, K.P. 1949. *Man the toolmaker*. London: The Trustee's of the British Museum.

Oakley, K.P. 1957. Tools Makyth Man. *Antiquity*. 31: 199-209.

Pagel, M. 2002. Modelling the evolution of continuously varying characters on phylogenetic trees, in N. MacLeod and P.L. Forey (eds) *Morphology, Shape and Phylogeny*: 269–286. CRC Press.

Panger, M.A., A.S. Brooks, B.G. Richmond and B. Wood. 2002. Older than the Oldowan? Rethinking the emergence of hominin tool use. *Evolutionary Anthropology* 11: 235–245.

Penfield, W. and T. Rasmussen. 1950. *The cerebral cortex of man: a clinical study of localization of function*. New York: The Macmillan Company.

Platt, J.R. 1964. Strong inference. *Science*. 146: 347-353.

Potts, R. 1996. Evolution and Climate Variability. *Science* 273: 922–923.

Raup, D.M. 1977. Stochastic Models in Evolutionary Palaeontology, in A. Hallam (ed.) *Developments in Palaeontology and Stratigraphy*: 59–78. Elsevier.

Rightmire, G.P. 1981. Patterns in the evolution of *Homo erectus*. *Paleobiology* 7: 241–246.

Roopnarine, P.D. 2001. The description and classification of evolutionary mode: a computational approach. *Paleobiology* 27: 446–465.

Sanz, C., Call, J. and D. Morgan. 2009. Design complexity in termite-fishing tools of chimpanzees (*Pan troglodytes*). *Biology Letters*: rsbl.2008.0786.

Shultz, S. and R.I.M. Dunbar. 2010. Species differences in executive function correlate with hippocampus volume and neocortex ratio across nonhuman primates. *Journal of Comparative Psychology* 124: 252–260.

Shultz, S., E. Nelson and R.I.M. Dunbar. 2012. Hominin cognitive evolution: identifying patterns and processes in the fossil and archaeological record. *Philosophical Transactions of the Royal Society B: Biological Sciences* 367: 2130–2140.

Smith, R.J. and B. Wood. 2017. The principles and practices of human evolution research: Are we asking questions that can be answered? *Comptes Rendus Paleovol*. 16: 670-679.

Spoor, F., P. Gunz, S. Neubauer, S. Stelzer, N. Scott, A. Kwekason and M.C. Dean. 2015. Reconstructed *Homo habilis* type OH 7 suggests deep-rooted species diversity in early *Homo*. *Nature* 519: 83.

Stout, D., E. Hecht, N. Khreisheh, B. Bradley and T. Chaminade. 2015. Cognitive Demands of Lower Paleolithic Toolmaking. *PLOS ONE* 10: e0121804.

Stout, D. and N. Khreisheh. 2015. Skill Learning and Human Brain Evolution: An Experimental Approach. *Cambridge Archaeological Journal* 25: 867–875.

Striedter, G.F. 2005. *Principles of Brain Evolution*. Sunderland, Mass.: Sinauer Associates.

Tobias, P.V. 1971. *The Brain in Hominid Evolution*. New York: Columbia University Press.

Vrba, E.S. 1996. Climate, Heterochrony, and Human Evolution. *Journal of Anthropological Research* 52: 1–28.

Washburn, S.L. 1959. Speculations on the Interrelations of the History of Tools and Biological Evolution. *Human Biology*. 31: 21-31.

Wood, B. and M. Collard. 1999. The Human Genus. *Science* 284: 65-71.

Andrew Du
Colorado State University
Department of Anthropology & Geography
1787 Campus Delivery
Fort Collins, CO80523
USA
Andrew.Du2@colostate.edu

Bernard Wood
Center for the Advanced Study of Human Paleobiology
800 22nd St NW
Suite 6000
Washington, District of Columbia 20052
USA
bernardawood@gmail.com

Australopithecus or *Homo*? The Postcranial Evidence

Robin H. Crompton

Abstract

The distinction between *Homo* and *Australopithecus* is becoming increasingly blurred. This paper reviews the fossil evidence pertaining to the distinction with respect to the postcranial skeleton. It asks whether *Australopithecus* is a valid genus from the locomotor perspective, given a lack of locomotor distinction between some material referred to Early African *Homo* and *Australopithecus*; and given that the earliest stone tool industry, the Lomekwian, considerably antedates the earliest *Homo*. In the postcranial skeleton, only loss of anterior iliac flare seems to suggest any possible clear distinction. Neither is there, it appears, clear evidence of a step-change in cranial capacity. Unless the gnathocranial evidence for validity of the genus *Australopithecus* is held to be overwhelming, serious thought should be given to sinking *Australopithecus* into *Homo*.

Main body

This review honours John's work on the first culturally formalized and transmitted technology of *Homo*, the Acheulian (to which I have had the pleasure of contributing), by reviewing the boundaries of *Homo* from a postcranial, rather than as more usual, gnathocranial, (see eg Schwartz and Tattersall 2015) perspective. While the Acheulian seems to be characterized by an almost perverse determination to produce the classic ovate handaxe whatever the material, suggesting strong isochrestic traditions and perhaps therefore a human-like mind, earlier technologies, the Oldupan and the more recently recognized Lomekwian are apparently not. While it has not been much considered whether repeated and consistent stone tool shapes might be lacking in the Oldupan or Lomekwian because of limitations in hand morphology of the hominin toolmaker (although Oldupan discussed in Susman [1991] and reviewed in de la Torre [2011]), Domalain *et al.* (2017) have attempted to argue that the hand of AL 288-1 *Australopithecus afarensis* at least could not have exerted enough force to make the Lomekwian tools. Both from a biomechanical and motor control perspective, this seems *prima facie* unlikely given that the function of both hominin hands and hominin feet, comprised of many joints crossed by many ligaments, tendons and muscles, are almost certainly characterised by very high degrees of

'functional abundance' ('functional redundancy' as was [Bernstein 1967]). Recent finds show that *Au. afarensis,* at least as represented by AL-288-1, is postcranially rather atypical of contemporary australopiths. This is witnessed particularly by the Woranso-Mille (Haile-Selassie *et al.* 2010) and Sterkfontein Member 2 (see eg Clarke 1998, 1999. 2002, 2003; Clarke and Tobias, 1995) hominins and the Laetoli S footprint trails (Masao *et al.* 2016), which indicate that other australopiths were less obviously distinct from ourselves in stature and proportions. Thus it seems that a review of the postcranial evidence for the boundaries of *Homo* is timely and appropriate.

George Gaylord Simpson (1963) admonished the palaeanthropological community for poor taxonomic practice, which he noted most of its members recognised: but only in others. Little appears to have changed since then. Simpson noted that while species can readily be defined as groups of actually or potentially interbreeding animals, taxa above the species level are more difficult to define. The situation is worse today since we now know that in nature species do sometimes interbreed and produce fertile offspring, which has led to alternative definitions of species being proffered, such as the 'recognition concept' of Paterson (1981). In practice, today, most workers would accept that genera are taxonomic groups of animals diagnosed by distinct phylogeny, morphology and adaptation. Each genus has a type species, to which other potential members of that genus should be compared. That for our own genus, *Homo* is *Homo sapiens*, ourselves. Unhelpfully, the diagnosis for *Homo sapiens* is 'know thyself' (*nosce te ipsum*) (Linnaeus, 1735 reviewed in Schwartz and Tattersall, 2015). While the first Javan *Homo erectus* were assigned their own genus, *Pithecanthropus* by Dubois (1896), and Dart (1926, 1953) named *Australopithecus africanus* for the Taungs skull, australopiths were sunk into *Homo* by Mayr (1950) on the basis of a shared adaptation to upright bipedality. Robinson (1954, page 182) notes that Mayr argues that '*man's great ecological diversity*' explains '*this puzzling trait of the hominid stock to stop speciating in spite of its eminent evolutionary success*'. Robinson (1954) however returned to a genus-level distinction of australopiths: '*one cannot agree with Mayr that in the early stages of euhominid development intraspecific variability was even greater than at the present time. The australopithecines had not yet reached the artefact manufacturing stage and were, on the whole, rather small brained.'* (Robinson 1954, page 183).

The authoritative, most formal and most detailed early taxonomic work on relationships within Hominoidea was that of Le Gros Clarke (1955) which remains one of the most cited.

He diagnosed Hominidae thus:

(only the first, locomotor features are given here: the full diagnosis is provided in Supplementary Material 1)

'—a subsidiary radiation of the Hominoidea distinguished from the Pongidae by the following evolutionary trends; progressive skeletal modifications in adaptation to erect bipedalism, shown particularly in a proportionate lengthening of the lower extremity, and changes in the proportions and morphological details of the pelvis, femur, and pedal skeleton related to mechanical requirements of erect posture and gait and to the muscular development associated therewith; preservation of well developed pollex; ultimate loss of opposability of hallux; (Le Gros Clark 1955, p 110)

and distinguishes them from Family Pongidae, in which he groups all living great apes apart from ourselves, which he diagnoses thus (and see Supplementary Materials 2 for the full diagnosis):

'—a subsidiary radiation of the Hominoidea distinguished from the Hominidae by the following evolutionary trends: progressive skeletal modifications in adaptation to arboreal brachiation, shown particularly in a proportionate lengthening of the upper extremity as a whole and of its different segments; acquisition of a strong opposable hallux and modification of morphological details of limb bones for increased mobility and for the muscular developments related to brachiation; tendency to relative reduction of pollex; pelvis retaining the main proportions characteristic of quadrupedal mammals' (Le Gros Clarke, 1955, pp. 111-112)

Interestingly, within Hominidae, Le Gros Clarke diagnoses *Homo* firstly and mostly on gnathocranial features (see Supplementary Material 3) but notes:

'limb skeleton adapted for a fully erect posture and gait.' (Le Gros Clarke 1955, p. 79).

Finally, he diagnoses *Australopithecus* again firstly on gnathocranial features, but notes :

'the limb skeleton (so far as it is known) conforming in its main features to the hominid type but differing from Homo *in a number of details, such as the forward prolongation of the region of the anterior superior spine of the ilium and a relatively small sacro-iliac surface, the relatively low position (in some individuals) of the ischial tuberosity, the marked forward prolongation of the intercondylar notch of the femur, and the medial extension of the head of the talus.'* (Le Gros Clarke 1955, p 156)

Thus, all taxa below the superfamily Hominoidea were diagnosed in part by locomotor features, and these were leading features in his diagnoses of both Hominidae and Pongidae, where progressive adaptations for erect bipedalism in hominids were contrasted with progressive adaptations for suspensory (brachiating) locomotion in pongids, foreshadowing the recent evidence that suspensory adaptations are recent and independent adapations in the great ape lineage (see *eg.* Almécija *et al.* 2007) Locomotor features are less prominent in diagnoses for *Homo* and *Australopithecus*, and, despite Robinson's (1954) use of artefact manufacture to distinguish *Homo* from *Australopithecus*, toolmaking does not enter into Le Gros Clarke's diagnosis. Only his (now untenable, see *eg.* Miguel and Henneberg 2001) limitation of brain-size to 450-700 cc in *Australopithecus*, versus a mean in excess of 1,100 cc in *Homo*, reflects cognitive features. It is notable that Le Gros Clarke (1955) commented on a progressive tendency to reduce the taxonomic separation of *Homo* from (other) apes, and the likelihood that this would continue.

The discovery of OH-5 ('*Zinjanthropus*') in Bed I at Olduvai in 1959 overshadowed the slightly earlier discovery of isolated teeth from one much more gracile individual, OH-4. Leakey (1959) was in no doubt about declaring that *Zinjanthropus* was the maker of the Olduwan industry. However, in 1959-1960 further teeth and skull fragments of another individual (OH-6); mandible, upper molars, parietal bones and hand bones of a further individual (OH-7); a partial proximal foot (OH-8) and juvenile cranial fragments (OH-14) were discovered in Bed 1. These suggested the existence of a far more gracile, larger brained human ancestor with postcanine teeth considerably smaller and less complex than those of OH-5, paenocontemporaneous with the robust '*Zinjanthropus*' (OH-5). In 1964, Leakey, Tobias and Napier placed these fossils within *Homo*, naming a new species *Homo habilis* (apparently at Robinson's suggestion) on the basis of the apparent manual dexterity suggested by the OH-7 hand, OH-7 being declared as the type specimen. They accepted Le Gros Clarke's (1955) diagnosis of Hominidae, but in order to include the Olduvai Bed 1 material, had to revise that for *Homo,* reducing minimal brain capacity below 1100 cc, down to as little as 600 cc. They now reattributed the Oldupan technology of Bed I to *Homo habilis* rather than to 'Zinjanthropus' (Tobias, 1965). Robinson [1965] argued that *Homo habilis* was a mixture of *Australopithecus* like material from Bed I and *Homo erectus* like material (eg OH-13) from Bed II, but in 1971, the discovery of a far more intact cranium, OH-24, near the bed of Bed I (Leakey et al 1971), countered this argument. Since then, the existence of a valid species, *Homo habilis,* has been generally accepted. The 1964 revised diagnosis of *Homo* by Leakey, Tobias and Napier is as follows (note that in their diagnosis gnathocranial features [for which see Supplementary Material 5] are given *after* locomotor features, suggesting these authors, *unlike* Le Gros Clarke, regard them as *less* important in the diagnosis of *Homo*:

> '*A genus of the Hominidae with the following characters: the structure of the pelvic girdle and of the hind-limb skeleton is adapted to habitual erect posture and bipedal gait; the fore-limb is shorter than the hind-limb; the pollex is well developed and fully opposable and the hand is capable not only of a power grip but of, at the least, a simple and usually well developed precision grip*' (Leakey, Tobias and Napier 1964, pages 7-8).

Note that tool-making *per se* does not appear in the diagnosis: in this respect Leakey, Tobias and Napier follow Le Gros Clarke (1955). The recent stricture of Schwartz and Tattersall (2015) suggesting that the Leakey, Tobias and Napier re-diagnosis was over-influenced by Oakley's (1949) concept of 'Man the Toolmaker' thus seems a little excessive.

Habitual bipedal gait, often regarded as suggested by the OH-8 foot, is far more central to the *Homo habilis* diagnosis, and a purported contrast between 'habitual' (*Homo*) and 'facultative' (*Australopithecus*, other apes) bipedalism has remained commonplace until quite recently (*eg.* Harcourt-Smith, 2007). An obvious difficulty with the revised diagnosis is that the only evidence pertinent to habitual bipedalism in the holotype of *Homo habilis* at that time was the OH-8 foot, which was not associated with other skeletal evidence, and indeed there was no possible evidence of relative hindlimb and forelimb length in the area at that time. It is further questionable whether hindlimb length does exceed forelimb length in all sub recent and modern human populations, although formal studies still appear to be yet unperformed. But a much more recent discovery at Olduvai (Johansen et al. 1987) generally assigned to *Homo habilis*, that of the highly fragmentary OH-62, suggests that body proportions were more similar to some australopiths than to other *Homo* (Richmond et al. 2002). Further, Wood (1974) argues that the geometry of the OH-8 talus resembles the TM

1517 *Paranthropus robustus* talus from Swartkrans and that OH-8 should be assigned to the East African robust australopith, *Australopithecus boisei*, that to the Olduvai 'Zinjanthropus'.

The OH-8 foot has remained a key source of information on the locomotion of early human ancestors, in addition to the Sterkfontein Sts 14 partial axial skeleton, partial pelvis, and partial femur; and robust and gracile australopith material from other sites in or near the Vredefort craton/Cradle of Humankind zone. It is relevant to note that this material was mostly referred to *Homo africanus* by Robinson in his (1972) monograph, and regarded as that of an efficient upright biped, although he argued that the robust asutralopiths were less well adapted to upright bipedalism and substantially arboreal. Prior to the discovery of the 30% complete skeleton of AL 288-1, this material stood as our best source on the postcranial skeleton of australopiths.

Whether it belongs to *Homo habilis* or *Australopithecus* (*Zinjanthropus*) boisei, the OH-8 foot, which lacks the anterior part due to carnivore or scavenger action, has been interpreted in very different ways (discussed in Crompton et al. 2008), either as essentially human-like, with transverse and longitudinal arches *etc.* (Day and Napier 1964) or as in part 'ape'-like, with no medial arch (Kidd *et al.* 1996). Others describe the foot morphology as a 'mosaic' of adaptive features, but suggest the medial arch was weight-bearing (Harcourt-Smith and Aiello, 2004). There have been many discussions or claims made with respect to the arches of the foot within palaeoanthropology (*eg.* Ward *et al.*, 2011) but few rigorous investigations of their functional/ biomechanical significance. In fact, since two-thirds of us produce mid-foot pressures in excess of the clinical threshold for flat-foot bilaterally within 5 minutes of treadmill walking (Bates *et al.* 2013), it is clear that a strong medial arch is far from being a defining characteristic of the healthy *H. sapiens* foot. High intraspecific variability is inevitable in the function of the extremities as a consequence of the high numbers of joints, ligaments, tendons and muscles, and hence high functional abundance *sensu* Bernstein 1967.

The functional significance of the AL-288-1 skeleton of *Au. afarensis* has been the subject of even more dispute than the OH-8 foot, largely because of its combination of long forelimbs with generally human-like knee morphology. Some regarded the latter as more significant and conclude that it was as a functionally efficient terrestrial biped (*eg.* Latimer, 1991). Others regard the forelimb morphology as signalling arboreality. The latter supposedly would have compromised its terrestrial bipedalism, leading to a 'bent-hip, bent-knee' gait (*eg.* Susman *et al.* 1984 and Susman and Stern, 1991). We have shown, using computer simulation, that AL 288-1 was, on the contrary, a rather effective fully upright biped (Crompton *et al.* 1998, 2003; Sellers *et al.* 2003, 2004, 2005). Lordkipanze's (2007) use of upright bipedalism to define *Homo* (even alongside 'moderately large' [p. 47] brains is thus completely unsustainable. Footprints, representing a record of interaction between the feet and ground sediments, might be expected to provide a secondary source of information on function. We have performed a topographical statistical analysis of all 22 usable records in the Laetoli G-1 trail attributed to *Au. afarensis* and together with forwards dynamic modelling based on the proportions of AL 288-1, this confirms that the tracks were made by an effective fully upright biped (Crompton *et al.* 2012). Further confirmation flows from statistical tests by Raichlen and Gordon (2017) which also included some data from the newly discovered Laetoli S trail. Together they obviate the contrary claim of Hatala and colleagues (2016), who selected just 5 of the 22 usable prints on which to base their stdy and conclusions.

As the *Au. afarensis* hypodigm has increased, it has become evident that AL 288-1 was a particularly small individual. Her limb proportions are beyond the range of even the smallest living humans, which do tend to have shorter hindlimbs than Western populations, as do many populations with relatively poor nutrition, but (Jungers and Baab, 2009) appear a close match for those of *H. floresiensis*: the so-called 'hobbits', from as little as 50-60 Kyr ago (Sutikna *et al.* 2016). Note that not only AL 288-1 but also Sts 14 are small individuals (Jungers and Baab, 2009). The most complete postcranium within the generally accepted hypodigm of *Homo habilis*, the very fragmented and incomplete OH-62, is also particularly small (Jungers and Baab 2009) and also has long forelimbs compared to hindlimbs, like AL 288-1. This has been interpreted as indicating a more arboreal habitus (*cf.* AL 288-1). If the 3.66 Ma Laetoli footprint trails were made by *Au. afarensis*, as seems likely, then the S trail indicates that the makers of G1, G2 and S had a great range of body size, comparable with our own (Masao *et al.* 2016). Our knowledge of the postcranial skeleton of australopiths has recently been much enlarged, by the discovery of two skeletons, on the one hand from Woranso-Mille in Ethiopia (Haile-Selassie *et al.* 2010), and the other, far more complete, from Sterkfontein Member 2. The Woranso-Mille skeleton, KSD-VP-1/1 is fragmentary, but might have had limbs of subequal size, although a foot from a different locality (Haile-Selassie *et al.*, 2012), shows evidence of greater hallucal divergence than in AL 288-1. The Woranso-Mille KSD-VP-1/1 skeleton is that of a male, short, but probably well within the modern Western human height range, and dates to about 3.4 Ma, similar to the age of AL 288-1. The Sterkfontein StW573 female skeleton, from Member 2, dated to around 3.67 Ma (Granger *et al.*, 2015) is also

probably within the modern human height range. The limbs of this by far most complete of all early hominins unquestionably show (Clarke, 2002) that the forelimb is shorter than the hindlimb, the first time this is known to be the case in human evolution (Crompton *et al.* 2018); 2.17 Ma. earlier than the previously known earliest occurrence of this phenomenon in *Homo ergaster*, KNM WT-15000. The hallux may be relatively divergent (Clarke and Tobias, 1995), although quite likely within the modern human functional range (Crompton *et al.* 2018) and the hand capable of a strong grasp (Clarke 1999) if deep soft precision grips possibly limited by a long apical ridge on the trapezium (Crompton *et al.* 2018). Both KSD-VP-1/1 and StW573 are described as substantially arboreal upright bipeds (Haile-Selassie *et al.* 2010; Clarke, 2003, Crompton *et al.* 2018).

As suggested by Clarke (2013), the hypodigm of *Homo habilis* may be better represented by *Homo rudolfensis* KNM ER-1470 than by OH-62 (taking into consideration also the possibly associated ER-1470 femora and ER-1500 incomplete tibia and radius, which if from the same skeleton might tend to suggest subequal limb proportions). The earliest generally accepted *Homo habilis*, as recognised by gnathocranial evidence, currently dates to 2.8 Ma, at Ledi Geraru (Villmoare *et al.* 2015). Cut-marks on bone, providing evidence for meat-butchering (de Heinzelin *et al.* 1999), were found near a cranium attributed to *Au. garhi* from to 2.5 Ma. (BOU-VP- 12/130, Asfaw et al. 1999). Also nearby, a partial humeral shaft, partial radius and ulna, and partial femur (BOU-12/1) were discovered . The humerofemoral index resembles that of *Homo ergaster*, KNM WT-15000, but (if correctly reconstructed) the proportion of the humeral length to ulnar length is orang-utan-like (Richmond *et al.* 2002). All this material, even Ledi Geraru (Villmoare *et al.* 2015), is at least 0.5 Ma younger than the 3.3 Ma Lomekwi 3 stone tools from West Turkana (Harmand *et al.* 2015). Clearly, the hominins who made this industry -- and, if they have been correctly interpreted, the cut marks on bone at Dikika (3.39 Ma, McPherron et al., 2010) -- had hands which could exert sufficient *controlled* force to utilize stone tools, and make them, even if they were relatively informal.

We have seen that *Au. afarensis*' limb morphology and proportions were not necessarily typical of other early hominins of the time, such as the Woranso-Mille and Sterkfontein member 2 hominins. The latter, though not associated with (and not claimed to have made) stone tools, had a powerful manual grip which might have been *exaptive* for stone tool manufacture (Clarke, 1999). Thus, the argument of Domalain et al. (2017), based on modelling of *Au. afarensis* AL 288-1 (*c.* 3.4 Ma) hand morphology, that the proportions of the hypothenar and fifth-digit bones would have prevented sufficient controlled force to be exerted, although ingenious, has little broad significance. We have noted that human foot function is highly plastic. A high number of joints, ligaments, tendons and muscles in both the hands and feet signals 'functional abundance' (Bernstein, 1967) and a great number of possible locomotor and manipulative strategies to achieve any required action, underlying 'variation without variation' (Bernstein, 1967) step-to-step and grasp-by-grasp. Bernstein's Dynamic Systems Theory is now accepted as the basis of motor control and motor skills learning in mainstream biomechanics, as the theory of neurobiological degeneracy (see eg. Van Emmerick and van Wegen, 2002, for a readily accessible review), and urgently needs to be taken on board by the palaeoanthropological community (reviewed in McClymont, 2017).

Thus, some early hominins, at 3.67-3.4 Ma, unquestionably had postcranial morphology appropriate for both effective fully upright bipedal walking on the ground, and stone-tool manufacture, even if there is no evidence that they had a stone-tool industry, as is the case for StW 573. They combined this with a capacity for arboreal climbing, which should also be unsurprising, given clear evidence that humans retain sufficient locomotor plasticity (Venkataraman *et al.,* 2013) to engage in arboreal foraging (see *eg.* Kraft *et al.* 2014) irrespective of body build.

A change in abdomino-thoracic shape between *Australopithecus* and *Homo erectus sensu lato* might however be signalled by change in pelvic form (reduction of the anterior iliac flare which does seem very common in australopiths, including StW 573). However, to the best of my knowledge, there is to date no evidence that this change is in place in material referred to, or referable to *Homo habilis* (such as *Homo rudolfensis, eg.* KNM ER 1470, Clarke, 2013). Indeed, the reconstructed biacromial and bi-iliac dimensions of a second *Au prometheus* pelvis from Sterkfontein, StW 431, are near-identical, suggesting strongly that in this species at least, in contradistinction to *Au. afarensis* AL-288-1 ' Lucy' the thorax was barrel-shaped, like our own, not conical, like hers (Crompton *et al.* 2018).

It has been argued that environmental instability may have been a driver in evolution of early *Homo* (Antón *et al.* 2002, Antón 2003, Antón *et al.* 2014) but the step-changes in brain size we traditionally associate with *Homo* do not commence until some 1.8 Ma, at the time of *Homo ergaster* (Schultz *et al.* 2012). Moreover, the reality of step changes is challenged both by the high levels of variation in estimates of cranial capacity by method, and by the good fit of cranial capacity to a smooth exponential curve from 3.3 Mya to the present (De Miguel and Henneberg, 2001). Thus, dietary shifts, and gnathocranial responses to them, may have been

slow in impacting brain size via energetic , social or other routes, and ultimately impacting cultural norms for technology (and see Wood and Collard 1999) Brain-size is simply not a reliable diagnostic for a separation of *Homo* and *Australopithecus*, as the case of the *c.* 400 *cc.* brainsize of *H. floresiensis* (Jungers and Baab, 2009) reminds us.

Wood and Collard (1999, 2001) and Wood and Lonergan (2008); have further acknowledged that a family-level distinction between *Homo, Australopithecus* and *Ardipithecus* (and perhaps also by extension *Orrorin* and *Sahelanthropus*) as Hominidae and *Pan* as a member of Pongidae is unjustifiable.

Further the ancient separation of orang-utans from the African apes (>13 Ma) and the *c.* 10.5 Ma separation of gorillas from the human/chimpanzee clade (reviewed in Crompton et al. 2008, 2010 and Crompton, 2015) versus the 5-8 Ma separation of human and chimpanzee clades (Bradley, 2008) makes a pongid family, opposed to a hominid family, untenable. If we accept a hominid family including all living great apes (including *Homo*) and opposed to a hylobatid family for the lesser apes (gibbons, siamangs etc), a division which would be supported by the molecular clock, then a tripartite division of living great apes and their fossil relatives into Ponginae (orang-utans); Gorillinae (gorillas) and Homininae (humans and chimpanzees) is reasonable (if not completely consistent with splitting patterns, as it fails to recognize a probable African ape clade: humans, chimpanzees and gorillas and their fossil relatives). Homininae would require to be split into two tribes, Hominini and Panini, (Wood and Collard, 1999, 2001; Wood and Lonergan, 2008) as adopted in this paper. *Orrorin* (Senut et al. 2001 and Senut 2006) and *Sahelanthropus* (Brunet et al. 2002) sit very close to that division, but the weight of evidence places them into Hominini. *Oreopithecus* (Moyà-Solà et al. 1999; Rook et al. 1999, Köhler and Moyà-Solà 2003,) also seems to fall into Hominini, but its age, *circa* 7-9 Ma, is at the upper end of the molecular dating range for the chimpanzee-human split (Bradley 2008). But there is now abundant evidence that knucklewalking is recently derived (Dainton and Macho, 1999; Kivell and Schmitt 2009), independently in *Pan* and *Gorilla* and *Pan* secondarily specialized in forelimb morphology (Drapeau and Ward, 2007) and forelimb/hindlimb ratios remarkably closely matched for effective high-speed quadrupedalism, (Isler *et al.,* 2006). Postcranially at least, the common ancestor of *Pan* and *Homo* was thus likely more like *Homo* (perhaps excluding lower limb elongation) than like *Pan*, and the common great-ape ancestor probably an arboreal biped (Crompton *et al.* 2008, 2010 and Crompton *et al* 2016), which used its forelimbs for hand-assisted bipedality (*Thorpe et al.* 2007) as the recent discovery of *Danuvius* (Böhme *et al.* 2019) clearly indicates.

Great apes are generally highly plastic in muscular capacity and locomotor performance (Myatt et al. 2011, 2012). As we have argued, and extending the MacLatchy and colleagues (2000) and Harrison (2002) concept of 'crown hominoids' the ancient (18-22 Ma) orthogrady of *Morotopithecus* (MacLatchy et al. 2000), and the lack of suspensory adaptations in *Pierolapithecus* (12.5-13 Ma, Moyà-Solà et al. 2004, 2005; Almécija *et al.* 2009) indicate that whenever the Hylobatidae first became suspensory the common *hominid* ancestor (in the new sense, including all great apes and their ancestors) was an arboreal biped and climber (see Almécija et al. 2007), which the discovery of *Danuvius* confirms. Le Gros Clark's (1955) diagnoses of the Pongidae and Hominidae are no longer tenable. Further, given the abundant evidence, reviewed briefly here, that with respect to the locomotor system, early hominins were highly variable, and even taking *Homo floresiensis* and present locomotor plasticity within *Homo sapiens* into account, probably more variable than modern and recent *Homo*, it seems that Mayr (1950) was right in suggesting sinking *Australopithecus* into *Homo*. Only gnathocranial changes suggest an *Australopithecus/Homo* distinction. Some, such as Schwartz and Tattersall (2015), who oppose the 'adaptationist' interpretation of evolution genus *Homo,* expressed in Antón *et al.* (2014), seem to hold that they are the prime consideration, presumably as teeth (rather less the cranium) are not plastic in life, possibly slow to adapt, and hence good phylogenetic indicators. However, Schwartz and Tattersall's (2015) review does not really leave the impression that there are indeed clear taxonomic markers in hominin gnathocranial morphology. Thus, following Mayr (1950) and more recently Groves (2001) perhaps it is now time for *Australopithecus* to be sunk into *Homo. Homo* was not the first effective upright biped and not the first toolmaker. Neither locomotion nor -- since a precision grip may first have appeared in *Oreopithecus* (Moyà-Solà et al, 1999), not *Homo habilis* -- manipulative/tool-making capabilities serve to distinguish genus *Homo* from genus *Australopithecus*. Whether development and evolution of cultural norms later affected or indeed, even effected, speciation within genus *Homo* remains to be determined (and see also Wood and Collard 1999).

Acknowledgements

I thank The Leverhulme Trust in particular for continued support of my human evolution research, currently via a Leverhulme Emeritus Fellowship, I have also been funded in this area by The Natural Environment Research Council, The Biotechnology and Biological Sciences Research Council, The Engineering and Physics Research Council and The Medical Research Council. I thank Juliet McClymont and Sarah Elton, for help with references and for stimulating conversations, and two anonymous reviewers for their support and detailed checking of my text. I particularly salute my friend and colleague Ron Clarke whose patient and determined excavation of *Au. prometheus* StW 573 over 20 years in the face of the impatience and hostility of less scrupulous palaeoanthropologists promises to revolutionize our understanding of early human ancestors. This paper is dedicated to the memory of my close friend and colleague Mr. Russell Savage and our many happy evenings in 'Gait Lab 2' during which many of the ideas in this paper were first conceived.

References

Almécija, A., Alba, D.M., Moyà-Solà, S., Köhler, M. 2007. Orang-like manual adaptations in the fossil hominoid *Hispanopithecus laietanus*: first steps towards great ape suspensory behaviours. *Proc R Soc B* 274, 2375–2384.

Almécija, A., Alba, D.M., Moyà-Solà, S., Köhler, M. 2009. Pierolapithecus and the functional morphology of Miocene ape hand phalanges: paleobiological and evolutionary implications. *J Hum Evol* 57, 284–297.

Antón, S.C., Leonard, W.R., Robertson, M.L. 2002. An ecomorphological model of the initial hominid dispersal from Africa. *J Hum Evol* 43, 773–785.

Antón, S.C. 2003. Natural history of *Homo erectus. Yrbk Phys Anthropol* 46, 126–170.

Antón, S.C., Potts, R., Aiello, L.C. 2014. Evolution of early *Homo:* An integrated biological perspective 345, 6192, 45-52

Asfaw, B., White, T.D., Lovejoy, C.O., Latimer, B., Simpson, S., Suwa, G. 1999. *Australopithecus garhi*: a new species of early hominid from Ethiopia. *Science* 284, 629–635.

Bates, K.T., Collins, D., Savage, R., McClymont, J., Webster, E., Pataky, T.C., D'Aout, K., Sellers, W.I., Bennett, M.R., Crompton, R.H. 2013. The evolution of compliance in the human lateral mid-foot. *Proc R Soc B,* 280 (1769), p. 20131818 1-7

Bernstein, N. 1967. *The co-ordination and regulation of movements.* Oxford: Pergamon.

Böhme, M., Spassov, N., Fuss, J., Tröscher, A., Deane, A.S., Prieto, J., Kirscher, U., Lechner, D., Begun, D.R. 2019. A new Miocene ape and locomotion in the ancestor of great apes and humans. *Nature* 575 489-491.

Bradley, B.J. 2008. Reconstructing phylogenies and phenotypes: a molecular view of human evolution. *J Anat* 212, 337–353.

Brunet, M., Guy, F., Pilbeam, D., Mackaye, H.T., Likius, A., Ahounta, D., Beauvilain, A., Blondel, C., Bocherens, H., Boisserie, J.R., De Bonis, L. 2002. A new hominid from the Upper Miocene of Chad, Central Africa. *Nature* 418, 145–151.

Clarke, R.J. 1998. First ever discovery of a well-preserved skull and associated skeleton of *Australopithecus. S Afr J Sci* 94, 460–463.

Clarke R.J. 1999. Discovery of a complete arm and hand of the 3.3 million-year-old *Australopithecus* skeleton from Sterkfontein. *S Afr J Sci* 95, 477–480.

Clarke, R.J. 2002. Newly revealed information on the Sterkfontein Member 2 *Australopithecus* skeleton from Sterkfontein. *S Afr J Sci* 98, 523–526.

Clarke, R.J. 2003. Bipedalism and arboreality in *Australopithecus. Cour Forsch-Inst Senckenberg* 243, 79–83.

Clarke, R. 2013. *Australopithecus from Sterkfontein Caves, South Africa*. In: The Paleobiology of Australopithecus, Vertebrate Paleobiology and Paleoanthropology. (eds Reed, K.E., Fleagle, J.G., Leakey, R.E.F.), pp. 105–123, Dordrecht, Heidelberg: Springer Science+Business Media doi: 10.1007/978-94-007-5919-0_7.

Clarke, R.J., Tobias, P.V. 1995. Sterkfontein Member 2 foot bones of the oldest South African hominid. *Science* 269, 521–524.

Crompton, R.H. 2016. The hominins: a very conservative tribe? Last common ancestors, plasticity and ecomorphology in Hominidae. Or, What's in a name? *J Anat* 228(4),686-699.

Crompton, R.H., Li, Y., Wang, W., Günther, M.M., Savage, R. 1998. The mechanical effectiveness of erect and 'bent-hip, bent-knee' bipedal walking in *Australopithecus afarensis. J Hum Evol* 35, 55–74.

Crompton, R.H., Thorpe, S.K.S., Wang, W., Li, Y., Payne, R., Savage, R., Carey, T., Aerts, P., Van Elsacker, L., Hofstetter, A., Günther, M. The biomechanical evolution of erect bipedality. *Cour Forsch- Inst Senckenberg* 243, 115–126.

Crompton, R.H., Vereecke, E.E., Thorpe, S.K.S. 2008. Locomotion and posture from the last common hominoid ancestor to fully modern hominins, with special reference to the last common panin-hominin ancestor. *J Anat* 212, 501–543.

Crompton, R.H., Sellers, W.I., Thorpe, S.K.S. 2010. Arboreality, Terrestriality and Bipedalism. *Phil Trans R Soc Lond B* 365, 3301–3314.

Crompton, R.H., Pataky, T.C., Savage, R., D'Août, K., Bennett, M.R., Day, M.H., Bates, K., Morse, S., Sellers, W.I. 2012. Human-like external function of the foot, and fully upright gait, confirmed in the 3.66 million year old Laetoli hominin footprints by topographic statistics, experimental footprint-formation and computer simulation. *J R Soc Interface* 9, 707–719.

Crompton, R.H., McClymont, J., Heaton, J., Pickering, T., Sellers, W.I., Thorpe, S.K.S., Pataky, T., Stratford, D., Carlson, K., Jashashvili, T., Beaudet, A., Elton, S., Bruxelles, L., Goh, C., Kuman, K., Clarke, R.J. 2018. Ecomorphology of the *Australopithecus prometheus* skeleton, StW573 3.67 Ma, from Sterkfontein Caves, South Africa. BioRxiv: http://dx.doi.org/10.1101/481556.

Dainton, M., Macho, G.A. 1999. Did knuckle walking evolve twice? *J Hum Evol* 36, 171–194.

Dart, R. 1926. Taungs and its significance. *Nat Hist* 115, 875.

Dart, R. 1953. The predatory transition from Ape to Man. *Int Anthropol Linguistic Rev* 1, 201–217.

Day, M.H., Napier, J.R. 1964. Fossil foot bones. *Nature* 201, 969–970.

DeMiguel, C., Henneberg, M. 2001. Variation in hominid brain size: How much is due to method? *HOMO* Vol. 52/1 3–58

de la Torre, I. 2011. The origins of stone tool technology in Africa: a historical perspective. *Phil Trans Roy Soc B* 366, 1028-1037.

de Heinzelin, J., Clark, J.D., White, T., Hart, W., Renne, P., WoldeGabriel, G., Beyene, Y., Vrba, E. 1999. Environment and behavior of 2.5-million-year-old Bouri hominids *Science* 284, 5414, 625-629.

Domalain, M., Bertin, A., Daver, G., 2017. Was *Australopithecus afarensis* able to make the Lomekwian stone tools? Towards a realistic biomechanical simulation of hand force capability in fossil hominins and new insights on the role of the fifth digit *C. R. Palevol* 16 572–584

Drapeau, M.S.M., Ward, C.V. 2007. Forelimb segment length proportions in extant hominoids and *Australopithecus afarensis. Am J Phys Anthropol* 132, 327–343.

Dubois, E. 1896. On *Pithecanthropus Erectus*: A Transitional form Between Man and the Apes *J. Anthrop. Inst. Great Britain and Ireland* 25, 240-255

Granger, D.E., Gibbon, R.J., Kuman, K., et al. 2015. New cosmogenic burial ages for Sterkfontein Member 2 *Australopithecus* and Member 5 Oldowan. *Nature* 522, 85–88.

Groves, C. 2001. *Primate Taxonomy*. Washington: The Smithsonian Institution

Haile-Selassie, Y., Latimer, B.M., Alene, M., Deino, A.L., Gibert, L., Melillo, S.M., Saylor, B.Z., Scott, G.R., Lovejoy, C.O. 2010. An early *Australopithecus afarensis* postcranium from Woranso-Mille, Ethiopia. *Proc Natl Acad Sci USA* 107, 27, 12121-12126.

Haile-Selassie, Y., Saylor, B.Z., Deino, A., Levin, N.E., Alene, M., Latimer, B.M. 2012. A new hominin foot from Ethiopia shows multiple Pliocene bipedal adaptations. *Nature* 483, 565–569.

Harcourt-Smith, W.E.H., Aiello, L.C. 2004. Fossils, feet and the evolution of human bipedal locomotion. *J Anat* 204, 403–416.

Harcourt-Smith, W.E.H. 2007. The Origins of Bipedal Locomotion. *Handbook of Palaeoanthropology* (Hencke W, Tattersall I eds.) Berlin: Springer.

Harmand, S., Lewis, J.E., Feibel, C.S., Lepre, C.J., Prat, S., Lenobe, A., Boës, X., Quinn, R.L., Brenet, M., Arroyo, A., Taylor, N., Clément, C., Daver, G., Brugal, J-P., Leakey, L., Mortlock, R.A., Wright, J.D., Lokorodi, S., Kirwa, C., Kent, D.V., Roche, H. 2015. 3.3-million-year-old stone tools from Lomekwi 3, West Turkana, Kenya *Nature* 521, 310-314.

Harrison, T. 2002. Late Oligocene to Middle Miocene catarrhines from Afro-Arabia. In *The Primate Fossil Record* (ed. Hartwig, W.C.), pp. 311–338. Cambridge: Cambridge University Press.

Harrison, T., Rook, L. 1997. Enigmatic anthropoid or misunderstood ape? The phylogenetic status of *Oreopithecus bambolii* reconsidered. In: *Function,*

Phylogeny, and Fossils – Miocene Hominoid Evolution and Adaptations (Begun, D.R., Ward, C.V., Rose, M.D. eds), pp. 327–362. New York: Plenum Press.

Hatala, K.G., Demes, B., Richmond, B.G. 2016. Laetoli footprints reveal bipedal gait biomechanics different from those of modern humans and chimpanzees. *Proc. R. Soc. B* 283: 20160235 1-8. http://dx.doi.org/10.1098/rspb.2016.0235

Isler, K., Payne, R.C., Günther, M.M., Thorpe, S.K., Li, Y., Savage, R., Crompton, R.H. 2006. Inertial properties of hominoid limb segments. *J Anat* 209, 201–218.

Johansen, D.C., Masao, F.T., Eck, G.G., White, T.D., Walter, R.C., Kimbel, W.G., Asfaw, B., Manega, P., Ndessokia, P., Suwa, G. 1987. New partial skeleton of *Homo habilis* from Olduvai Gorge, Tanzania *Nature* 327, 205-209.

Jungers, W.L. 2009. Interlimb proportions in humans and fossil hominins: variability and scaling. In: *The First Humans: Origins of the Genus Homo* (F.E. Grine, R. E. Leakey and J.G. Fleagle, eds .Springer: Dordrecht), pp. 93–98.

Jungers, W., Baab, K. 2009. The geometry of hobbits: *Homo floresiensis* and human evolution. *Significance* , December. 159-164

Kidd, R.S., O'Higgins, P.O., Oxnard, C.E. 1996. The OH-8 foot: a reappraisal of the hindfoot utilizing a multivariate analysis. *J. Hum. Evol.* 31, 269–291.

Kivell, T.L., Schmitt, D. 2009. Independent evolution of knucklewalking in African apes shows that humans did not evolve from a knuckle-walking ancestor. *Proc Natl Acad Sci USA* 106, 14241–14246.

Köhler, M., Moyà-Solà, S. 2003. Understanding the enigmatic ape *Oreopithecus bambolii. Cour Forsch-Inst Senckenberg* 243, 111–123

Kraft, T.S., Venkataraman, V.V., Dominy, N.J. 2014. A natural history of human tree climbing. *J Hum Evol* 71, 105–118.

Latimer, B. 1991. Locomotor adaptations in *Australopithecus afarensis*: the issue of arboreality. In: *Origine(s) de la Bipedie Chez les Hominides* (eds Senut, B., Coppens, Y.), pp. 169–176. Paris: CNRS.

Leakey, L.S.B. 1959. A New Fossil Skull from Olduvai. *Nature* 4685 491-493.

Leakey, L.S.B., Tobias, P.V., Napier, J.R. 1964. A New Species of the Genus *Homo* from Olduvai Gorge. *Nature, 4927*, 7-9.

Leakey, M.D., Clarke, R.J., Leakey, L.S.B. 1971. New hominid skull from bed I, Olduvai Gorge, Tanzania *Nature* 232, 308-312.

Le Gros Clark, W.E. 1955. The Fossil Evidence for Human Evolution. Chicago: University of Chicago Press.

Linnaeus, C. 1735. *Systema naturae per regna tria naturae, secundum classes, ordines, genera, species cum characteribus, differentiis, synonymis, locis* (Laurentii Salvii, Stockholm).

Lordkipanze, D. 2007. The History of Early *Homo.* In: On Human Nature (Tibayrenc, M., Ayala, F.J. eds.) Amsterdam: Elsevier pp. 47-54

Mayr, E. 1950. Taxonomic categories in fossil hominids. *Cold Spring Harbor Symposia on Quantitative Biology* 15, 109-118.

MacLatchy, L., Gebo, D., Kityo, R., Pilbeam, D. 2000. Postcranial functional morphology of *Morotopithecus bishopi*, with implications for the evolution of modern ape locomotion. *J Hum Evol* 39, 159–183.

Masao, F.T., Ichumbaki, E.B., Cherin, M., Barili, A., Boschian, G., Iurino, D.A., Menconero, S., Moggi-Cecchi, J., Manzi, G. 2016. New footprints from Laetoli (Tanzania)provide evidence for marked body size variation in early hominins *eLife* 5:e19568. DOI: 10.7554/eLife.19568 1-29.

McPherron, S.P., Alemseged, Z,, Marean. C.W., Wynn, J.G., Reed, D., Geraads, D., Bobe, R,, Bereat, H.A. 2010. Evidence for stone-tool-assisted consumption of animal tissues before 3.39 million years ago at Dikika, Ethopia *Nature* 466, 857-860.

McClymont, J. 2016. *Foot Pressure Variability and Locomotor Plasticity in Hominins* PhD Thesis, The University of Liverpool .

Moyà-Solà, S., Köhler, M., Rook, L. 1999. Evidence of a hominid like precision grip capability in the hand of the Miocene ape *Oreopithecus. Proc Natl Acad Sci USA* 96, 313–317.

Moyà-Solà, S., , Köhler, M., Alba, D.M., Casanovas-Vilar, I., Galindo, J. 2004. *Pierolapithecus catalaunicus*, a new Middle Miocene great ape from Spain. *Science* 306, 1339–1344.

Moyà-Solà, S., Köhler, M., Alba, DM., Casanovas-Vilar, I., Galindo, J. 2005. Response to comment on 'Pierolapithecus catalaunicus, a new Middle Miocene great ape from Spain'. *Science* 308, 203d.

Myatt, J.P., Crompton, R.H., Thorpe, S.K.S. 2011. Hindlimb muscle architecture in non-human great apes and a comparison of methods for analyzing inter-species variation. *J Anat* 219, 150–166.

Myatt, J.P., Crompton, R.H., Payne-Davis, R.C., Vereecke, E.E., Isler, K., Savage, R., D'Août, K., Günther, M.M., Thorpe, S.K.S. 2012. Functional adaptations in the forelimb muscles of nonhuman great apes. *J Anat* 220, 13–28.

Oakley, K. 1949. *Man the Toolmaker* London: British Museum (Natural History),

Paterson, H.E.H. 1981. The Continuing Search for the Unknown and Unknowable: A Critique of Contemporary Ideas on Speciation. *S Afr J Sci* 77 113-119.

Raichlen, D.A., Gordon, A.D. 2017. Interpretation of footprints from Site S confirms human-like bipedal biomechanics in Laetoli hominins *J Hum Evol.* 107, 134-138

Richmond, B.G., Aiello, L.C., Wood, B.A. 2002. Early hominin limb proportions. *J Hum Evol* 43, 529–548.

Robinson, J.T. 1954. The Genera and Species of the Australopithecinae. *Am J Phys Anthropol* 12, 181.

Robinson, J.T. 1965. *Homo habilis* and the Australopithecines. *Nature* 205, 121.

Robinson, J.T. 1972. *Early Hominid Posture and Locomotion.* London: The University of Chicago Press.

Rook, L., Bondioli, L., Köhler, M., Moyà-Solà, S., Macchiarelli, R. 1999. *Oreopithecus* was a bipedal ape after all: evidence from the iliac cancellous architecture. *Proc Natl Acad Sci USA* 96, 8795–8799.

Schultz, S., Nelson, E., Dunbar, R.I.M. 2012. Hominin cognitive evolution: identifying patterns and processes in the fossil and archaeological records. *Phil Trans Roy Soc Lond B* 367, 2130-2140.

Schwartz, J.H., Tattersall, I. 2015. Defining the genus *Homo*. *Science* 349, 6251, 931-932.

Sellers, W.I., Dennis, L.A., Crompton, R.H. 2003. Predicting the metabolic energy costs of bipedalism using evolutionary robotics. *J Exp Biol* 206, 1127–1136.

Sellers, W.I., Dennis, L.A., Wang, W., Crompton, R.H. 2004. Evaluating alternative gait strategies using evolutionary robotics. *J Anat* 204, 343–351.

Sellers, W.I., Cain, G., Wang, W.J., Crompton, R.H. 2005. Stride lengths, speed and energy costs in walking of *Australopithecus afarensis*: using evolutionary robotics to predict locomotion of early human ancestors. *J Roy Soc Interface* 2, 431–442.

Senut, B., Pickford, M., Gommery, D., Mein, P., Cheboi, K., Coppens, Y. 2001. First hominid from the Miocene (Lukeino formation, Kenya). *CR Acad Sci Paris* Ser IIA 332, 137–144.

Senut, B. 2006. Arboreal origin of bipedalism. In: Human Origins and Environmental Backgrounds (eds Ishida, H., Tuttle, R., Pickford, M., Ogihara, N., Nakatsukasa, M.), pp. 199–208. Berlin: Springer.

Simpson, G.G. 1965. The Meaning of Taxonomic Statements. In: Classification and Human Evolution (Washburn, S. ed.) Chicago: Aldine pp 1-31,

Susman, R.L. 1991. Who Made the Oldowan Tools? Fossil Evidence for Tool Behavior in Plio-Pleistocene Hominids. *J. Anthropological Research*, 47, 2, 129-151

Susman, R.L., Stern, J.T. 1991. Locomotor behavior of early hominids: epistemology and fossil evidence. In: Origine(s) de la Bipedie Chez les Hominides (Senut, B., Coppens, Y. eds), pp. 121–131. Paris: CNRS.

Susman, R.L., Stern, J.T., Jungers, W. 1984. Arboreality and bipedality in the Hadar hominids. *Folia Primatol* 43, 113–156.

Sutikna, T., Tocheri, M.W., Morwood, M.J., Saptomo, E.W., Jatmiko, D., Awe, R., Wasisto, S., Westaway, K.E., Aubert, M., Li, B., Zhao, J-X., Storey, M., Alloway, B.V., Morley, M.W., Meijer, H.J.M., van den Bergh, G., Grün, R., Dosseto, A., Brumm, A., Jungers, W.L., Roberts, R.G. 2016. Revised stratigraphy and chronology for *Homo floresiensis* at Liang Bua in Indonesia *Nature* 532, 366–369.

Thorpe, S.K., Holder, R.L., Crompton, R.H., 2007. Origin of human bipedalism as an adaptation for locomotion on flexible branches. *Science,* 316 5829, 1328-1331.

Tobias, P.V. 1965. *Australopithecus, Homo Habilis,* Tool-Using and Tool-Making *S. Afr. Arch. Bull.* 20, 80, 167-192

van Emmerik, R., van Wegen, E. 2002. On the functional aspects of variability in postural control. *Exerc. Sport Sci. Rev.,* 30, 177-183.

Venkataraman, V.V., DeSilva, J.M., Dominy, N.J. 2013. Phenotypic Plasticity of Climbing-Related Traits in the Ankle Joint of Great Apes and Rainforest Hunter- Gatherers. *Human Biol.* 85, 1, 309-328.

Villmoare, B., Kimbel, W.H., Seyoum, C., Campisano, C.J., DiMaggio, E., Rowan, J., Braun, D.R., Arrowsmith, J.R., Reed, K.E. 2015. Early *Homo* at 2.8 Ma from Ledi-Geraru, Afar, Ethiopia *Sciencexpress* 10.1126/science. aaa1343, 1352-1355.

Walker, A., Leakey, R.E. 1993b. The postcranial bones. In The Nariokotome *Homo erectus* Skeleton (Walker, A., Leakey, R.E. eds.), pp. 95–160. Cambridge, MA: Harvard University Press.

Ward, C.V., Kimbel, W.H., Johanson, D.C. 2011. Complete fourth metatarsal and arches in the foot of *Australopithecus afarensis*. *Science* 331, 750–753.

Wood, B.A. 1974. A *Homo* talus from East Rudolf, Kenya. *J. Anat* 117, 203–204.

Wood, B.A., Collard, M. 1999. Is *Homo* defined by culture? *Proc. British Academy* 99: 11-23.

Wood, B.A. Collard, M. 2001. The Meaning of *Homo*. *Ludus Vitalis* 9, 15 63-74.

Wood, B.A., Lonergan, N. 2008. The hominin fossil record: taxa, grades and clades. *J Anat* 212, 354–376.

Supplementary Material: full Le Gros Clarke and Leakey, Tobias and Napier diagnoses

1) Hominidae

'—a subsidiary radiation of the Hominoidea **distinguished from the Pongidae by the following evolutionary trends; progressive skeletal modifications in adaptation to erect bipedalism, shown particularly in a proportionate lengthening of the lower extremity, and changes in the proportions and morphological details of the** *pelvis, femur, and pedal skeleton related to mechanical requirements of erect posture and gait and to the muscular development associated therewith; preservation of well developed pollex; ultimate loss of opposability of hallux; increasing flexion of basicranial axis associated with increasing cranial height; relative displacement forward of the occipital condyles; restriction of nuchal area of occipital squama, associated with low position of inion; consistent and early ontogenetic development of a pyramidal mastoid process; reduction of subnasal prognathism, with ultimate early disappearance (by fusion) of facial component of premaxilla;*

diminution of canines to a spatulate form, interlocking slightly or not at all and showing no pronounced sexual dimorphism; disappearance of diastemata; replacement of sectorial first lower premolars by bicuspid teeth (with later secondary reduction of lingual cusp) ; alteration in occlusal relationships, so that all the teeth tend to become worn down to a relatively flat even surface at an early stage of attrition; development of an evenly rounded dental arcade; marked tendency in later stages of evolution to a reduction in size of the molar teeth; progressive acceleration in the replacement of deciduous teeth in relation to the eruption of permanent molars; progressive "molarization" of first deciduous molar; marked and rapid expansion (in some of the terminal products of the hominid sequence of evolution) of the cranial capacity, associated with reduction in size of jaws and area of attachment of masticatory muscles and the development of a mental eminence.' (Le Gros Clark 1955, p 110)

2) Pongidae

*'—a subsidiary radiation of the Hominoidea **distinguished from the Hominidae by the following evolutionary trends: progressive skeletal modifications in adaptation to arboreal brachiation, shown particularly in a proportionate lengthening of the upper extremity as a whole and of its different segments; acquisition of a strong opposable hallux and modification of morphological details of limb bones for increased mobility and for the muscular developments related to brachiation; tendency to relative reduction of pollex; pelvis retaining the main proportions characteristic of quadrupedal mammals;** marked prognathism, with late retention of facial component of premaxilla and sloping symphysis; development (in larger species) of massive jaws associated with strong muscular ridges on the skull; nuchal area of occiput becoming extensive, with relatively high position of the inion; occipital condyles retaining a backward position well behind the level of the auditory apertures; only a limited degree of flexion of basicranial axis associated with maintenance of low cranial height; cranial capacity showing no marked tendency to expansion; progressive hypertrophy of incisors with widening of symphysial region of mandible and ultimate formation of "simian shelf"; enlargement of strong conical canines interlocking in diastemata and showing distinct sexual dimorphism; accentuated sectorialization of first lower premolar with development of strong anterior root; post canine teeth preserving a parallel or slightly divergent alignment in relatively straight rows; first deciduous molar retaining a predominantly unicuspid form; no acceleration in eruption of permanent canine.'* (Le Gros Clarke, 1955, pp. 111-112)

3) Homo

A genus of the family Hominidae, distinguished mainly by a large cranial capacity with a mean value of more than 1,100 cc. , supra-orbital ridges variably developed, becoming secondarily much enlarged to form a massive torus in the *species* H. neanderthalensis, *and showing considerable reduction in* H. sapiens; *facial skeleton orthognathous or moderately prognathous; occipital condyles situated approximately at the middle of the cranial length; temporal ridges variable in their height on the cranial wall, but never reaching the mid-line; mental eminence well marked in* H. sapiens *but feeble or absent in* H. neanderthalensis; *dental arcade evenly rounded, with no diastema; first lower premolar bicuspid with a much reduced lingual cusp; molar teeth rather variable in size, with a relative reduction of the last molar; canines relatively small, with no overlapping after the initial stages of wear;* **limb skeleton adapted for a fully erect posture and gait.***' (Le Gros Clarke 1955, p. 79.

4) Australopithecus

*'a genus of the Hominidae distinguished by the following characters : relatively small cranial capacity, ranging from about 450 to about 700 cc. ; strongly built supra-orbital ridges; a tendency in individuals of larger varieties for the formation of a low sagittal crest in the frontoparietal region of the vertex of the skull (but not associated with a high nuchal crest) ; occipital condyles well behind the mid-point of the cranial length but on a transverse level with the auditory apertures; nuchal area of occiput restricted, as in Homo; consistent development (in immature as well as mature skulls) of a pyramidal mastoid process of typical hominid form and relationships; mandibular fossa constructed on the hominid pattern but in some individuals showing a pronounced development of the postglenoid process; massive jaws, showing considerable individual variation in respect of absolute size; mental eminence absent or slightly indicated; symphysial surface relatively straight and approaching the vertical; dental arcade parabolic in form with no diastema; spatulate canines wearing down flat from the tip only; relatively large premolars and molars; anterior lower premolar bicuspid with subequal cusps; pronounced molarization of first deciduous molar; progressive increase in size of permanent lower molars from first to third; **the limb skeleton (so far as it is known) conforming in its main features to the hominid type but differing from Homo in a number of details, such as the forward prolongation of the region of the anterior superior spine of the ilium and a relatively small sacro-iliac surface, the relatively low position (in some individuals) of the ischial tuberosity, the marked forward prolongation of the intercondylar notch of the femur, and the medial extension of the head of the talus.***'* (Le Gros Clarke 1955, p 156)

5) **Homo** *according to Leakey, Tobias and Napier (1964)*

- '**A genus of the Hominidae with the following characters: the structure of the pelvic girdle and of the hind-limb skeleton is adapted to habitual erect posture and bipedal gait; the fore-limb is shorter than the hind-limb; the pollex is well developed and fully opposable and the hand is capable not only of a power grip but of, at the least, a simple and usually well developed precision grip;**

the cranial capacity is very variable but is, on the average, larger than the range of capacities of members of the genus Australopithecus, *although the lower part of the range of capacities in the genus* Homo *overlaps with the upper part of the range in* Australopithecus; *the capacity is (on the average) large relative to body-size and ranges from about 600 c.c. in earlier forms to more than 1,600 c.c.; the muscular ridges on the cranium range from very strongly marked to virtually imperceptible, but the temporal crests or lines never reach the midline; the frontal region of the cranium is without undue post-orbital constriction (such as is common in members of the genus* Australopithecus; *the supraorbital region of the frontal bone is very variable, ranging from a very salient, supra-orbital torus to a complete lack of any supra-orbital projection and a smooth brow region; the facial skeleton varies from moderately prognathous to orthognathous, but it is not concave (or dished) as is common in members of the* Australopithecinae; *the anterior symphyseal contour varies from a marked retreat to a forward slope, while the bony chin may be entirely lacking, or may vary from slight to very strongly developed mental trigone; the dental arcade is evenly rounded with no diastema in most members of the genus; the first lower premolar is clearly bicuspid with variably developed lingual cusp'.* (Leakey, Tobias and Napier 1964 pages 7-8).

Robin H. Crompton
Leverhulme Emeritus Fellow
School of Archaeology, Classics and Egypotology
University of Liverpool
and
Institute of Ageing and Chronic Disease
Aintree University NHS Trust
robinhuwcrompton@gmail.com

Evolutionary Diversity and Adaptation in Early *Homo*

Alan Bilsborough and Bernard Wood

Introduction

Our knowledge of the East African early hominin fossil record is mostly a product of discoveries made in the last half century, a period that is only a decade or so longer than John Gowlett's involvement in the study of that region's prehistory. This paper focuses on the hominin fossil record referred to early *Homo*, and considers changes in its interpretation over that period. These changes reflect additions to the fossil record, developments in evolutionary theory and debates about how phylogeny and adaptation should be reflected in taxonomy. We therefore provide a brief historical summary of the relevant discoveries before reviewing current views on the material assigned to early *Homo*.

Prior to the early 1960s most interpretations of human evolution were founded on a severely typological approach to the hominin fossil record, resulting in candelabra-like phylogenies with multiple lineages which, given the paucity of the fossil record, in some cases included taxa represented by only a single specimen. However, a paradigm shift occurred from around the mid-twentieth century onwards when the 'New Synthesis' (Huxley 1942), which successfully married population genetics with Darwinian mechanisms, began to influence areas of biological anthropology, including palaeoanthropology. In particular, studies exposed and documented the substantial amount of intraspecific variation in *Homo sapiens* serving as the 'raw material' for selection and hence evolution. The realization of how much variation was subsumed within modern humans led to a marked reduction in the number of both taxa and lineages recognised in the hominin fossil record and a correspondingly greater emphasis on anagenetic change. Bernard Campbell (1962; 1965) Theodosius Dobzhansky (1962; 1963), Ernst Mayr (1950; 1963) and Sherwood Washburn (1950; 1963) were particularly influential voices in promoting a 'lumping' taxonomy for hominins while Simpson (1953; 1963) offered broader paleontological perspectives. In retrospect it is also evident that — so far as the study of human variability was concerned — there was also a definite socio-political dimension to this shift of evolutionary perspective in the sense that it was also a response to racist interpretations of human diversity.

At the same time field studies had drawn attention to the importance of ecological perspectives and the need to consider hominins within the context of their habitats and as part of the broader mammalian community. Howell and Bourlière (1963) provided a pioneering overview, with others focusing on the distinctive nature of the human ecological niche, and the role of culture in defining that niche (Oakley, 1949). Some, following Gauss's Exclusion Principle, argued that the acquisition of culture made the hominin niche (not just that of modern humans) so broad that there was no ecological 'space' for two sympatric species, hence only one culture-bearing species could exist at any one time (Wolpoff 1971). Because this interpretation relegated what previously had been regarded as species-level differences to within-species variation, the upshot of this thinking was the replacement of taxic diversity and polyphyletic phylogenies with a single, anagenetic, hominin lineage. Within this new framework *Homo erectus*, which had evolved from *Australopithecus africa*nus, was viewed as the most primitive member of the *Homo* genus. There was also a general, albeit informal, agreement that a 'cerebral rubicon' of 750 cm^3 (Keith 1948) represented the *Homo* threshold with respect to endocranial volume, with the smallest then known *H. erectus* braincase at 775 cm^3 and the largest gorilla at just under 700 cm^3.

Homo habilis

In 1964 Leakey, Tobias and Napier proposed a new species – *Homo habilis* – together with a revised definition of the genus *Homo*. They took these steps after assessing cranial, dental and some postcranial specimens (e.g., OH 7, 8, 13, 16, 48) recovered from Bed I and Lower Bed II at Olduvai Gorge. In their diagnosis of the new taxon they emphasised dental size and proportions (relatively large incisors and canines; relatively small cheek teeth with the lower ones, especially the premolars, narrow and mesiodistally elongated) and neurocranial morphology (endocranial volume larger than *Australopithecus*, overlapping with smaller *H. erectus*, and with slight to strongly marked cranial ridging, but lacking the latter's pronounced bony reinforcement). They also stressed that the hand bones of OH 7 indicated stout, curved fingers capable of powerful flexion and an opposable thumb, and they interpreted the well-developed lateral and transverse arches and adducted hallux of the OH 8 foot as adaptations for terrestrial bipedal locomotion. The species name *habilis* ('able, handy, mentally skillful'), which was suggested by Raymond Dart, was considered highly appropriate given the co-occurrence of early stone tools at Olduvai and the emphasis placed at that time on tool-making as a distinctively human (*sensu Homo*) trait.

Leakey *et al.* (1964) were strongly criticised for the creation of the new taxon. Several authors e.g., (Campbell 1964; 1967; Clark 1964; 1967; Oakley and Campbell 1964; Robinson 1965; 1966) claimed that there was insufficient morphological (and so taxonomic) space to accommodate a new species between *A. africanus* and *H. erectus*. Central to these critics' argument was the assumption of an ancestor-descendant relationship between these two taxa (see above), with the earlier evidence for *H. habilis* being more like its presumed precursor, *A. africanus*, and the temporally-later fossils more like *H. erectus*. The recovery in 1968 of a relatively complete, albeit crushed, cranium (OH 24) from the DK 1 site (M. D. Leakey *et al.* 1971) low down in the section at Olduvai, negated the argument for a temporal cline in morphology, and for many it confirmed the reality of *H. habilis* as a hominin more advanced than *Australopithecus*, but smaller- brained and with a more lightly constructed cranium than *H. erectus*. Details of the fossils' contexts, the stratigraphy of Beds I and II, and the associated archaeological record are provided in Leakey (1971). Recent refinements to the Bed I/Lower Bed II stratigraphy and dating (Deino, 2012) place the Olduvai *H. habilis* fossils in the interval 1.877-1.65 Ma.

The 1970s saw a further expansion of the early *Homo* fossil record with new discoveries in South Africa. Ron Clarke's reconstruction of the SK 847 partial cranium from Swartkrans Member 1 (<2.2 Ma) resulted in a *Homo*-like fossil cranium (Clarke *et al.* 1970) that has since been variously attributed to *H. habilis* (Howell 1978; Grine *et al.* 1993), *H. erectus* (Tobias 1991), *Homo gautengensis* (Curnoe 2010), or an unnamed *Homo* species comparable to, but distinct from, the East African taxa (Clark and Howell 1972; Grine *et al.* 1993). Another specimen, Stw 53, this time from Sterkfontein (Hughes and Tobias, 1977), which was originally thought to be from the tool-bearing Member 5 but is now interpreted as coming from an infill that may be as old as c.2.6 Ma (Pickering and Kramers, 2010), was added to the *H. habilis* hypodigm (Tobias 1991). However, others have argued that it belongs either in a different *Homo* species (Grine *et al.* 1996; Curnoe 2010), or to *A. africanus* (Kuman and Clarke 2000).

Koobi Fora and *Homo habilis* sensu lato

The main additions to the 1970's early *Homo* record were those made by Richard Leakey and colleagues from localities at Koobi Fora east of Lake Turkana - formerly Lake Rudolf, hence the prefix 'ER'(Leakey *et al.* 1978). Researchers refrained from specific attribution

until full descriptions and comparative studies were completed, so these fossils were initially assigned by their discoverers at the genus level only to either *Australopithecus* or *Homo*. However, this policy did not deter others from creating new taxa to accommodate the hominins recovered from Koobi Fora, including *Homo ergaster* for the KNM-ER 992 mandible matched with the KNM-ER 1805 and ER 1813 crania (Groves and Mazák 1975), *Pithecanthropus rudolfensis* (Alexeev 1986) for the KNM-ER 1470 cranium, and *Homo microcranous* (Ferguson 1995) for the KNM-ER 1813 cranium. Others clustered the finds without necessarily formalising the resulting groupings as taxa, with some schemes (e.g., Stringer 1986) dividing the Olduvai finds into subsets that matched separate Koobi Fora clusters.

Opinions rapidly coalesced around two interpretations. One was that the Koobi Fora early *Homo* finds could be included with the Olduvai evidence, plus other fossils, within a more inclusive interpretation of *H. habilis* sensu stricto that we label *H. habilis* sensu lato. An outstanding example of this 'broad' interpretation of *H. habilis* was Howell's treatment of the Olduvai and Koobi Fora material in his detailed survey of African hominins (Howell 1978). The other interpretation was that the extent and pattern of morphological diversity in the Koobi Fora evidence of early *Homo* indicated multiple species, of which *H. habilis* sensu stricto was one. In other words, there was widespread — although not universal - agreement (Chamberlain 1987; Chamberlain and Wood 1987) that the *H. habilis* hypodigm included at least some of the Koobi Fora early *Homo* specimens; the issue was whether other *Homo* species were also represented in that site sample. With hindsight it seems likely that advocates of a 'broad' interpretation of *H. habilis* continued to be influenced by a phyletic framework in which a single clade, subject to necessarily anagenetic processes , was the dominant evolutionary paradigm for early *Homo*. In this model taxa within the genus are considered to be arbitrary chronospecies.

Early *Homo* postcranial evidence

Olduvai Gorge had yielded several postcranial fossils from the same levels as cranial evidence of *H. habilis* sensu stricto. Although Koobi Fora had yielded many postcranial fossils, most were unassociated with cranial remains, thus leaving open the possibility that these individuals belonged to other taxa such as *Paranthropus boisei*. What was needed were postcranial bones that were part of an associated skeleton that included *H. habilis*-like cranial remains. In 1987 Johanson *et al.* assigned an incomplete and much abraded associated skeleton from Olduvai (OH 62) to *H. habilis* (Johanson *et al.* 1987). Following this announcement Leakey *et al.* (1989) drew attention to the earlier discovery at Koobi Fora of a larger, but less complete and also abraded partial skeleton (KNM- ER 3735), which they suggested was a male to OH 62's female. The OH 62 associated skeleton was assigned to *H. habilis* in part because of its location in the section, but also because of the ways its palate resembled that of Stw 53. But if Stw 53 is, in fact, *Australopithecus* (see above), then OH 62's assignment to *H. habilis* is in doubt. This is important since the Olduvai and Koobi Fora associated skeletons provide the only information on body size and proportions in *H. habilis*, with some claiming that it had distinctively primitive limb proportions (Hartwig-Scherer and Martin 1991; but see Haeusler and McHenry 2004; 2007) that contrasted markedly with other, broadly contemporary, fossil evidence that we discuss in the next section.

In addition to evidence of *H. habilis*, East Turkana sites yielded cranial and postcranial fossils of broadly *H. erectus*-like morphology but lacking some derived traits seen in Asian fossils of that species (e.g., KNM -ER 3733, KNM- ER 3883). Wood (1991) associated the crania with the KNM-ER 992 mandible, which Groves and Mazak (1975) had assigned to *H. ergaster*, so that these new cranial specimens were incorporated in the latter hypodigm. The taxon *H. ergaster* has gained wide but far from universal support , with other workers preferring to view these East African specimens as early examples of *H. erectus*. Because of the lack of consensus regarding *H. ergaster* from hereon we refer to this material as Early African *H. erectus* (EAHE). Most of our understanding of the EAHE postcranium comes from the KNM-WT 15000 specimen from Nariokotome, West Turkana (Brown *et al.* 1985; Walker and Leakey 1993). This virtually complete skeleton has proved exceptionally influential in forming views on the significance of EAHE morphology and, by inference, its niche and adaptations. Ward *et al.* (2015) have recently published an associated skeleton from Koobi Fora, KNM-ER 5881,that appears to depart from the morphology exemplified by KNM-WT 15000.

Older (2.5 Ma) postcranial fossils from Bouri (Ethiopia) lack direct association with cranial material of *A. garhi* known from elsewhere in the region (Asfaw *et al.* 1999) and which displays a mosaic of cranio-dental affinities with both *Australopithecus* and *Paranthropus*. Given their spatial and temporal proximity it is reasonable to also assign the postcrania to *A. garhi*. Irrespective of species identity the Bouri postcrania provide independent evidence of at least one basal Pleistocene hominin species that combined an elongated hindlimb with a powerful forelimb.

Cladistics and early *Homo* diversity

The development of mainframe computing in the 1960s facilitated the use of multivariate approaches for studying primate morphology and character covariation, but most studies were concerned with exploring functional affinities rather than phylogenetic relationships, and few studies focused on early *Homo*.

However by the mid-1980s the availability in English of Hennig's *Phylogenetic Systematics* (Hennig 1966), together with the introduction of progressively more powerful desk top computers and the development of bespoke software like MacClade, prompted researchers to pursue cladistic approaches to the study of the increasingly diverse early hominin fossil record, and thus initiated a consequent shift towards generating groupings based on hypotheses about shared derived traits rather than overall (phenetic) affinity.

These cladistics analyses resulted in differing interpretations of the early *Homo* hypodigm, with multiple specimen clusters reinforcing polyphyletic interpretations of the material. The outcomes of such analyses differed in detail, if only because they used different outgroups, different OTUs, different trait lists (e.g., dental versus non-dental characters), and different ways to score those traits (Wood and Baker 2011). Many of these studies made no *a priori* allowance for character inter-correlation, so for studies based on multiple characters from restricted anatomical regions there was the real possibility that the same feature was being incorporated into the analysis several times over, resulting in inadvertent differential weightings of traits.

Attempts to track temporal trends in early *Homo* were also bedeviled by changes in the age of the KBS Tuff — a crucial marker for several important *Homo* specimens. Early results suggested that the tuff dated to 2.61 Ma and so fossils such as KNM- ER 1470 from below that level were thought to be up to 3 Ma. This implied that a large-brained *Homo* taxon antedated what were then interpreted as smaller- brained *H. habilis* from Olduvai and Koobi Fora by >1Ma. Subsequent fieldwork and further dating showed the KBS tuff to be c.1.9 Ma and the early *Homo* fossils to be broadly contemporary (McDougall and Brown, 2006). If the hypodigm of early *Homo* sampled a single species, then that species was thus comparatively variable in brain size, facial structure, and in jaw and dental dimensions. Alternatively, if the hypodigm of early *Homo* sampled more than one species not only did it imply split(s) in the *Homo* lineage prior to 2 Ma at the latest, but it also raised questions of sympatry and niche differentiation. See Bilsborough (1992) and Wood (1987) for summaries of differing interpretations of early *Homo* diversity around that time.

The beginning of the 1990s saw the publication of two major descriptive accounts and associated interpretations of early *Homo*. These were Tobias' long-awaited *magnum opus* on the Olduvai *H. habilis* cranial fossils (Tobias, 1991), and Wood's description and taxonomic assignment of the Koobi Fora hominins (Wood 1991; 1992). The end of the decade saw an attempt to justify and clarify the criteria for diagnosing genera in general, and the genus *Homo* in particular (Wood and

Collard 1999). During the same decade the recovery of the Hadar AL 666-1a palate extended fossil evidence for the genus back to c.2.4 Ma (Kimbel *et al.* 1997) while the finds from Hata Bouri (Asfaw *et al.* 1999) revealed that an extended hindlimb – crucial for energetically efficient bipedalism — had evolved in at least one hominin species by 2.5 Mya (see above). The 1990s also saw the first discoveries at Dmanisi, Georgia (see below).

By the 1990s most workers had abandoned the notion of a single taxon, *H. habilis sensu lato*, as a plausible explanation for early *Homo* variability in favour of taxic diversity. The majority view was accordingly one of multiple *Homo* species: a relatively primitive *H. habilis*, an enigmatic *H. rudolfensis*, and a terrestrially-adapted, postcranially essentially modern human-like, early African *H. erectus*.

21st Century Discoveries

Since the millennium further discoveries within Africa and beyond have, perhaps inevitably, eroded the distinctiveness of these early *Homo* taxa and accordingly blurred the above picture. A maxilla from Olduvai (OH 65) has been argued by its discoverers (Blumenschine *et al.* 2003) to both resemble the palate of KNM-ER 1470 and provide a good match for the OH 7 mandible. This challenged interpretations that saw dental and gnathic contrasts between *H. rudolfensis* and *H. habilis*, and implied that the former should be subsumed within the latter as *H. habilis sensu lato*. But since the major component of intraspecific variation in such an inclusive species is likely to have been sexual dimorphism this assignment posed its own problems, for the larger, presumed-male, individuals of *H. habilis sensu lato* would have anterior teeth no larger than those of the smaller, presumed-female, individuals, and, unusually, the face of the former would be flatter than the more prognathic faces of the presumed-females. Not surprisingly, most workers have been unconvinced by this interpretation. Clarke (2012) agrees that OH 7 and OH 65 (and so KNM-ER 1470) are a good match and represent *H. habilis* at Olduvai and Koobi Fora. However, he considers that smaller crania from these two localities (OH 13, OH 24, KNM-ER 1813) represent a derived version of *A. africanus* (*A. cf. africanus*). This interpretation has also not gained wide support, and in any event a subsequent study by Fred Spoor and colleagues (Spoor *et al.* 2015) refuted the claimed similarities between OH 7 and OH 65, suggesting that the latter, and the AL 666-1 maxilla from Hadar, show resemblances to early *H. erectus*.

Spoor *et al.* (2015) have also reported a new digital reconstruction of OH 7, incorporating missing bone spalls in the symphysis and correcting the distortion of the corpus. Their analysis suggests that the primitive symphyseal morphology, sub-parallel dental arcade and long post-canine tooth rows of OH 7 distinguish it from other early *Homo* specimens in the sense that it resembles *A. afarensis* and some mandibles attributed to early *Homo* (KNM-ER 1802 and OH 13), but differs markedly from others such as early *H. erectus* from Africa and Georgia and from *H. rudolfensis* (KNM-ER 1482 and KNM-ER 60000) to an extent comparable to that between *Pan* and *Gorilla*. Virtual and manual reconstructions of a palate to match the OH 7 mandible show that the former reconstruction most closely resembles *A. afarensis* palates while the latter is most like KNM-ER 1813, but only to an extent comparable to the extreme limits of intraspecific shape differences between pairs of extant hominids (humans and apes), and differs from other early *Homo* to an even greater degree.

Spoor *et al.* (2015) also used a newly reconstructed biparietal arch and a more extensive comparative data set relating biparietal and total endocranial volumes, to derive new estimated values of 729-824 cm^3 for OH 7's total endocranial volume, which are larger than those generated previously e.g., Tobias (1991) - 647 cm^3; Holloway *et al.* (2004) - 687 cm^3. The endocranial volume of OH 13 (650 cm^3) can easily be accommodated along with the newly revised OH 7 values within a single species (*H. habilis*) based on modern human ranges, whereas, given its endocranial capacity of 509 cm^3, KNM-ER 1813's position as a conspecific with OH 7 is marginal. It can only be accommodated within the same taxon if the *minimal* estimated values for OH 7 are used as the *upper* limit for the estimated species range.

Moreover, additional finds at Koobi Fora have extended the specimen base for *H. rudolfensis*, confirming the distinctive nature of its dental proportions and gnathic form (Leakey *et al.* 2012). Spoor *et al.* (2015) suggested that a sub-adult face including the maxilla (KNM-ER 62000), resembles that of KNM-ER 1470 (except for its overall size), and along with an almost complete mandible (KNM-ER 60000) and a mandible fragment (KNM-ER 62003) provides further information on jaw and tooth morphology . Together with earlier cranial (KNM-ER 1470, 3732) and mandibular (KNM-ER 1482, 1801) discoveries the recent finds now provide a larger, but still small, sample for *H. rudolfensis*. Stratigraphic and chronometric refinements indicate a time range of 2.09-1.78 Ma for these fossils.

Other cranial remains from Koobi Fora (e.g., KNM-ER 42700) have a brain size comparable to smaller *H. habilis sensu stricto* specimens such as KNM-ER 1813, but morphological features such as mid-sagittal thickening and an incipient angular torus that are reminiscent of *H. erectus* (Spoor *et al.* 2007), so further blurring the distinction between these two taxa.

Direct support for the multiple, as opposed to single species, framework for early *Homo* diversity around

2 Ma is provided by the AL 666-1 palate (see above) and the recently recovered Ledi-Geraru mandible, (Villmoare *et al.* 2015) both of which potentially extend the chronology for early *Homo* back to the late Pliocene. Despite its age (2.3 Ma) Spoor *et al.* (2015) suggest that the Hadar palate is appreciably more derived than the virtual palate of OH 7, and it also shows suggestive resemblances to Dmanisi skull 5 in palatal form and dento- alveolar features (Lordkipanidze *et al.* 2013) (see below). This opens the possibility of considerable antiquity for the EAHE lineage (or a similar, otherwise unknown clade) and an even earlier origin for the *H. habilis* lineage. The 2.8 Ma morphologically primitive LD 350-1 partial mandible (which is unassigned at the species level), if it does belong to *Homo*, would push the first-appearance date of the genus back to close to 3 Ma (Villmoare *et al.* 2015).

Dmanisi and its implications

The exceptional finds from Dmanisi, Georgia, have important implications for the functional morphology, zoogeography, population dynamics, toolmaking and cognitive behaviour of early *Homo*. The finds, which consist of cranial and postcranial fossils and Oldowan tools from levels dated around 1.81-1.77 Ma, constitute what is presently the earliest confirmed evidence for hominins beyond Africa, although Zhu *et al.* (2018) claim that stone tools from the Loess Plateau of China antedate the Dmanisi evidence. The fossils recovered from Dmanisi are unweathered and remarkably complete, pointing to little or no transport before being covered, although carnivore tooth marks and cut marks – as well as the tools themselves – indicate hominin and other predator involvement in their accumulation.

The hominin crania are highly variable, with marked contrasts in face and jaw size (Bräuer and Schultz 1996; Rightmire *et al.* 2006; Lordkipanidze *et al.* 2013). Endocranial volumes are modest and the site sample includes smaller crania (545-650 cm^3) that overlap in brain size with *H. habilis sensu stricto*, as well as crania that fall within the lower part of the EAHE range (730-c.780 cm^3). One individual (D3444/D3900) had lost virtually all its teeth well before death, and researchers have interpreted this as implying some form of co-operation and supportive behaviour within the group (Lordkipanidze *et al.* 2005). Another skull (D4500/D2600) combines a very large face and jaws with a small (545 cm^3) braincase, effectively straddling the *Australopithecus/Homo* boundary and uniting in a single individual features of all three early *Homo* species (Lordkipanidze *et al.* 2013; Schwartz *et al.* 2014; Zollikofer *et al.* 2014).

There are postcranial remains of an adolescent (also represented by one of the skulls), and of one large and two small adults (Lordkipanidze *et al.* 2007). The

individuals were relatively short (c.1.5 m) and small-bodied (c.40-50 kg), but with hindlimb proportions like those of modern humans. The shoulder joint was inclined more superiorly and less laterally than in *H. sapiens*, and the relatively short humerus lacked the torsion seen in modern humans. The foot bones (Pontzer *et al.* 2010) suggest that, while pedal morphology and that of the hindlimb overall are consistent with that of a functionally effective biped, the feet of the Dmanisi hominins could have provided a less stable platform than was the case in EAHE, and in modern humans

The Dmanisi fossils pre-date undoubted EAHE in East Africa by 0.1-0.2 Ma, and in several respects they appear more primitive than KNM-WT 15000 and other EAHE specimens. Among the more primitive features are the small endocranial volumes, the large, strongly built face and jaws, and postcranial features such as the reduced humeral torsion, thoracic and lumbar vertebral shapes and femoral shaft proportions. The 1.9 Ma KNM-ER 5881 right femur (Ward *et al.* 2015) resembles the Dmanisi femora in its shaft proportions and together with the associated left hip bone may represent a pre- EAHE *Homo* population antecedent to that genus' expansion into Asia. However, the specimen also resembles the OH 62 (*H. habilis*) femur, or it may belong to *H. rudolfensis*, so that specific assignment is currently not possible. KNM-ER 5881's iliac and femoral morphology does, however, contrast with that of EAHE, which undermines previous assumptions based on KNM-WT 15000 that its essentially modern, terrestrially adapted, postcranial morphology extended back to the earliest evidence of an *H. erectus*-like cranial morphology.

Whether an early, proto-EAHE originated in East Africa and subsequently dispersed out to the Levant and Caucasus, with African populations then evolving into full blown EAHE, or whether the latter evolved in western Asia from an earlier, more primitive form, and subsequently re-colonised East Africa by back migration is a moot point. Both views have strong proponents who can cite the movement of other mammal species in either direction to support their case (Wood and Turner 1995; Wood 2011).

Because the Dmanisi hominin fossils show marked variability, some have argued the site samples more than one species (Schwartz 2000; Skinner *et al.* 2006; Schwartz *et al.* 2014). But for many the Dmanisi sample represents the closest approximation known to a Lower Pleistocene palaeodeme *sensu* Howell (1999). Indeed, the combination of features seen in the Dmanisi hominins, and the overlap of some traits with other early *Homo* taxa, has led some to argue that the morphological variation apparent at Dmanisi has implications for the recognition of other species. In one of the more extreme examples of this taxonomic nihilism Lordkipanidze *et al.* (2013) argue that all early *Homo* species should be subsumed within *H. erectus.*

Lordikipanidze *et al.* (2013) modelled the variation expected in traditional cranial and mandibular linear variables taken from palaeodemes of multiple overlapping early *Homo* species and compared it with that displayed in modern hominid comparators – a single, geographically widespread, sample of modern humans, three subspecies of *Pan troglodytes* and *Pan paniscus* – and with samples of other fossil *Homo*. They argued that cranial variation in the Dmanisi site sample is as great as the differences among the species recognized within early *Homo*, and so they suggest this effectively refutes the hypothesis of taxic diversity within early *Homo*. Their conclusion is that the morphological variation seen among the hypodigms of *H. habilis sensu stricto, H. rudolfensis,* and EAHE, as well as the Dmanisi fossils, is consistent with variation in a single evolving species, which, because of priority would be *H. erectus.*

However, the Lordkipanidze *et al.* study can be criticized on several grounds. First, there is empirical evidence that it lacks discriminatory power. For example, their analysis of fossil and modern comparator crania, which emphasises facial size and prognathism, fails to differentiate the Dmanisi sample from both *Au. africanus* (Sts 5 and Sts 71) and *P. paniscus*, as well as from early *Homo*. In fact the widest spread is that within early *Homo*, such that its morphospace, with limits represented by KNM-ER 1805 and OH 24, encompasses members of four *Pan* demes (*P.t. verus, P.t. troglodytes, P. t. schweinfurthii,* and *P. paniscus*), plus *Au. africanus* and three of the four Dmanisi crania. Only the pathological D3444 (Skull 4), which displays marked alveolar bone resorption, lies outside the early *Homo* morphospace. Lordkipanidze *et al.'s* Fig. 4 shows that the geographically restricted subspecific chimpanzee demes each displays cranial variability comparable to that of an inter-continental (Africa, Australia, America) sample of *H. sapiens* composed of multiple demes. Moreover, in their analysis all of the Middle and Upper Pleistocene and extant *Homo* species (*H. erectus, H. heidelbergensis, H. neanderthalensis* and *H. sapiens*) exhibit less diversity than that displayed by the small early *Homo* sample, considered by Lordkipanidze *et al.* to represent inter-demic contrasts in a single evolving clade. In fact, early *Homo* displays more variation than all the post-*H. erectus Homo* taxa combined.

Lordkipanidze and colleagues also claim their analysis sorts specimens on the basis of differences in cranial shape associated with grade shifts in neurocranial size between species. However, the discriminatory power of their analysis within *Homo* is again poor, such that

the Dmanisi crania associate more closely with African and Javan *H. erectus*, Middle and Upper Pleistocene specimens (Kabwe, Steinheim, Neanderthals), and even with some modern individuals, rather than with early *Homo*. All this evidence suggests that the methods Lordikipanidze *et al.* used to capture cranio-facial form and brain size do not usefully discriminate among hominin fossils.

Technical issues aside, while we acknowledge the marked variation displayed by the Dmanisi sample, we fail to understand why this must be extrapolated to other evidence of early *Homo*, and why all specimens need be accommodated within a single polymorphic lineage. In particular, we do not accept Lordkipanidze *et al.'s* (2013) premise that palaeospecies' interspecific differences necessarily associate with restricted intraspecific variability (their Fig 3B). In fact, there appear to be good reasons for supposing that the contrary is more likely to occur, especially with species displaying significant sexual dimorphism and given that population numbers are a major determinant of expressed (phenotypic) intraspecific variability. All other things being equal, a species comprising many individuals is likely to display more variation than one of similar morphology with few individuals. It is also more likely to be recovered and formally recognised: see Wood and Boyle (2016) for a review of the influence of differential representation in the fossil record on taxic recognition.

In the absence of genetic barriers or geographic isolation we find it difficult to envisage how differing morphologies could become sufficiently frequent and stable enough to register as discrete, repeated signatures in the fossil record. We believe that what is needed for identifying interspecific contrasts associated with multiple species is a consistent pattern of regular differences, rather than heightened variability *per se.* Despite the variation it subsumes, it is noteworthy that the Dmanisi sample does not incorporate any individuals with the characteristic facial and gnathic morphology of *H. rudolfensis*. We therefore continue to interpret this as a discrete species, distinct from other early *Homo*, and we think it probable that *H. habilis* and early *H. erectus* also represent distinct species, albeit with overlapping morphologies. Anton *et al.* (2014) similarly argue that the early *Homo* fossil record represents species diversity rather than intraspecific variability, identifying early African *H. erectus* and two informal morphological clusters assembled around the relatively complete crania KNM- ER 1470 and KNM- ER 1813 which partly – but not entirely - correspond to *H. rudolfensis* and *H. habilis.*

Most recently the Dmanisi team have rowed back somewhat from their extreme monophyletic interpretation of early *Homo* (Rightmire *et al.* 2017). In further morphometric analyses the Dmanisi specimens cluster together with *H. habilis sensu stricto* (OH 24 and KNM-ER 1813) and some (but not all) EAHE, but not with later African or Asian *H. erectus*, which form separate clusters. The best preserved *H. rudolfensis* cranium (KNM-ER 1470) is isolated from the Dmanisi and early African *Homo* clusters. Rightmire *et al.* (2017) acknowledge contrasts, especially in facial morphology, between the Dmanisi fossils and KNM-ER 1470 and, given the latter's isolation, provisionally recognize *H. rudolfensis* as a distinct species of early *Homo*. In contrast, based on their morphometric results they view *H. habilis sensu stricto* and *H. erectus* as chronospecies, with the Dmanisi fossils sampling an intermediate population in this continuum.

The time ranges of *H. habilis sensu stricto* (2.1-1.65 Ma) and *H. rudolfensis* (2.1-1.78 Ma) overlap with the estimated age of the Dmanisi hominins (1.85-1.77 Ma). The Dmanisi fossils and *H. rudolfensis* antedate the time span of most of the hypodigm of EAHE (1.65- 1.43 Ma), but if the provenance of an *H. erectus*-like isolated occipital bone, KNM-ER 2598, is correctly interpreted, then its estimated age (1.9-1.88 Ma) means that both *H. habilis sensu stricto* and *H. rudolfensis*, overlap with the first appearance data for EAHE. If the Hadar 666-1 maxilla belongs to EAHE, then the case for temporal overlap would be even stronger. But if we leave the Hadar maxilla aside, the existing time ranges are compatible with the notion that *H. habilis sensu stricto* and *H. rudolfensis* represent discrete sympatric clades, either of which, given the chronology, and leaving aside the morphological evidence, might be ancestral to EAHE. However, we agree with Rightmire *et al.*'s assessment that the apparently derived facial, gnathic and dental morphology of *H. rudolfensis* make it an unlikely EAHE ancestor. Given the comparative lack of derived features in *H. habilis sensu stricto*, along with the facial, gnathic and dental similarities between the latter and EAHE, a stronger case can be made for it being that ancestor.

It is unclear whether EAHE's appearance involved a speciation event, or was the outcome of an anagenetic process. Rightmire *et al.* (2017) note the former possibility, but discount it in favour of the latter. According to our reading of their paper, they doubt the persistence of sympatric species with similar dietary niches based on carcass exploitation in a restricted region such as a Rift Valley basin, and so prefer anagenesis to account for EAHE's origin. However, although the evidence used to identify the dietary niches of early hominins is sparse, what little evidence that does exist suggests that there was a significant difference in the stable isotope signatures of *H. habilis sensu stricto* and EAHE (Patterson *et al.* 2019). We therefore find it comparably difficult

to envisage multiple morphs persisting in the same region as intra-specific variants – whether demes or larger groupings – without the isolating barriers that promote speciation and which act as a ratchet to fix locally originating morphologies and their underlying gene combinations, so that they persist long enough to be registered as discrete, repeated signatures in the fossil record (Futuyma 1987). By contrast, without such barriers intraspecific adaptive features, while locally prevalent for a time, would in due course be diluted and 'washed out' by migration and gene flow. We accordingly take the view that the observed pattern of morphological variation within early *Homo* is more plausibly interpreted as that among individuals belonging to separate species rather than variation among individuals belonging to a single polymorphic species.

Habitat diversity and early *Homo* evolution

The anagenetic evolution of EAHE from *H. habilis sensu stricto* would have required a pulse of very rapid morphological change given that the geologically most recent fossils attributed to the latter taxon include small, lightly-built, crania such as KNM-ER 1813 and OH 13 that are effectively contemporary with larger, more heavily constructed, *H. erectus* crania such as KNM-ER 3733, KNM-ER 3883, and KNM-WT 15000. An alternative interpretation would be to propose a speciation event prior to 2 Ma, given that first appearance dates (FADs) of most taxa are likely to get older with further discoveries, with one daughter species (*H. habilis sensu stricto*) persisting for upwards of 0.5 Ma, and with the other daughter species (*H. erectus*) displaying more morphological change, perhaps as a result of strong directional selection associated with the exploitation of a wider range of environments and resources. The Dmanisi fossils would then represent an early extra-African expansion of the EAHE clade before the establishment of its full character set which, on current dating evidence, would be sometime after 1.7 Ma. In this interpretation EAHE and its Dmanisi representatives would be reproductively isolated from *H. habilis sensu stricto,* and any common traits would be interpreted as hypothetical plesiomorphies, not conspecific markers, as in Rightmire *et al.*'s preferred model. Interestingly, recent developments indicate that later human evolution (<1.0 Ma) was probably appreciably more speciose than had hitherto been suspected (Wood 2017a; Wood and Boyle 2016) with consequent implications for the taxic diversity represented by the earlier *Homo* evidence which, as yet, is represented by a much sparser fossil record.

There is evidence in East Africa for multiple, relatively short-lived, climatic changes superimposed upon a longer term drying trend resulting in environmental instability and heterogeneity that could well have catalyzed the evolution of morphological and behavioural adaptive responses (Anton *et al.* 2014). In addition Fortelius *et al.* (2016) have suggested that the Turkana basin was significantly drier than other areas of East Africa, and that it might have acted as a source for dry-adapted species that subsequently expanded elsewhere, but this interpretation has been challenged by Robinson *et al.* (2017). While climate and habitat changes undoubtedly influenced hominin communities they do not allow us to determine their evolutionary responses in detail or help decide whether an anagenetic or cladogenetic response is the more plausible.

However, irrespective of process, climatic instability and environmental fragmentation doubtless resulted in demic diversification, a feature also expected of the presumably small and thinly spread bands involved in the initial phase of early *Homo* expansion beyond Africa. These circumstances would allow a significant role for stochastic factors in influencing group characteristics, resulting in emergent differentiated local and regional trends without necessarily leading to a reconstruction of the total gene pool at the species level (Bilsborough 1999). Such evolutionary processes may underlie the regional contrasts between later

African and Asian *H. erectus*, and those of north and south east Asia (Bilsborough 2005). They also raise the distinct possibility that the Dmanisi group may well not be representative of other extra-African early *Homo* populations.

Conclusion

Here we sound a cautionary note: All of the above needs to be read with the equivalent of a series of 'health warnings.' First, evolutionary scenarios *sensu* Eldredge (1979; Eldredge and Tattersall 1975) are out of fashion, and while Greene (2017) makes a cogent case for their judicious use, we agree with him that any evolutionary scenario, no matter how carefully it is assembled, should be 'regarded skeptically'. Second, we are guilty of falling into the trap of discussing the taxonomic and systematic implications of the fossil evidence for early *Homo* as if we had access to all of the relevant evidence. Nearly seven decades ago, no less a figure than Robert Broom wrote that 'within 20 years it is likely that we shall have the complete history of pre-man in South Africa, and the main facts of the evolution which led from a higher Primate to man' (Broom, 1950, p. 12). With the enormous benefit of hindsight, we know

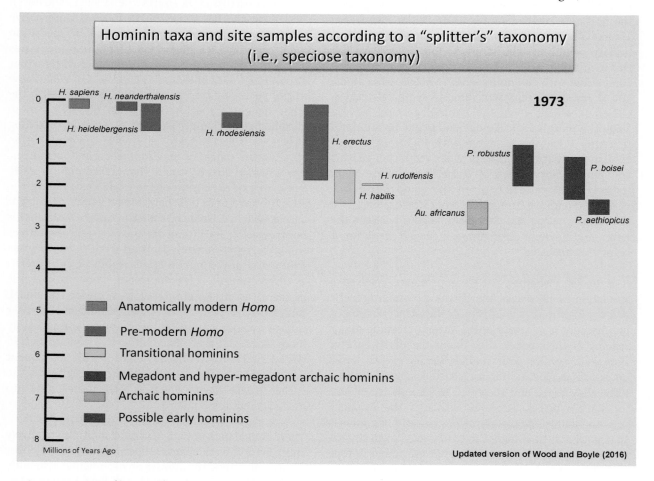

Figure 1. Hominin ('hominid') taxa recognized in 1973 on the basis of a relatively speciose interpretation of the fossil record and a resulting splitters' taxonomy.

that Broom's optimism was misplaced, and despite the best efforts of those who devote their lives to field work, there is no reason to think that in 2020 we are aware of all of the early hominin taxa that ever existed. For example, in 1973, which was the year of John Gowlett's first full field season at Kilombe (Gowlett 1978), a relatively speciose interpretation of what was then called the 'hominid' fossil record would have recognized eleven species (Figure 1). In 2020, especially if you include enigmatic site samples, and even if you adopt a more lumping philosophy (Wood and Boyle, 2016), that number is much higher. It is even higher under a speciose interpretation (Figure 2).

There is also the knotty and persistent problem of how to define *Homo*. Where, if at all, should we draw the line between *Homo* and its precursor genus? Nearly twenty years ago one of us suggested that *Homo* needed to be both a grade and clade (Wood and Collard 1999). In a recent paper Kimbel and Villmoore (2017) argued that recent evidence has blurred the distinction between the adaptations of *H. habilis sensu stricto* and taxa such as *Australopithecus afarensis*. They wrote that ' a fresh look at brain size, hand morphology and earliest technology suggest that a number of key *Homo* attributes may

already have been present in generalized species of *Australopithecus*, and that adaptive distinctions in *Homo* are simply amplifications or extensions of ancient hominin trends.' (Kimbel and Villmoare 2017). Although its discoverers claim that *Homo naledi* holds the secret to the origin of *Homo* (Berger *et al.* 2015; Berger *et al.* 2017), its puzzling mix of primitive and derived morphology, and its age (Dirks *et al.* 2017) suggest that it is a relict taxon that has little to say about the origin of our genus (Wood 2017b).

Over the last half century the fossil record for hominin evolution has documented ever increasing taxic diversity, and the early *Homo* sub-set of that record is no exception. Interpretations of the hominin record have shifted to a corresponding extent: at the outset of the period the default position was an anagenetic one; polyphyletic proponents had to marshal an overwhelming case for serious consideration. Now the position has shifted, with cladogenetic interpretations of hominins generally, and of early *Homo* in particular, attracting majority support. Controversies still persist, and disentangling intra- and inter-specific components of the observed variability remains a thorny issue.

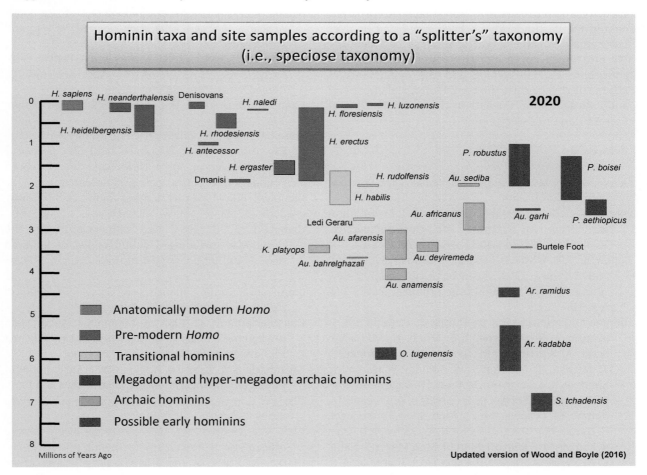

Figure 2. Hominin taxa recognized in 2020 on the basis of a relatively speciose interpretation of the fossil record and a resulting splitters' taxonomy.

Irrespective of phyletic perspective, there is now a greater awareness of the processes and potential drivers of evolution, and an increasing knowledge of hominin habitats (e.g., Patterson *et al.* 2017a, 2017b). In addition to longer term climate and biotic trends, there is an increasing awareness of the complexity of habitats even within a single lake basin. Climate change affects environments and habitats, and thus influences faunal communities that include hominins (Patterson *et al.* 2017a). Although some researchers have succumbed to the temptation to link particular environmental episodes with speciation and extinction 'events' within the hominin clade (Anton *et al.* 2014; Maslin *et al.* 2014), this is as perilous as confidently generating evolutionary scenarios. First and last appearance dates generated from a sparse early hominin fossil record recovered from sites that sample less than 3% of the land surface of Africa are unlikely to reflect the timing and location of hominin speciation and extinction events. For a host of reasons, the next half-century is likely to see changing interpretations of early *Homo* at least as dramatic as those of the last fifty years.

Acknowledgements

We are grateful to the institutions that have employed us and to the funders who have supported us over our combined careers that span close to a century. We also much appreciate the editors' invitation to contribute to this volume and the opportunity it affords us to recognize the contributions of our friend and colleague John Gowlett to the understanding of East African prehistory.

References

Alexeev, V. P. 1986. *The Origin of the Human Race.* Moscow: Progress Publishers

Antón, S. C., R. Potts and L. C. Aiello 2014. Evolution of early *Homo*: An integrated biological perspective. *Science* 345: 6192 (doi:10.1126/science.1236828)

Asfaw, B., T. D. White, O. Lovejoy, B. Latimer, S. Simpson and G. Suwa 1999. *Australopithecus garhi*: a new species of early hominid from Ethiopia. *Science* 284: 629-635.

Berger L., J. Hawks, D. J de Ruiter, S. E. Churchill, P. Schmid, L. K. Delezene, T. L. Kivell, H. M. Garvin Scott, A. Williams, J. M. DeSilva, M. M. Skinner, C. M. Musiba, N. Cameron, T. W. Holliday, W. Harcourt-Smith, R. R. Ackermann, M. Bastir, B. Bogin, D. Bolter, J. Brophy, Z. D. Cofran, K. A. Congdon, A. S. Deane, M. Dembo, M. Drapeau, M. C. Elliott, E. M. Feuerriegel, D. Garcia-Martinez, D. J. Green, A. Gurtov, J. D. Irish A. Kruger, M. F. Laird, D. Marchi, M. R. Meyer, S. Nalla, E. W. Negash, C. M. Orr, D. Radovcic, L.Schroeder, J. E. Scott, Z. Throckmorton, M. W. Tocheri, C. VanSickle, C. S. Walker, P. Wei and B. Zipfel 2015. *Homo naledi*, a new species of the genus *Homo* from the Dinaledi Chamber, South Africa. *eLife* 2015;4:e09560 DOI: 10.7554/eLife.09560

Berger, L. R., J. Hawks, P. H. Dirks, M. Elliott and E. M. Roberts 2017. *Homo naledi* and Pleistocene hominin evolution in subequatorial Africa. *eLife* 2017;6:e24234 DOI: 10.7554/eLife.24234

Bilsborough, A. 1992. *Human Evolution*. London: Blackie Academic & Professional

Bilsborough, A. 1999. Contingency, patterning and species in hominid evolution, in J. Bintliff (ed.) *Structure and Contingency: Evolutionary Processes in Life and Human Society:* 43-101. London: Leicester Univ. Press

Bilsborough, A. 2005. *Homo erectus* revisited: aspects of affinity and diversity in a Pleistocene hominin species. *Anthropologie* 43: 129-158.

Blumenschine, R.J., C. R. Peters, F.T. Masao, R.J. Clarke, A. L. Deino, R.L. Hay, C. C. Swisher, I.G. Stanistreet, G.M. Ashley, L.J. McHenry, and N. E. Sikes 2003. Late Pliocene *Homo* and hominid land use from western Olduvai Gorge, Tanzania. *Science* 299: 1217-1221.

Bräuer, G., and M. Schultz 1996. The morphological affinities of the Plio-Pleistocene mandible from Dmanisi, Georgia. *Journal of Human Evolution* 30: 445-481

Broom, R. 1950. The genera and species of the South African ape-men. *American Journal of Physical Anthropology* 8: 1-14

Brown, F. H., J. Harris, R.E. F. Leakey and A. Walker 1985. Early *Homo erectus* skeleton from West Lake Turkana, Kenya. *Nature* 316: 788-792

Campbell, B. G. 1962. The systematics of Man. *Nature* 194: 225-232

Campbell, B. G. 1963. Quantitative Taxonomy and Human Evolution, in S. L. Washburn (ed.) *Classification & Human Evolution:* 50-74 New York: Wenner–Gren Foundation for Anthropological Research

Campbell, B.G. 1964 Just another 'man-ape'? *Discovery* 25(6) : 37-38

Campbell, B. G. 1965. The nomenclature of the Hominidae including a definitive list of hominid taxa. *Royal Anthropological Institute Occasional Paper* 22: 1-34

Campbell, B. G. 1967. *Human evolution; an introduction to man's adaptations.* London: Heinemann

Chamberlain, A. T. 1987. A taxonomic review and phylogenetic analysis of *Homo habilis*. Unpublished PhD dissertation, University of Liverpool

Chamberlain, A. T., and B. A. Wood 1987. Early hominid phylogeny. *Journal of Human Evolution* 16: 119-133.

Clark, W. E. Le Gros 1964. The Evolution of Man. *Discovery* 25: 49

Clark, W. E. Le Gros 1967. *Man -apes or Ape-men? The story of discoveries in Africa.* New York: Holt, Rinehart

Clarke, R.J. 2012. A *Homo habilis* maxilla and other newly-discovered hominid fossils from Olduvai Gorge, Tanzania. *Journal of Human Evolution* 63: 418-428.

Clarke, R. J. and F.C. Howell 1972. Affinities of the Swartkrans 847 hominid cranium. *American Journal of Physical Anthropology* 37: 319–335

Clarke, R. J., F. C. Howell and C.K. Brain 1970. More Evidence of an Advanced Hominid at Swartkrans. *Nature* 225: 1112 - 1119 doi:10.1038/2251112a0

Curnoe, D. and P.V. Tobias 2006. Description, new reconstruction, comparative anatomy, and classification of the Sterkfontein Stw 53 cranium, with discussions about the taxonomy of other southern African early *Homo* remains. *Journal of Human Evolution* 50: 36-77.

Curnoe, D. 2010. A review of early *Homo* in southern Africa focusing on cranial, mandibular and dental remains, with the description of a new species (*Homo gautengensis* sp. nov.). *Homo - Journal of Comparative Human Biology* 61: 151-177.

Deino, A. L. 2012. 40 Ar/39 Ar dating of Bed I, Olduvai Gorge, Tanzania, and the chronology of early Pleistocene climate change. *Journal of Human Evolution* 63: 251-273

Dirks, P. H., E. M Roberts, H. Hilbert-Wolf, J. D. Kramers, J. Hawks, A. Dosseto, M. Duval, M. Elliott, M. Evans, R. Grün, J.Hellstrom, A.R. Herries, R. Joannes-Boyau, T. V. Makhubela, C. J. Placzek, J.Robbins, C. Spandler, J. Wiersma, J. Woodhead and L. R. Berger. The age of *Homo naledi* and associated sediments in the Rising Star Cave, South Africa. *eLife*, 2017; 6 DOI: 10.7554/eLife.24231

Dobzhansky, T. 1962. *Mankind Evolving: The Evolution of the Human Species*. New Haven and London: Yale University Press

Dobzhansky, T. 1963. Genetic entities in hominid evolution, in S. L. Washburn (ed.) *Classification and Human Evolution*: 347-363. New York: Wenner–Gren Foundation for Anthropological Research

Eldredge, N. 1979. Cladism and common sense, in J. Cracraft and N. Eldredge (eds.) *Phylogenetic Analysis and Paleontology*: 165- 198. New York: Columbia University Press, New York.

Eldredge, N. and I. Tattersall 1975. Evolutionary models, phylogenetic reconstruction, and another look at hominid phylogeny, in F. S. Szalay (ed.) *Approaches to Primate Paleobiology; Contributions to Primatology*: 5:218—42. Basel: Karger

Ferguson, W. W. 1995. A new species of the genus *Homo* (Primates: Hominidae) from the Plio/Pleistocene of Koobi Fora, in Kenya. *Primates 36:* 69-89.

Fortelius M., I. Žliobaitė, F. Kaya, F. Bibi, R. Bobe, L. Leakey, M.G. Leakey, D. Patterson , J. Rannikko and L. Werdelin 2016. An ecometric analysis of the fossil mammal record of the Turkana Basin. *Philosophical Transactions of the Royal Society B* 371: 1698 DOI: 10.1098/rstb.2015.0232

Futuyma, D. J. 1987. On the Role of Species in Anagenesis. *The American Naturalist* 130: 465-473

Gowlett, J. A. J. 1978. Kilombe - an Acheulian site complex in Kenya, in W. W. Bishop (ed.) *Geological Background to Fossil Man*: 337-360. Edinburgh: Scottish Academic Press.

Greene, H. W. 2017. Evolutionary Scenarios and Primate Natural History. *The American Naturalist* 190: 69-86. doi: 10.1086/692830.

Grine, F. E., B. Demes, W. L. Jungers and T. M. Cole 1993. Taxonomic affinity of the early *Homo* cranium from Swartkrans, South Africa. *American Journal of Physical Anthropology* 92: 411- 426.

Grine, F. E., W. L. Jungers and J. Schultz 1996. Phenetic affinities among early *Homo* crania from East and South Africa. *Journal of Human Evolution* 30: 189-225.

Groves, C. P. and V. Mazak 1975. An approach to the taxonomy of the Hominidae: gracile Villafranchian hominids of Africa. *Casopis pro mineralogii a geologii* 20: 225-247.

Haeusler, M. and H. M. McHenry 2004. Body proportions of *Homo habilis* reviewed. *Journal of Human Evolution* 46: 433-465.

Haeusler, M. and H. M. McHenry 2007. Evolutionary reversals of limb proportions in early hominids? Evidence from KNM-ER 3735 (*Homo habilis*). *Journal of Human Evolution* 53: 383-405.

Hartwig-Scherer, S., and R. D. Martin 1991. Was 'Lucy' more human than her 'child'? Observations on early hominid postcranial skeletons. *Journal of Human Evolution* 21: 439-449.

Hennig, W. 1966. *Phylogenetic Systematics*. Urbana: Univ. of Illinois Press.

Holloway, R.L., D. C. Broadfield and M. S. T. Yuan 2004. *The Human Fossil Record Volume 3: Brain Endocasts- the Paleoneurological Evidence*. Hoboken (N.J): Wiley.

Howell, F. C. 1999. Paleo-Demes, Species Clades, and Extinctions in the Pleistocene Hominin Record. *Journal of Anthropological Research* 55:191-243

Howell, F. C. 1978. Hominidae, in V. J. Maglio and H. S. B. Cooke (eds.) *Evolution of African Mammals*: 154-248. Cambridge (Mass): Harvard University Press

Howell, F. C. and F. Bourlière (eds.) 1963. *African Ecology and Human Evolution*. New York: Wenner–Gren Foundation for Anthropological Research

Hughes, A. R., and P. V. Tobias 1977. A fossil skull probably of the genus *Homo* from Sterkfontein, Transvaal. *Nature* 265: 310-312.

Huxley, J. 1942. *Evolution: The Modern Synthesis*. London: Allen & Unwin.

Johanson, D. C., F. T. Masao, G. G. Eck, T. D. White, R. C. Walter, W. H. Kimbel, B. Asfaw, P. Manega, P. Ndessokia and G. Suwa 1987. New partial skeleton of *Homo habilis* from Olduvai Gorge, Tanzania. *Nature* 327: 205-209.

Keith, A. 1948. *A New Theory of Human Evolution*. London: Watts

Kimbel, W.H., D. C. Johanson and Y. Rak 1997. Systematic assessment of a maxilla of *Homo* from Hadar, Ethiopia. *American Journal of Physical Anthropology* 103:235-62

Kimbel, W.H. and B. Villmoare 2016. From *Australopithecus* to *Homo*: the transition that wasn't. *Philosophical Transactions of the Royal Society B Biological Sciences*. 371:1698 DOI: 10.1098/rstb.2015.0248

Kuman, K., and R. J. Clarke 2000. Stratigraphy, artefact industries and hominid associations for Sterkfontein, Member 5. *Journal of Human Evolution* 38: 827-847.

Leakey, L.S.B. 1966. *Homo habilis, Homo erectus* and the Australopithecines. *Nature* 209: 1279-1281

Leakey, L.S.B., P. V. Tobias and J. R. Napier 1964. A new species of the genus *Homo* from Olduvai Gorge. *Nature* 202: 7-9

Leakey, M. D. 1969. Recent discoveries of hominid remains at Olduvai Gorge, Tanzania. *Nature* 223: 756

Leakey, M. D. 1971. *Olduvai Gorge: Volume 3. Excavations in Beds I & II 1960-1963.* Cambridge: Cambridge University Press

Leakey, M. D., R. J. Clarke and L. S. B. Leakey 1971. New hominid skull from Bed I, Olduvai Gorge, Tanzania. *Nature* 232: 308-312

Leakey, M. G., F. M. Spoor, C. Dean, C. S. Feibel, S. C. Antón, C. Kiarie, and L.N. Leakey 2012. New fossils from Koobi Fora in northern Kenya confirm taxonomic diversity in early *Homo*. *Nature* 488: 201–204

Leakey, R.E., M. G. Leakey and A. K. Behrensmeyer 1978. The Hominid Catalogue, in M. G. Leakey and R. E. Leakey (eds) *Koobi Fora Research Project Volume 1: The Fossil Hominids and an introduction to their Context 1968-1974:* 86-182. Oxford: Clarendon Press

Leakey, R. E. F., A. Walker, C.V. Ward and H. M. Grausz 1989. A partial skeleton from the Upper Burgi Member of the Koobi Fora Formation East Lake Turkana, in G. Giacobini (ed.) *Proceedings of the 2nd International Congress on Human Paleontology:* 167-173. Milan: Jaca

Lordkipanidze D., T. Jashashvili, A. Vekua, M.S. Ponce de Leon, C. P. E. Zollikofer, G. P. Rightmire, H. Pontzer, R. Ferring, O. Oms, M. Tappen, M. Bukhsianidze, J. Agusti, R. Kahlke, G. Kiladze, B. Martinez-Navarro, A. Mouskhelishvili, M. Nioradze and L. Rook 2007. Postcranial evidence from early *Homo* from Dmanisi, Georgia. *Nature* 449: 305-310

Lordkipanidze D., A. Vekua, R. Ferring, G. P. Rightmire, J. Agusti, G. Kiladze, A. Muskhelishvili, M. Nioradze, M. S. Ponce de León, M. Tappen and C.P.E. Zollikofer 2005. Anthropology: The earliest toothless hominin skull. *Nature* 434: 717-718

Lordkipanidze, D., M.S. Ponce de León, A. Margvelashvili, Y. Rak, G. P. Rightmire, A. Vekua and C.P.E. Zollikofer 2013. A Complete Skull from Dmanisi,Georgia, and the Evolutionary Biology of Early *Homo*. *Science* 342: 326-331

Maslin, M. A., C. M. Brierley, A. M. Milnera, J. Shultz, M.H. Trauth and K.E. Wilson 2014. East African climate pulses and early human evolution. *Quaternary Science Reviews* 101: 1-17

Mayr, E. 1950. Taxonomic categories in fossil hominids, in M. Demerec (org.) *Origin and Evolution of Man* (Cold Spring Harbor Symposia on Quantitative Biology 15): 109-118 doi:10.1101/SQB.1950.015.01.013

Mayr, E. 1963. The taxonomic evaluation of fossil hominids, in S. L. Washburn (ed.) *Classification &*

Human Evolution: 332-346. New York: Wenner –Gren Foundation for Anthropological Research

McDougall, I. and F. H. Brown 2006. Precise ^{40}Ar/^{39}Ar geochronology for the upper Koobi Fora Formation, Turkana Basin, northern Kenya. *Journal of the Geological Society* 163: 205-220.

Oakley, K. P. 1949. *Man the Tool-Maker*. London: British Museum

Oakley, K. P. and B. G. Campbell 1964. Newly described Olduvai hominid. *Nature* 202: 732

Patterson, D.B., D.R. Braun, A.K.Behrensmeyer S.Merritt, I. Žliobaite, B. A. Wood,M. Fortelius, and R. Bobe, 2017a. Ecosystem evolution and hominin paleobiology at East Turkana, northern Kenya between 2.0 and 1.4 Ma. *Palaeogeography, Palaeoclimatology, Palaeoecology* 481: 1-13.

Patterson, D.B., D. R. Braun, A.K. Behrensmeyer, S. B. Lehmann, S. Merritt, J.S. Reeves, B.A. Wood, and R. Bobe 2017b. Landscape scale heterogeneity in the East Turkana ecosystem during the Okote Member (1.56–1.38 Ma). *Journal of Human Evolution* 112: 148-161.

Patterson, D.B., Braun, D.R., Allen, K., Barr, W.A., Behrensmeyer, A.K., Biernat, M., Lehmann, S.B., Maddox, T., Manthi, F.K., Merritt, S.R., Morris, S.E., O'Brien, K., Reeves, J.S., Wood, B.A. and Bobe, R. 2019. Comparative isotopic evidence from East Turkana suggests a dietary shift between early Homo and Homo erectus. *Nature Ecology and Evolution* 3: 1048-1056.

Pickering, R. and J. D. Kramers 2010. Re-appraisal of the stratigraphy and determination of new U-Pb dates for the Sterkfontein hominin site, South Africa. *Journal of Human Evolution* 59: 70-86.

Pontzer, H., C. Rolian, G. P. Rightmire, T. Jashashvili, M. S. Ponce de León, D. Lordkipanidze and C. P. E. Zollikofer 2010. Locomotor anatomy and biomechanics of the Dmanisi hominins. *Journal of Human Evolution* 58: 492-504

Rightmire, G. P., D. Lordkipanidze and A. Vekua 2006. Anatomical descriptions, comparative studies and evolutionary significance of the hominin skulls from Dmanisi, Republic of Georgia. *Journal of Human Evolution* 50: 115-141

Rightmire, G. P., M. S. Ponce de León, D. Lordkipanidze, A. Margvelashvili and C. P. E. Zollikofer 2017. Skull 5 from Dmanisi: Descriptive anatomy, comparative studies, and evolutionary significance. *Journal of Human Evolution* 104: 50–79

Robinson, J.T. 1965. *Homo 'habilis'* and the australopithecines. *Nature* 205: 121-124.

Robinson, J. T. 1966. The distinctiveness of *Homo habilis*. *Nature* 209: 957-60.

Robinson, J. R., J. Rowan, C. J. Campisano, J.G. Wynn and K. Reed 2017. Late Pliocene environmental change during the transition from *Australopithecus* to *Homo*. *Nature Ecology & Evolution* 15:1(6):159. doi: 10.1038/s41559-017-0159.

Schwartz, J. H. 2000. Taxonomy of the Dmanisi crania. *Science* 289: 55-6

Schwartz, J.H., I. Tattersall, and Z. Chi 2014. Comment on 'A complete skull from Dmanisi, Georgia, and the evolutionary biology of early *Homo*'. *Science* 344: 360-360.

Simpson, G. G. 1953 *The Major Features of Evolution.* New York: Columbia

Simpson, G. G. 1963 The Meaning of taxonomic statements, in S. L. Washburn (ed.) *Classification and Human Evolution:* 1-31. New York: Wenner–Gren Foundation for Anthropological Research

Skinner, M. M., A. D. Gordon and N. J. Collard 2006. Mandibular size and shape variation in the hominins at Dmanisi, Republic of Georgia. *Journal of Human Evolution* 51: 36 - 49

Spoor F., M. G. Leakey, P.N. Gathogo, F. Brown, S. C. Antón, I. McDougall, C. Kiarie, F. K. Manthi and L. N. Leakey 2007. Implications of new early *Homo* fossils from Ileret, east of Lake Turkana, Kenya. *Nature* 448: 688-691

Spoor, F., P. Gunz, S. Neubauer, S. Stelzer, N. Scott, A. Kwekason and M. C. Dean 2015. Reconstructed *Homo habilis* type OH 7 suggests deep-rooted species diversity in early *Homo. Nature* 519: 83-86

Stringer, C. B. 1986. The credibility of *Homo habilis,* in B. Wood, I. Martin, and P. Andrews (eds) *Major Topics in Primate and Human Evolution:* 266-294. Cambridge: Cambridge University Press.

Tobias, P. V. 1991. *Olduvai Gorge IV. The skulls endocasts and teeth of* Homo habilis. Cambridge: Cambridge University Press

Villmoare B., W. H.Kimbel, C. Seyoum, C. R. Campisano, E. R. DiMaggio, J. Rowan, D. R. Braun, J. R. Arrowsmith and K. E. Reed 2015. Early *Homo* at 2.8 Ma from Ledi-Geraru, Afar, Ethiopia. *Science* 347: 1352-1355

Walker, A., and R.E. F. Leakey (eds) 1993. *The Nariokotome* Homo erectus *skeleton.* Cambridge, Mass: Harvard University Press

Ward, C. V., C.S. Feibel, A.S. Hammond and L.N. Leakey 2015. Associated ilium and femur from Koobi Fora, Kenya, and postcranial diversity in early *Homo. Journal of Human Evolution* 81: 48-67

Washburn, S. L. 1963. (ed.) *Classification and Human Evolution:* New York: Wenner–Gren Foundation for Anthropological Research

Washburn, S. L. 1950 The analysis of primate evolution with particular reference to the origin of man. M.Demerec (org.) *Origin and Evolution of Man* (Cold Spring Harbor Symposia on Quantitative Biology 15): 67-78 doi:10.1101/SQB.1950.015.01.013

Wolpoff, M. H. 1971. Competitive exclusion among Lower Pleistocene hominids: the single species hypothesis. *Man* 6: 601-614

Wood, B. A. 1987. Who is the 'real' *Homo habilis? Nature* 327: 187-188

Wood, B. A. 1991. *Koobi Fora Research Project IV: Hominid Cranial Remains.* Oxford: Clarendon Press

Wood, B. A. 1992. Origin and evolution of the genus *Homo.* Nature 355: 783-790.

Wood, B. A. 2011. Did early *Homo* migrate 'out of' or 'in to' Africa? PNAS 108: 10375-10376 doi: 10.1073/pnas.1107724108

Wood, B. A. 2017a. Origins(s) of Modern Humans. *Current Biology* 27: R767-9. doi.org/10.1016/j.cub.2017.06.052

Wood, B. A. 2017b. Chalk and cheese 2.0. *Journal of Human Evolution.* doi.org/10.1016/j.jhevol.2017.08.010

Wood B. A. and J. L. Baker 2011. Evolution in the genus *Homo. Annual Review of Ecology, Evolution and Systematics* 42:47-69

Wood B. A. and K. E. Boyle 2016. Hominin taxic diversity: Fact or fantasy? *American Journal of Physical Anthropology 159 (Suppl 61):* 37-78. doi: 10.1002/ajpa.22902

Wood, B. A. and M. Collard 1999. The Human Genus. *Science,* 284: 65-71.

Wood, B. A. and A. Turner 1995. Out of Africa and into Asia. *Nature* 378:239-240

Zollikofer, C. P., M.S. Ponce de León, A. Margvelashvili, G. P. Rightmire and D. Lordkipanidze 2014. Response to Comment on 'A Complete Skull From Dmanisi, Georgia, and the Evolutionary Biology of Early *Homo.' Science* 344: 360.

Alan Bilsborough
Durham University
Evolutionary Anthropology Research Group
Department of Anthropology
Dawson Building
Durham DH1 3LE
UK
alan.bilsborough@durham.ac.uk

Bernard Wood
George Washington University
Center for the Advanced Study of Human Paleobiology
800 22nd St NW
Suite 6000
Washington, District of Columbia 20052
USA
bernardawood@gmail.com

Rift Dynamics and Archaeological Sites:
Acheulean Land Use in Geologically Unstable Settings

Simon Kübler, Geoff Bailey, Stephen Rucina, Maud Devès and Geoffrey C.P. King

Introduction

Our aim in this chapter is to examine the reconstruction of physical landscapes in rift settings and their relevance to archaeological interpretation and to reflect on the challenges and opportunities of such research, with particular reference to the Kenyan sector of the East African Rift. We focus on the mapping of physical landforms and how they change in relation to variations in geology, topography, hydrology, soils and sediments, their relationship to the dynamic underlying processes of tectonic activity in different sorts of geological and geodynamic settings, and the implications of these relationships for the surface distribution of plant and animal resources and physical features of potential human significance.

We take it as axiomatic that investigating the location of archaeological or human-fossil sites and site distributions in relation to their physical surroundings at many different geographical scales provides a valuable source of information about the types of landscapes that were of particular significance to their human occupants, how they were used and what this tells us about broader processes of human evolution, adaptation and dispersal.

Our geographical scale is that of the site catchment (the area within a 10–20 km radius) and the wider region beyond. We distinguish this from other landscape approaches, particularly those applied in Africa, which look at a much more localised scale of tens of metres to kilometres (Potts et al. 1999; Blumenschine and Peters 1998; Blumenschine et al. 2012; Bunn et al. 1980; Isaac 1981; Kingston 2007; Kroll 1994; Stern 1994).

At the same time, we acknowledge a major problem in pursuing such goals, especially when dealing with the earlier time ranges of human evolution. The first generation of site catchment studies applied to the prehistoric period took the present-day landscape as a starting point for evaluation, incorporating evidence of changes as opportunity offered but generally treating them as relatively minor impediments to interpretation (Bailey 2005; Higgs and Vita-Finzi 1972; Vita-Finzi and Higgs 1970). However, as one goes further back in time, and especially in geologically active regions such as rifts and plate margins, such an assumption is of doubtful validity; on the contrary the opposite assumption may hold sway – that so much has changed that any attempt at reconstruction is doomed to failure. The latter assumption is often allied to the belief that no useful statements can be made about the determinants of archaeological or fossil occurrences other than the banal – sites are associated with food and water; or that a more detailed investigation is hopeless – discovery of sites is the result of fortuitous conditions of preservation and exposure resulting from geological change that are beyond further investigation. It is probably these two reasons above all others – the reliance on an assumption of little or no physical change that may be unwarranted, or a belief that so much has changed that detailed investigation is impossible – that account for widespread scepticism and lack of progress in catchment-scale studies as applied to the earlier time ranges of the archaeological record.

We show here that it is possible to steer between these extremes, and that the key to doing so is an understanding of the underlying geodynamic processes that determine the nature of land forms. We have elsewhere shown how tectonically active environments can create and maintain features of long-term benefit to human land use. This can be achieved through the creation and rejuvenation of a complex topography and hydrology that traps water in lake basins, creates spring lines along fault boundaries, sharpens topographic features of tactical advantage in avoiding predators and accessing prey, maintains variations in relief over short distances that juxtapose different plant and animal communities within a relatively small area, protects against climatic extremes, renews the fertility of soils and shapes patterns of soil erosion and sedimentation (Bailey and King 2011, Bailey et al. 2011, Deves et al. 2014, 2015; King and Bailey 2010, Kübler et al. 2015, 2016; Reynolds et al. 2011; Winder et al. 2013, 2014). Here we expand on this theme. We review the variety of geodynamic processes that need to be taken into account when undertaking landscape reconstruction, and examine specific archaeological examples.

The study of Earth geodynamics and active tectonics is a rapidly evolving field, with many different sorts of interacting processes. The Earth's crust, or lithosphere, is highly unstable and irregular in its surface morphology, and there is no place on Earth that is immune from these instabilities. These processes include the effects of convection currents in the Earth's interior associated with plate motions and plumes, and isostatic movements associated with differential loading and unloading of mountain and water masses. The scale and rate of change varies enormously across the globe and through time, but there is no such thing as a static landscape. Even geodynamic processes that operate very slowly and over very long periods of time (millions of years) can give rise to short-term phenomena such as volcanic eruptions and earthquakes with immediate human impact, or in their cumulative effect sustain physical features advantageous to human land use. Paradoxically, the processes that provide long-term human benefit are also those that make for difficulties of landscape reconstruction. A static landscape would in time undergo erosion and removal of topographic features, depletion of soil nutrients and environmental degradation of little human benefit.

Of course, geodynamic processes can also damage or destroy. They also play an important role in the taphonomic effects that determine the burial, exposure, preservation and discovery of archaeological and fossil material. From every point of view, an understanding of their effects on human landscapes is indispensable to archaeological interpretation, and we do not minimise the complexities involved in unravelling their impact. There is much that is still unresolved in understanding the geodynamic history of the lithosphere and how the various processes interact, and we outline these issues here.

For examples of landscape reconstruction, we draw on early Acheulean localities in Africa and the Levant but with a particular focus on the central and southern Kenyan Rift. We emphasise this archaeological period and region not only because it provides a good illustration of our approach, but because it has also been the focus of John Gowlett's own field research, beginning with his PhD research at Kariandusi in the early 1970s (Figure 1, Figure 2), and subsequently at Chesowanja and Kilombe (Gowlett 1991, 1999; Gowlett et al. 2015). We present this chapter both as a tribute to John's fieldwork in the region and also more generally as a contribution to debates about archaeological and fossil materials in their landscape setting.

Figure 1. The site of Kariandusi looking southeast. The tin-roofed structure visible at the top of the cliff is the open-air museum built over the area excavated by John Gowlett in 1974. The white deposits represent lake sediments accumulated when lake level was higher than the present. They are quarried today for their diatomites, which have many applications in building and agriculture. Photo by Geoff Bailey, February 2012

Figure 2. Left: view of section and artefacts in situ on John Gowlett's excavation at Kariandusi. Right: the sign at Kariandusi, marking the location of the Gowlett excavation. Photos by Geoff Bailey, February 2012.

Earth geodynamics

Plates

Ever since the acceptance of plate tectonic theory in the 1960s, plate motions driven by ocean-spreading have dominated understanding of geodynamic processes. According to plate-tectonic theory, the lithosphere (the rigid ~100km-thick outer crust of the Earth's surface) is broken up into seven major plates, and these plates are in continuous motion relative to each other, sliding over the more viscous material of the underlying mantle where rocks are in a more fluid or semi-plastic state as a result of heat and pressure. These plates are composed either of continental crust including blocks of geologically very ancient rocks known as cratons, or of relatively young oceanic crust composed primarily of basaltic lava, or of continental crust with a piece of oceanic crust attached.

The primary driving force for these plate motions is the emergence of fresh lava at mid-ocean ridges, the best known being the mid-Atlantic ridge. Here basaltic lava erupts from relatively shallow sources in the mantle along cracks or seams in the lithosphere, resulting in the creation of a ridge formed by volcanoes, a divergent plate boundary with slabs of oceanic crust moving in opposite directions, and the progressive opening up of an ocean basin. As the basaltic lava spreads, it becomes cooler and denser, eventually sinking back into the mantle. Slabs of oceanic crust can also slide against each other resulting in transform boundaries, or collide with continental crust creating subduction boundaries where the denser oceanic material slides underneath the edge of the lighter continental crust. These different types of boundaries are associated with narrow belts of intense earthquake and volcanic activity, and are often associated with large-scale uplift and subsidence. These processes, which create and destroy oceanic crust, have resulted in the almost complete renewal of the ocean floor during the past 65 million years.

Subduction systems do not develop in continents because the continental lithosphere is not dense enough to sink back into the mantle. Instead the crust becomes squashed and thickened by convergent plate motions, with the result that continental deformation can spread over 1000s of kilometres to form the Earth's major mountain belts, for example the Alpine-Himalayan belt that marks the collision of the African, Arabian and Australian Plates with the Eurasian Plate. Earthquakes are widely distributed over such regions, hosted by complex and continuously evolving fault systems. Transform faults in the form of major strike-slip faults also occur in continental lithosphere, where blocks of continental crust slide past each other, but these can be geometrically more complex than the features found in oceans.

In general, uplift and subsidence on land, and the consequent irregularity or roughness of the surface at many different scales can mostly be explained as a secondary consequence of horizontal compression due to plate motions acting on the boundaries of a deforming region and causing thickening by faulting and folding. However, Earth scientists are increasingly aware that there are other factors at work. No phenomenon has posed a greater challenge to our understanding of uplift and subsidence than the tectonic evolution of the African continent (Burke 1996, Burke and Gunnel 2008).

Plumes

Although a major part of the heat from the Earth's core that reaches the surface is carried by basalt erupting from shallow mantle sources at ocean ridges as part of the plate tectonic process, heat also reaches the surface in other ways. Prominent features, referred to as hot spots, are considered to be associated with hot plumes coming from deep sources (>1000km depth) at the core-mantle boundary of the Earth's interior, much deeper than the shallow source magmas of ocean ridges (Davies 1999). When a plume reaches the lower boundary of the lithosphere the plume head spreads out like a mushroom head to create a swelling of the Earth's surface with kilometre-scale vertical motion and 1000km horizontal extent, which can cause uplift, doming and rupture of the lithosphere. Uplifted regions such as Ethiopia, built up from a 4km-thick sequence of flood basalts that erupted at about 30 Ma, are thought to be the result of plume activity. These plumes are fixed relative to the motions of plates in the lithosphere. Where the overlying plate is relatively stationary, as is the case with the African plate, the effect of the plume head can lead to swelling and uplift, and this probably accounts for the elevation of the Ethiopian plateau, parts of the Kenyan Rift, and the interior of South Africa. Where the overlying plate is moving, it can result in a chain of volcanic activity, and this is thought to be the case for the Hawaiian Islands, resulting from a plate moving over a long-lived plume source.

This plume mode of deformation is not as well understood as the plate mode, mainly because it is difficult to observe, and it has in general and until recently been considered at best of minor importance in comparison with plate motions, and of relevance only to very early periods of Earth's geological history. However, views are changing as a result of new information, in particular tomographic images of heterogeneity in the deep mantle and satellite measurements of earth gravitational anomalies. Also, a growing number of surface phenomena that are not adequately explained by plate motions are increasingly being referred to the effects of the plume mode and to landscape effects that can manifest on shorter (Myr to kyr) geological time-scales (Bunge and Glasmacher 2017; Friedrich et al. 2017; Guillocheau et al. 2017).

Rifts

Rifts were originally described as Graben, first named for the Rhine Valley (Suess 1885) and were thought to be the result of a down-dropped piece of land bordered by parallel faults, creating a wedge-shaped block rather like the V-shaped keystone of an archway. This is now known to be a misconception, and rifts are recognised as resulting from crustal extension and thinning. This can result from mechanical effects associated with either mode of deformation (plate or plume), or from thermal expansion and uplift associated with lithospheric melting and volcanism in the plume mode. In all cases extension is accommodated by normal faulting with progressive uplift of the rift flanks and subsidence of the rift floor. It is this extensional process that allows the formation of new ocean floor, but not all rifts evolve into oceans.

Rifts can vary in scale from 100km or less (e.g., the Gulf of Corinth, Greece) to the >3000km structure that forms

the Rhone and Rhine valleys and the axis of the North Sea. The latter is thought to be the result of the nascent extension that finally formed the Atlantic Ocean, but it was abandoned as motion concentrated on other rifts to the west. Rifts can therefore be early features in the creation of new oceans, only to be abandoned, and these are sometimes referred to as Failed Rifts. The status of the East African Rift in this regard is not clear. Is it a new plate boundary in process of being formed that will ultimately become an ocean basin, or will it become quiescent? There is no agreement (e.g., King and King 1967).

To the north, the Red Sea basin is a new ocean that results from the continued widening of an extension of the early East African Rift. This was made possible by the opening of the Gulf of Aden and the creation of the Arabian Plate, which is now moving away from Africa and converging with the Eurasian Plate (Hubert-Ferrari et al. 2003). The boundary of this plate extends through the Levant as a strike-slip feature with compression and uplift in Lebanon and some extension to the north and south.

Rift mechanisms

Rifts are characterised by a single main fault (sometimes with closely spaced subparallel branches) and are normally asymmetric in cross section (Figure 3). The main fault can occur to the left of the rift axis, or to the right, and there is no general rule in this regard. Faulting on one side of the rift (the foot wall) dominates and shows upward flexure and uplift, while the other side of the rift (the hanging wall) shows downward flexure and antithetic faulting with no more than 10% of displacement compared to the main fault (Figure 3a).

The lithosphere associated with old cratons, as is the case throughout much of the East African Rift, is generally thicker than elsewhere, and the elastic properties associated with the lithosphere therefore extend to greater depth. Hence, the flexure associated with faulting has a greater lateral extent than would otherwise be the case, which is why the African Rift valleys are mostly wider than those in other parts of the world.

The main fault is commonly complex, composed of several subparallel faults, which create a ragged main scarp (Figure 3b). The faults associated with tension structures at the surface of the footwall create local narrow valleys, well-illustrated by the Delphi valley in the Gulf of Corinth (Armijo et al. 1996). Unlike many other rift valleys, volcanoes are common within the African Rift and are considered to be related to thinning of the crust associated with extension, which is thought to be greatest along the rift axis. However large volcanoes also appear on the footwall and are

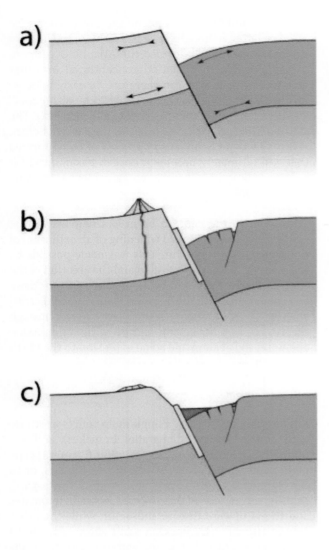

offset from the main rift axis, such as Mt. Kenya and Mt. Kilimanjaro. They are probably the result of extension at depth that creates tensional features aiding the rise of magma (Ellis and King 1991) (Figure 3b). In time, the effects of erosion smooth the visible part of the main fault making it a less obvious feature than some of the much smaller (antithetic) faults in the hanging wall. Sediments derived from slope erosion of the uplifted rift flanks accumulate on the rift floor where they can obscure the original morphology of the fault scarp in the valley bottom (Figure 3c).

The East African Rift

The East African Rift extends for about 1200km, from Ethiopia in the north to Mozambique and the borders of South Africa in the south, and is a composite structure with different segments of clearly identifiable rift broadly aligned on a north–south axis but with some branching and divergence (Figure 4, Figure 5). Rifting is thought to have been initiated with the massive flood-basalt eruptions that occurred in Ethiopia at 30 Ma. Propagation of the Rift appears to have proceeded generally from north to south exploiting weaknesses between cratonic blocks, splitting into an Eastern Rift passing south through Kenya and Tanzania, and a Western Rift curving around the border of the Democratic Republic of the Congo, before re-joining the Eastern Rift to propagate further south through Malawi. Further south again, the Rift begins to break up and spread out with structures developing in different directions towards Mozambique, the Zambesi and the Okavango Delta. The asymmetry associated with local rifting is clearly visible in a series of cross-sections taken at different points along the Rift (Figure 4).

Geology

Rock types include cratonic basement rocks, volcanic lavas such as basalt and trachyte, sedimentary and metamorphic rocks of various types, and Quaternary sediments (Figure 5b). Quaternary sediments are patchy and of limited extent, being confined to lake basins and valley bottoms. Volcanic rocks, on the other hand, are more common than in most other rifts, and are especially prominent in Ethiopia, Kenya and Tanzania.

Altitude and drainage

There are substantial variations of altitude of the rift floor, ranging from close to sea-level to nearly 2000m (Figure 5a). These differences of elevation are related to uplift from mantle heating in some regions associated with plumes. Considerable differences also occur at a more local scale within some parts of the Rift valley,

Figure 3. Diagrammatic sections showing normal faulting resulting from extension in the Rift setting. (a) Simplified diagram, with a single fault, demonstrating the asymmetric nature of displacement associated with faulting. The footwall, shown in green, is put into compression at depth and extension at the surface (shown by the arrows). The hanging wall, in yellow, shows compression at the surface, and extension at depth. (b) This shows that the main fault may be a series of faults – indicated by the extra slab. Also shown is the antithetic fault in the footwall and the small clean faults created by extension at the surface of the footwall. Extension of the hanging wall at depth allows the entry of magma, which is then able to force its way to the surface to create volcanoes in the hanging wall, counterintuitively, given that the surface is under compression. Mt. Kenya and Mt. Kilimanjaro are examples of volcanoes formed in this way on the hanging wall of the Rift. (c) This shows the effects of erosion from the Rift wall and accumulation of sediments in the trough formed at the base of the main fault. This gives the impression of a flat surface on the Rift floor and hence contributes to the mistaken belief that a whole block has dropped down.

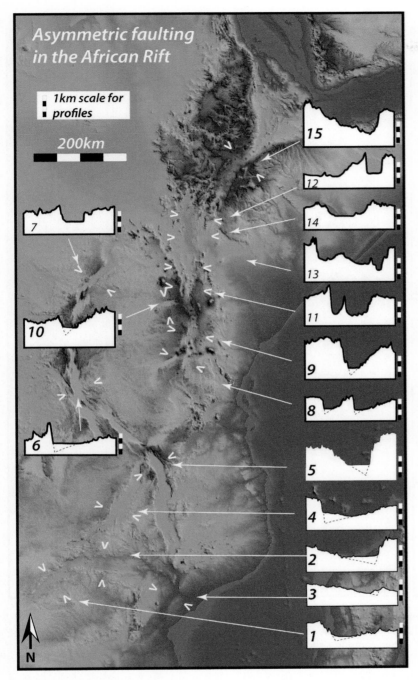

Figure 4. An overview of the main sectors of the East African rift, showing a series of cross-sectional profiles. Note that the main fault and the hanging wall may be on either the left side or the right side of a given rift segment.

according to global climate changes, especially changes in rainfall patterns.

Some of the depressions created by rifting are filled with lakes, the largest being Lake Turkana in northern Kenya, Lake Tanganyika in the Western Rift and Lake Malawi in the south (Figure 5a). The Turkana basin is very deep (>3km) but filled with sediment brought in from Ethiopia by the Omo River. Lake Turkana has at times been connected to the Indian Ocean and may in the past have drained west into the River Nile. Lake Victoria is a relatively shallow lake in a depression formed between the uplifted flanks of the Western and Eastern Rifts. Other lakes contained within the rift are fed by internal drainage from the surrounding slopes, with relatively thin sediment cover on the valley floor and in lake bottoms.

Lake levels of all Rift lakes have varied significantly during the Plio-Pleistocene, responding both to global changes in climate, particularly precipitation falling on the major catchments such as the Ethiopian highlands, and to tectonic movements.

Vegetation and relief

For much of the Rift, savannah (grassland and scrub with some trees) is the main vegetation type, but conditions vary locally between semi-arid desert and forest, largely as a function of altitude and local climate conditions (Figure 5c). The well-known savannah hypothesis, that reduction of forest cover and the need to seek food on the ground was the primary agent selecting for a bipedal gait has now been largely discredited, mainly because the earliest anatomical evidence of habitual bipedalism occurred in hominins such as the Australopithecines, who were still semi-arboreal and living in environments with extensive tree cover. However, the characterisation of the African Rift as a savannah environment is misleading in another way. Savannah is a type of vegetation and usually associated with an image of flat or gently rolling and smooth topography. In reality it is associated with a wide range of topographic conditions and land forms, ranging from basaltic lavas to heavily weathered basement and metamorphic rocks, and from flat

with differences in elevation of 1000m or more between the Rift floor and the crest of the major fault scarps, resulting from progressive uplift of the rift flanks by fault motions. As a result, there are considerable variations in climatic conditions, both regionally and locally, with semi-arid or desert conditions at low altitude, particularly towards the lower latitudes of the northern Rift, and higher rainfall and forest cover on the highest elevations (much of which has been cleared for agricultural purposes in recent times). The relative extent of these different climate zones has varied

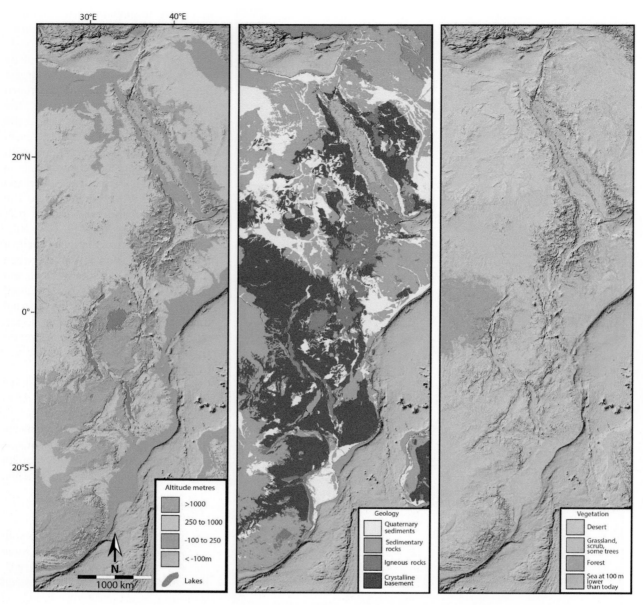

Figure 5. Simplified maps of the East African Rift and its northward extension to the Red Sea and the Levant. 5a, geology; 5b, altitude; 5c vegetation. Note that the coastline is placed at the –100m bathymetric contour. The amount of land between the modern coastline and the –100m contour is shown in Figure 5a by the dark green shading

valley bottoms to steep slopes, fault scarps and rough topography. Concentration on the term savannah has therefore tended to obscure the immense variations of topography and topographic roughness that occur within the savannah vegetation zone and how this may have influenced the evolution of the human body form (Reynolds et al. 2011; Winder et al. 2013).

The Red Sea and the Levant

Structures to the north of Ethiopia are often considered to be an extension of the African Rift. This is true for the 'Proto Red Sea', which extended from the Ethiopian rift into the Gulf of Suez during the Oligocene. However, at about 30Ma the Arabian Peninsula started to move

to the north resulting in the formation of the Gulf of Aden, extension in the Afar region of Ethiopia and the widening of the 'Proto Red Sea'. The original rift flanks formed land escarpments on either side of the Red Sea, with a more pronounced escarpment on the Arabian side (Figure 6).

The Gulf of Aden is typical of what may be expected in the early stages of the formation of an ocean, with a central spreading ridge associated with basaltic volcanism of mantle origin. The water depths are less than for an ocean over a fully formed ocean-ridge, probably because the basaltic magmas in this region are modified by lighter elements derived from the Ethiopian 'Hot Spot', and are in consequence less dense

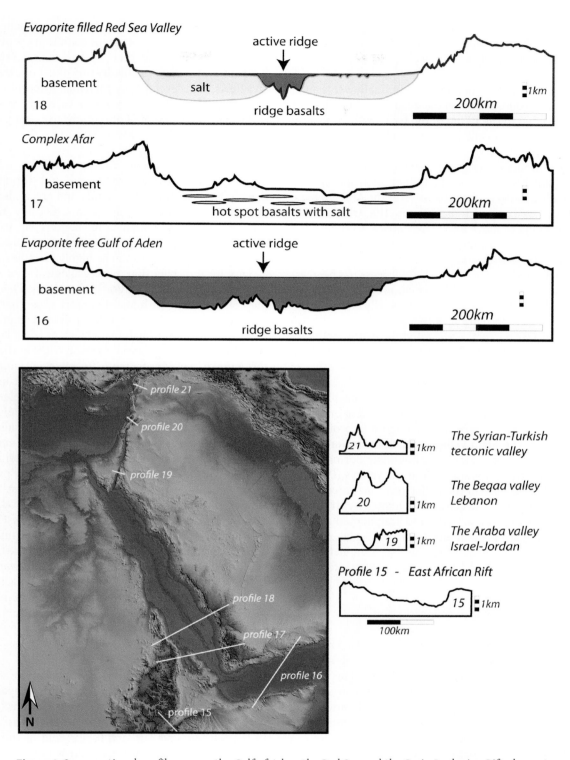

Figure 6. Cross-sectional profiles across the Gulf of Aden, the Red Sea and the Syrio-Jordanian Rift, drawn to the same vertical and horizontal scale as those shown in Figure 4. Profile 15 from the African Rift is shown for comparison. The Gulf of Aden and the Red Sea show a much wider opening than the East African Rift. In the Gulf of Aden a mid-ocean ridge is beginning to form in the centre (profile 16). In the Afar depression, the basalt from the Afar Hot Spot is mixed with patches of salt, producing material that is less dense than the ridge basalts in the Gulf and the Red Sea, putting most of the floor of the Afar rift above sea level (profile 17). The Red Sea is filled with massive deposits of salt (evaporite) formed during the Miocene when the basin was closed off from the oceans, but with a bit of central ridge poking through (profile 18). The main activity is near the centre of these depressions with less activity on the sides. The Levantine profiles show a much narrower opening, resulting from strike-slip movement with some extension. There is very little normal faulting, the main exception being the Haifa fault in Israel, which is not in the Rift zone proper.

than the basalts extruded from other mid-ocean ridges. Similarly, the basalt-dominated Afar is within the Hot Spot region and is above sea level.

In the Red Sea, water depths are even shallower than in the Gulf of Aden, except in the deep axial trough, most probably due to the filling of the basin with very large thicknesses of Miocene-age salt (Figure 6).

Today's Arabian topography and geology have been shaped by the opening of the Gulf of Aden and the Red Sea and the progressive separation of the Arabian plate from the African and Indian Plates. The Arabian Plate is continuing to move away from Africa, to slide along the large transform faults that define its western and eastern boundaries and to collide with the Eurasian continent to the north. The whole process has been associated with episodes of deformation and magmatism, with uplift of the rift shoulders, especially on the Arabian side. This deformation, with a possible contribution from the Afar hotspot, has created the major mountain chains observed along the coasts of the Red Sea and the Gulf of Aden, rising to about 3000m in the southwest corner of the Arabian Peninsula.

The Syrio-Jordanian valley produces very different profiles from the Red Sea, the Gulf of Aden, or any of the profiles in the African Rift (Figure 6, profiles 19, 20 and 21). The main valley is much narrower with a U-shape and little evidence for bounding normal faults considered characteristic of true rift valleys (Devès et al. 2011). This reflects the major tectonic processes at work in this region, which are dominated by strike-slip motion. Earthquake activity and fragmented oblique slip faulting are present along with extensive areas of volcanic activity, generating similar combinations of rough topography and lake-basins to those found in the African Rift.

Methods of landscape reconstruction

In our reconstructions, we place particular emphasis on topographic features such as faults formed by repeating earthquakes that form low barriers or cliffs, and on volcanic lava flows which can act in a similar way as impediments to animal movement. We do so on the assumption that large mammals are a major source of food and that minor barriers can create predictable pathways of animal movements and natural constrictions that facilitate interception and capture. These features can also act as traps for sediment and water, creating lake basins or spring lines. Faulting and volcanic activity are dynamic and episodic processes that can change the landscape setting. In some cases, these changes can be beneficial to human subsistence activity, rejuvenating and sustaining advantageous topographic features such as barriers, lakes and springs, and renewing soil nutrients through surface disturbance and erosion or addition of new material. In other cases, ongoing change can destroy previously advantageous conditions or obscure earlier landscapes and their archaeological remains, for example through the blanketing effect of extensive lava flows. We therefore pay particular attention to the dates of geological features where these are available and to the modelling of fault motions to re-set parts of the landscape to an earlier configuration.

We also pay particular attention to the edaphic properties of different rock and soil formations, namely the presence of minerals and trace elements important to animal health and especially to the growth of young animals. In regions of variable bedrock geology and geomorphological processes, these edaphic properties can be highly variable in their geographical distribution. Both ecological studies, for example of wildebeest migrations (Murray 1995), and our conversations with Masai herders show that these factors play an important role in animal migrations today and that the animals move towards advantageous conditions at the appropriate season without the need for human encouragement or intervention. Also, systematic measurements of the nutrient properties of soils and plants in the African context (Kübler et al. 2015, 2016) show a close relationship between their edaphic properties and the bedrock geology, allowing the use of bedrock geology as a proxy for mapping edaphic properties over larger areas. More general field observations and anecdotal evidence in other parts of the Near East and the Mediterranean suggest that a similar relationship holds elsewhere (Devès et al. 2014; Sturdy et al. 1997; Sturdy and Webley 1998). Finally, in a digital age, we make full use of satellite imagery, digital elevation models and computer software to map geologically and topographically significant variables and to construct visual images. This greatly facilitates our ability to place local features in the immediate vicinity of sites into their wider regional context.

The Kenyan Rift

Here we concentrate on just two sites, Kariandusi and Olorgesailie, in the central and southern Kenyan Rift, although we note that there are concentrations of sites with archaeological or fossil material to the north, in the Baringo and Turkana basins, and to the south across the Tanzanian border at Olduvai Gorge as well as sites such as Kilombe in the central Rift (Figure 7). We summarise key features of their surrounding regions, drawing on more detailed analyses in Kübler et al. (2015, 2016) which provide supporting detail.

Kariandusi

This site consists of two closely adjacent concentrations of stone tools including flakes and numerous Acheulean bifaces, the first investigated by Louis Leakey between

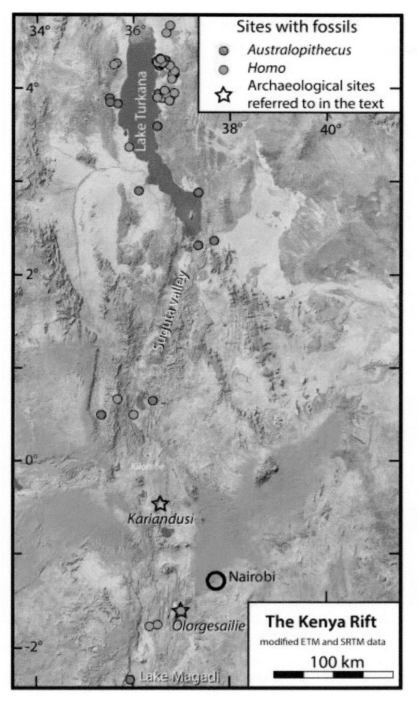

Figure 7. General location map of sites in the central and southern Kenyan Rift in relation to topography, showing position of Olorgesailie and Kariandusi. Other sites with hominin fossils are also shown. Data from the Paleobiology database (https://paleobiodb.org). After Kübler et al. 2016. The image is based on ETM+ legacy data. Topography from SRTM v4.1. Maps created by SK and GK using Adobe Illustrator CS 5.1, Adobe Photoshop CS 5.1, MAPublisher 9.4.0, and Global Mapper.

that filled a large part of the rift basin (Figure 8). Faunal remains are rare because of preservation conditions and consist only of some equid teeth. The material is bracketed between 0.73 and 0.98Ma and this corresponds to a period of high lake level some 80–90m above the present rift floor.

The region is a typical asymmetric rift with uplifted flanks reaching to 3000 to 4000m and a rift floor at about 1800m. Rifting is aligned on a NNW–SSE axis and started in the late Miocene, between 12 and 6Ma, with further spells of activity between 5.5 and 2.6Ma, when the 30km-wide Kinangop Plateau and the 40km-wide inner rift depression were formed. The inner rift was subsequently covered by trachytic, basaltic and rhyolitic lavas and tuffs and continues to be cut by normal faulting around a central axis. Thus, the major faults that define the topography of the flank region around the Kinangop Plateau were already in place when the site was occupied. However, the younger structures of the inner rift and much of the volcanic activity visible today post-date the occupation of the Kariandusi site and limit what can be said about the landscape features and edaphics of these areas at the time when the site was in use (Figure 9).

The other major change in physical conditions is variations in lake level, with evidence of at least three episodes of high lake levels associated with wetter climate over the past one million years: at ~1ma, 150–60ka, and 15–4ka (Bergner et al. 2009; Trauth et al. 2005), the first episode coinciding closely with the Acheulean occupation at Kariandusi.

The tectonic history has created a complex topography with extensive upland basins such as the Kinangop plateau bounded by faults that form steep cliffs and limited points of access for large mammals (Figure 8). Edaphic properties are also variable in their distribution, with the best conditions on some of the volcanics and on the lake sediments of the rift floor. The distribution of edaphically-rich soils in the wider region is patchy, with the best concentrations on the rift flank to the east of the site. Here, an extensive upland basin

1928 and 1931, the second by John Gowlett in 1974 (Gowlett & Crompton 1994; Leakey 1931; Shipton 2011) (Figure 2). The material was originally deposited on the banks of a small tributary of the Kariandusi River at a time when it drained into an expanded Lake Elmentaita

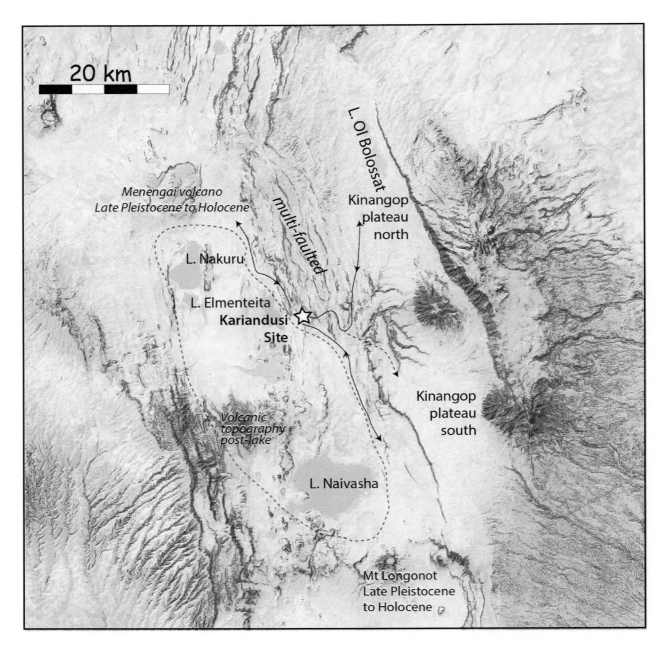

Figure 8. Morphology of the region including Lakes Nakuru, Elmenteita, Naivasha and Bolossat. Slopes greater than 15° are indicated by grey shading, and would have restricted movements of large animals to well defined areas and restricted pathways that converge on the Kariandusi location. Lakes Nakuru, Elmenteita and possibly Naivasha were joined to form a single lake, shown by the blue area enclosed by a dashed line, during periods of high lake level. Some earlier features (that pre-date the Kariandusi site) have been masked by later volcanism between Elmenteita and Naivasha and around Menengai volcano and Mt. Longonot (see Figure 9)

is enclosed by steep faults and easily accessible to herds of large animals only from the rift floor to the west and northwest (Figure 10). At the time when the site was occupied, much larger areas of the rift floor were submerged under an enlarged freshwater lake, and the lakeshore was at a higher level, forming a barrier to animal movements from west to east except through a narrow corridor close to Kariandusi (Figure 11).

Freshwater, raw material for making stone artefacts and good viewpoints for tracking animal movements were all locally available throughout the Pleistocene. However, the site appears to have been used only during the period when high lake levels gave the location strategic advantage in relation to topographic barriers to target large mammals during their seasonal migrations.

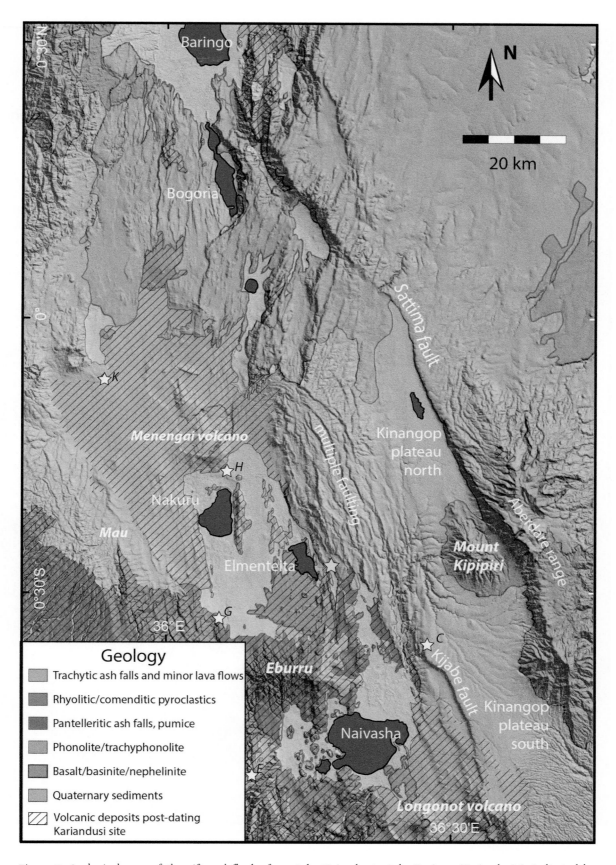

Figure 9. Geological map of the rift and flanks from Lake Naivasha to Lake Baringo. Kariandusi is indicated by a yellow star. Cross hatching indicates volcanic deposits c. 700 ka – postdating the Kariandusi site. Small stars indicate other sites. C: Cartwright's site, E: Ekapune Ya Munto site, G: Gamble's site, H: Hyrax Hill, K = Kilombe site (After Kübler et al. 2016).

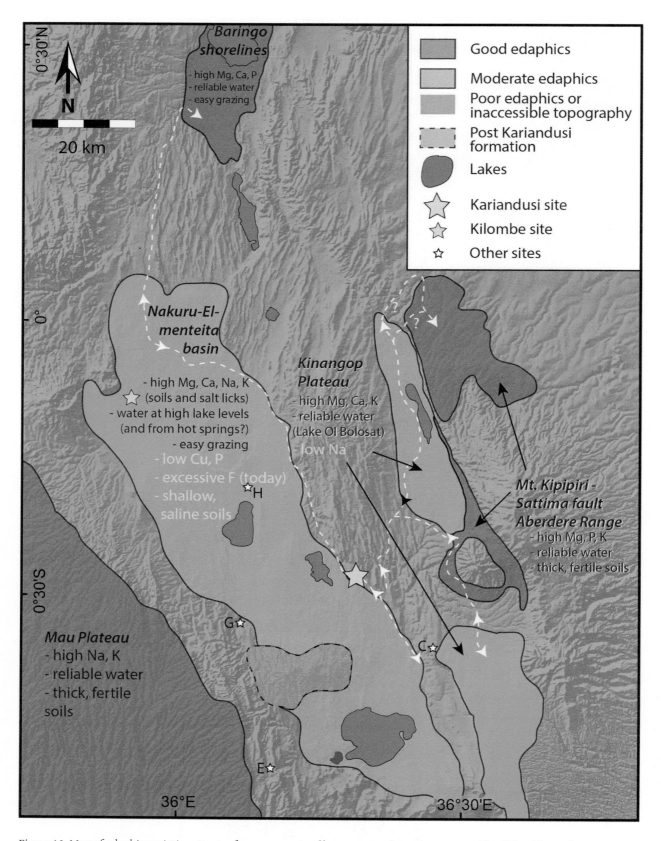

Good edaphics

Moderate edaphics

Poor edaphics or inaccessible topography

Post Kariandusi formation

Lakes

☆ Kariandusi site

☆ Kilombe site

☆ Other sites

20 km

Baringo shorelines
- high Mg, Ca, P
- reliable water
- easy grazing

Nakuru-El-menteita basin
- high Mg, Ca, Na, K (soils and salt licks)
- water at high lake levels (and from hot springs?)
- easy grazing
- low Cu, P
- excessive F (today)
- shallow, saline soils

Kinangop Plateau
- high Mg, Ca, K
- reliable water (Lake Ol Bolosat)
- low Na

Mt. Kipipiri - Sattima fault Aberdere Range
- high Mg, P, K
- reliable water
- thick, fertile soils

Mau Plateau
- high Na, K
- reliable water
- thick, fertile soils

H

G

C

E

36°E

36°30'E

Figure 10. Map of edaphic variation. Routes for movements of large mammals on the eastern side of the rift are shown. At very high lake levels, large animals would have to pass the Kariandusi site. After Kübler et al. 2016.

Olorgesailie

Olorgesailie lies in the centre of a 60km-wide rift floor adjacent to a palaeo-lake (Figure 1). Numerous Acheulean artefacts, fossil mammals including elephant, hippo, giant baboon, equids and bovids with evidence of carcase butchery, and part of a *Homo* skull are preserved in sediments spanning ~1.2 to <0.5Ma (Behrensmeyer et al. 2002; Isaac 1977; Potts 1989; Potts et al. 2004).

Trachyte flows laid down between 0.7 and 1.4Ma are the dominant rock type, together with Plio-Pleistocene basalts and tuffs (1.4–2.7Ma) associated with the volcanoes of Mt. Olorgesailie and Mt Esayeti. There are also small areas of sediment carried from the rift flanks, but uplift and back tilting prevent the entry of sediments from outside the main rift. There are many sub-parallel fault scarps, and these are nearly vertical in places, especially on the trachyte, which is particularly resistant to erosion, creating a complex topography that would have constrained east–west movements of large animals and humans (Figure 12).

Ongoing faulting and the partial collapse of the Olorgesailie caldera have created a modern landscape that has clearly changed during the past million years (Figure 13a). In particular tectonic motion on two north-south trending normal faults has resulted in tilting of the Legemunge lake beds that contain the archaeological material. Also, partial collapse of the Olorgesailie caldera has resulted in draining of the palaeolake. The geometry of the landscape past and present is defined by its tectonic structure, allowing reconstruction of palaeo-morphology by modelling fault displacements. By removing the effects of fault motion and making corrections for erosion and deposition of sediment, a palaeo-DEM of the landscape can be created as it would have appeared during the period when hominins were using the Olorgesailie locality (Figure 13b).

Modern analyses of macronutrients and trace elements from soils on a representative sample of lithological and sedimentary units demonstrate a consistent relationship between edaphic quality and the underlying regolith (Kübler et al. 2015). The trachytes, which dominate the region, have poor-quality soils; richer soils develop only on some other volcanic rocks and on sediments brought by rivers from the north and east. This is supported by interviews with Masai shepherd families who are well aware where animals must graze and browse to remain healthy.

In the past, soils close to the palaeolake would have provided an attractive focus for animal grazing and browsing, especially during the dry season. But this area is comparatively small, so that large herds would need to move to the more distant rift flanks. Routes of

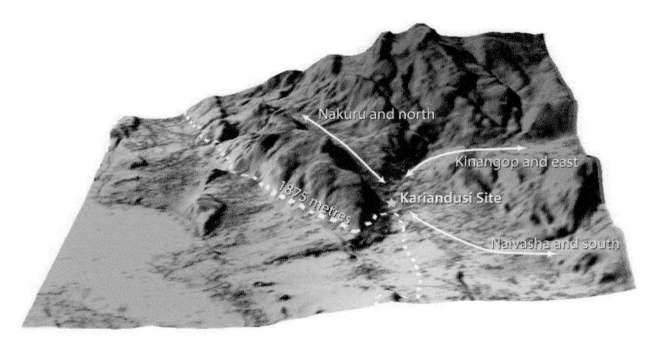

Figure 11. Oblique 3D view looking north showing the routes for large animals when the level of Lake Elmenteita is high. Lake levels of greater than 1875m completely block access along the lake-shore. For lower lake levels, movements along the lake shore would have been straightforward and the lake sediments would also have provided good grazing or browsing. At high lake-levels, animals moving to or from Nakuru and the north would have to pass along a valley bounded by fault scarps and cliffs. Kariandusi is ideally located to intercept animals crossing the Kariandusi River. The DEM used for this figure is a product of two World View 2 scenes processed with "Stereo Pipeline" software. Data made available through collaboration with Ryan Gold, USGS. After Kübler et al. 2016.

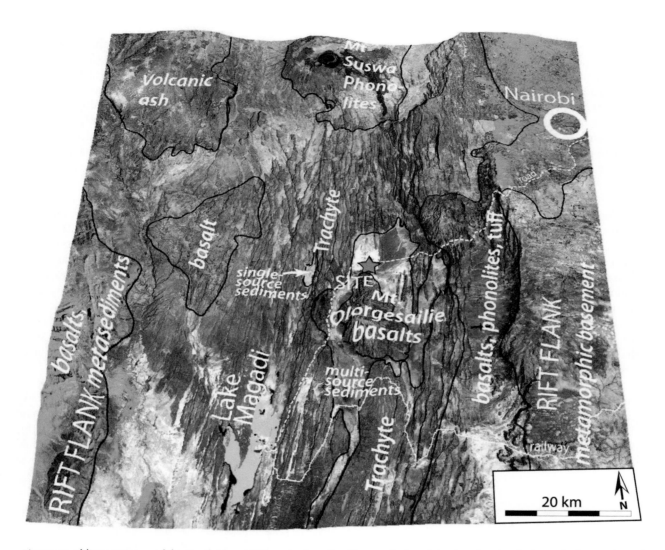

Figure 12. Oblique 3D view of the South Kenya Rift centred on the Olorgesailie hominin site, showing faults and other geological features. The region is heterogeneous with some areas well-vegetated and others with thin vegetation. Thin vegetation can result from soils that do not favour plant growth, but can also result from heavy grazing and browsing of favoured vegetation. The image is based on ETM+ legacy data. Topography from SRTM v4.1 data with a vertical exaggeration of ~8. A red star indicates the Olorgesailie site. Maps created by SK and GK using Adobe Illustrator CS 5.1, MAPublisher 9.4.0, Global Mapper 16, and ENVI 5.1.

animal migration to or from the flanks were greatly constrained by the numerous north–south fault scarps and would need to pass along a predictable route between the lake and the volcanic edifice (Figure 13b). It is on this route that the archaeological site is located.

We conclude that a key reason why the Olorgesailie site was attractive to hominins and repeatedly used over a long period is its proximity to a nearby area that has excellent edaphics in a wider region that is highly deficient, and that it controlled the only route allowing movement of large animals between east and west, facilitating trapping by ambush hunting. Proximity to a reliable source of drinking water, suitable volcanic stone for making artefacts, and good look-out points for monitoring animal movements are additional advantages of the Olorgesailie location, but are not sufficient to explain why the site remained a repeated

focus of human activity over such a long period. When the caldera collapsed and the lake was drained, the topographic constraints that favoured ambush hunting disappeared and occupation ceased.

The Southern Levant

Although the Jordanian Rift is a strike-slip structure rather than a rift, it nevertheless has many similar features to the East African Rift, with high levels of earthquake activity and faulting (Devès et al., 2011), volcanic activity with extensive basaltic lava flows, elongated valleys with lake-filled basins subject to variations in lake level, marked changes of elevation over relatively short distances, and a complex topography with numerous cliffs and deep, steep-sided valleys. The geology is dominated by limestone and basalt, which offer attractive environments for herbivores including

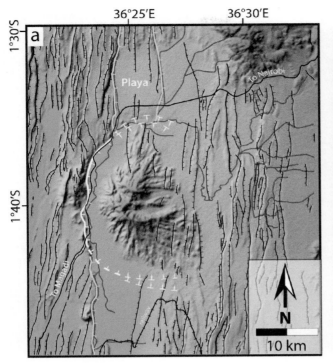 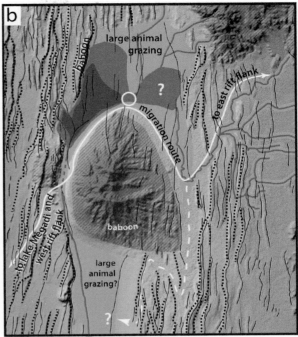

Figure 13. Digital elevation models (DEMs) of the Olorgesailie region. (a) present day topography. Prominent fault scarps are indicated in black and those indicated in yellow and white are used in the modelling of earlier topography in (b). The yellow faults are young faults, and post-date the lake and the period of hominin activity at Olorgesailie. The white faults result from caldera collapse. (b) PalaeoDEM of the Olorgesailie region. White circle indicates the position of the Olorgesailie site. The volcanic edifice of Mt. Olorgesailie was already in place and impeded drainage to the south, resulting in the formation of the lake. A possible late lower lake level is indicated in dark blue. Some faults in the trachytes do not cut basalts of the Olorgesailie edifice and therefore clearly pre-date it. Dotted lines show faults that formed barriers to animal movement and are thought to have existed when the site was used by hominins. The drainage system was limited by a barrier in the same place as hypothesized by Behrensmeyer and colleagues (Behrensmeyer et al. 2002), so the barrier could have been higher than the lake without fully blocking drainage from the lake. Likely grazing areas are indicated. Routes to the flanks of the Rift negotiable by large animals are shown. Fault scarps and the volcanic edifice would only be accessible to smaller and more agile animals. Images are based on SRTM v4.1 data. Maps created by GK using Adobe Illustrator CS 5.1, and Global Mapper 9.4.0. Landscape reconstruction and fault mechanisms are calculated with Almond 7.05 software (www.ipgp.jussieu.fr/ king) based on the program developed by Okada (1982)

good edaphic properties (Figure 14). It is also a region with a long history of archaeological and geological investigation and many Palaeolithic sites dating back to over 1 Ma (Enzel and Bar-Yosef 2017)

Gesher Banat Ya'aqov

One of the major sites of the Lower Palaeolithic in the region is the Acheulean site of Gesher Banot Ya'acov with thousands of stone artefacts including Acheulean bifaces, smaller flakes designed for hafting, evidence for the use of fire and repeated use over a period of about 150,000 years between about 0.7 and 0.85Ma (Alperson-Afil and Goren-Inbar 2016; Goren-Inbar et al. 2000; Rabinowich et al. 2012). The site is located near the edge of the palaeo-Hula Lake, and part of it remains waterlogged, with preservation of organic remains including wood, and evidence of plant foods. Animals exploited include elephants, hippo, rhino, gazelle, horse and bovids, with evidence of carcase butchery, and fish from the lake.

At the time of its occupation, the site was strategically located to intercept large mammals moving through one of the few available corridors for east–west movement across the Jordan valley (Figure 14, Figure 15). Immediately to the north were the margins of the expanded Hula Lake. To the south was the palaeo-Lake Lisan, with an intervening area dissected by rough topography and steep slopes. Like the African sites already discussed, the location of the site has a number of advantageous features including proximity to freshwater, volcanic material for artefact manufacture, and, in the case of Gesher Banot Ya'acov, additional food supplies of non-migratory animals such as wild boar, plants and fish. However, given that the large herbivores, particularly elephant and fallow deer, were the major support of the palaeoeconomy, the location of the site on one of the few crossing points from west to east, combined with the tactical opportunities for bringing down large prey, including miring along lake edges, is a prime factor in the importance of this site (Devès et al. 2014: 152–153).

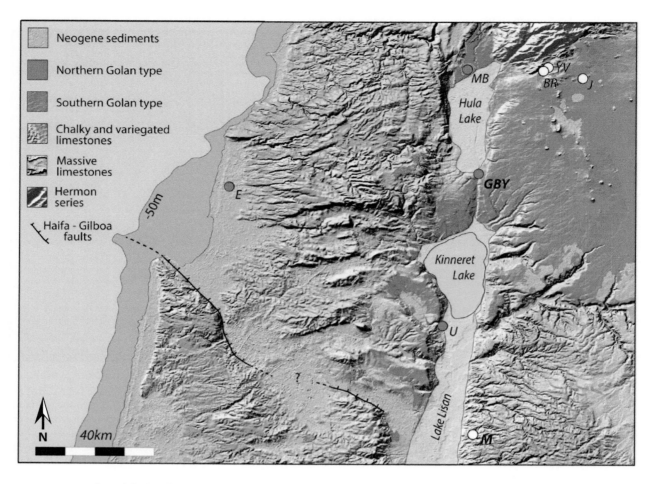

Figure 14. Map of simplified geology and topographic roughness in the central Jordanian Rift, showing the position of the palaeo-Lisan Lake and Lake Hula at their maximum extent. Red circles indicate Acheulean sites with fauna, white circles Acheulean sites without fauna. Steep slopes and rough topography render large areas inaccessible to large mammals such as elephants, with the smoothest terrain on the Golan Heights to the top right and on the Neogene sediments towards the coast. The Golan Heights also have the largest extent of edaphically rich conditions. Gesher Banat Ya'acov (GBY) is located on the only feasible pathway for large mammals moving between summer grazing territories on the Golan Heights and winter territories on the coast. Moreover the site is located on a narrow neck of land ideal for trapping large mammals as well as providing access to the resources of the nearby lake (see Figure 15). After Devès et al. 2014.

Conclusion

We draw two main conclusions from the above observations, the first is about the adaptive strategies implied by our landscape approach, the second is about our methods of landscape reconstruction in tectonically active environments and their feasibility and wider applicability.

Regarding adaptive strategies, all three sites that we have examined have a number of features in common. In the first place they are all sites which have accumulated large assemblages of stone artefacts, numbering thousands of specimens, with large cutting tools of the Acheulean tradition including bifaces and cleavers. Two of the three sites (Olorgesailie and Gesher Banot Ya'acov) also have large assemblages of large-mammal bones with evidence of on-site butchery of large animals such as elephants, hippos, bovids and

equids, while the third site hints at similar activity but is largely lacking in evidence apparently because of poor conditions of bone preservation. They are all 'large' sites in the sense that the quantities of material indicate either intensive activity in one place, or repeated visits over a long period, or most likely both. In relation to their wider landscape setting they are all located on corridors of movement that represent almost the only feasible pathway of travel for large mammals moving between different feeding territories. In relation to their immediate surroundings they are all located close to natural topographic barriers that would have greatly facilitated the trapping and capture of large and dangerous animals.

They are also all in lake-edge settings that would have provided abundant freshwater, perhaps an additional attractor for large mammals, and potential additional food supplies such as fish (with certain evidence of

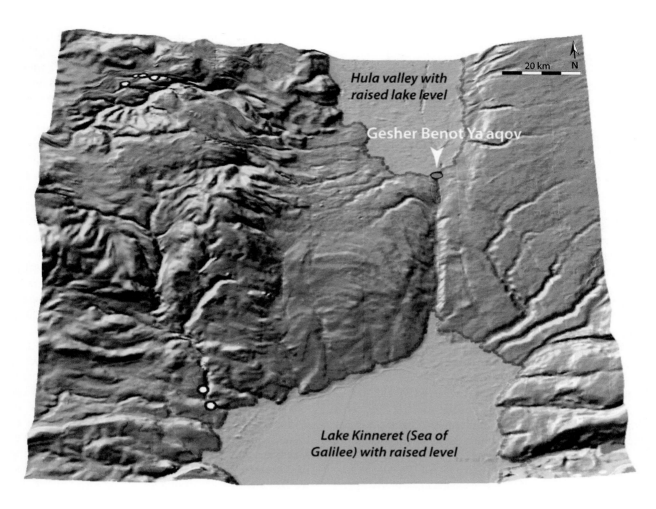

Figure 15. Oblique 3D view of the region around Gesher Banat Ya'aqov in relation to geology and topography at the time of its occupation. The image is based on etm+ legacy data. Topography from SRTM v4.1 data with a vertical exaggeration of ~8. Colour coding of geology as in Figure 14. Note the restricted corridor between the expanded Lake Hula to the north and the steep-sided Upper Jordan valley to the south. This forms a natural ambush for trapping large mammals moving between the grazing lands west of the Rift valley and the Golan Heights. After Devès et al. 2014.

their exploitation at Gesher Banot Ya'acov). Supplies of water and raw materials for stone tool manufacture, though offering added attractions, are widely available elsewhere. They do not in themselves uniquely determine why the sites we have discussed are located where they are or why they attracted so much activity over long periods. It is the added factor of their position in relation to the local and regional topography and the likely pattern of large-mammal movements that is significant. We infer from these features that their occupants were using the topography as an aid to ambush hunting.

Kariandusi is of particular interest in relation to the other sites, because it has the shortest time-depth of occupation, and one that appears to have coincided with a short-lived period of high lake level. Since the lake shore at its highest level was essential to creating the funnel that compelled animals to move through the Kariandusi trap, we conclude that the site was not

viable for regular visitation before or after that time because large mammals were able easily to bypass the Kariandusi location when lake level was lower. We recognize that other hypotheses could be advanced in this case, for example that Acheulean occupation in the Central Rift was only possible during wet climatic episodes, hence the association with high lake levels, or that proximity to the lake shore was important for other reasons such as potable water or fish. However, we reject the climatic hypothesis as implausible (the large mammal community has persisted in the region through periods of low lake level, as today). We also reject the hypothesis of proximity to lake resources as an insufficient factor for reasons explained above.

We conclude that the use of topography to ambush large mammals was a strategy widely practised by early *Homo* populations making Acheulean tools from at least 1.2Ma onwards, and one that contributed to their successful dispersal within and beyond Africa.

Regarding our methods of landscape reconstruction, we note that despite ongoing tectonic and volcanic activity, we have been able to show that many details of topography and landscape can be reconstructed both locally and over large areas with some confidence as they would have existed many hundreds of thousands of years ago. This includes not only physical features such as cliff lines and lake shores but also the edaphic properties of rocks and soils of importance to animal movements and health. Of course, some features have changed substantially, but they are also those features that are most amenable to mapping and dating: changing lake levels as inferred from lake sediments, volcanic lava flows, and changes in fault scarps as inferred from stratigraphic observations and modelling of fault motions.

Of course, questions arise as to whether the same methods of reconstruction can achieve similar results in other parts of the Rift, or whether similar features of dynamic landscapes are associated with major Acheulean sites elsewhere. We have noted that even within the central Kenyan Rift a large part of the area around the site of Kilombe is covered by volcanic lavas that postdate the occupation of the site and limit what can be said about the local landscape at that time. Other parts of the East African Rift have undergone more dramatic changes.

At Olduvai Gorge, one of the richest source of fossils and archaeological materials in Africa (Ashley et al. 2010), it is clear in general terms that the palaeo-Olduvai lake was created by damming of lava flows from the Ngorongoro volcano and that archaeological deposits are associated with lake-shore settings intimately associated with continued extension and normal faulting on the west side of the Rift together with associated volcanic activity. Variable edaphics in the wider region around Olduvai have probably also played a significant role in animal migrations (Murray 1995). However, subsequent fault motion created a gorge that cut through earlier sediments and destroyed many of the details of the original landscape setting. It is that erosion of course that has facilitated the exposure and discovery of archaeological and fossil material. This in its turn has given rise to the belief that rift tectonics destroys local environments and, through uplift and erosion, provides a sufficient explanation for the burial and subsequent exposure and discovery of early finds. However, our earlier examples demonstrate that this taphonomic factor cannot be universally true and is not a sufficient explanation for the occurrence of large sites.

The Lake Turkana region is another rich source of early finds and is different again. Although, this is one of the deepest parts of the East African Rift system, unlike other deep valleys (e.g. Lake Tanganyika) it is completely filled with Plio-Pleistocene sediments and contains only a shallow lake. On both sides of the lake and in the Omo valley these sediments contain hominin and other animal fossils. The sediments on the east side of the Lake, notably at Koobi Fora, are close to outcrops of basement rocks, basalt lava flows, small faults and river terraces. The relief is modest and at a small scale but includes features that hominins could have exploited for security and perhaps to facilitate access to prey. However, we currently lack the information to reconstruct the earlier environments in sufficient detail to test alternative hypotheses of hominin exploitation. Further north in the Afar, many early sites were associated with the active rift, a highly dynamic setting where a great deal has changed, and we have previously suggested that the original setting of the early sites can only be understood at best by analogy with the area where the present-day Awash River enters the currently active rift (Bailey et al. 2011: 265–267).

South Africa lies outside the Rift zone proper, yet it is another region rich in early finds of human fossils and Oldowan and Acheulean stone assemblages over a wide time range. Here too there is evidence of geological disturbances that create features similar to those we have described in the Rift. In the region of Makapansgat and Taung, minor rifting and faulting has created uplifted areas next to down-dropped wetlands with river drainage cutting into fault scarps to form a steep-sided gorge (Bailey et al. 2011). In the Cradle of Humankind region, faulting cross-cuts older geological structures and disturbed river profiles are evidence of more recent activity. However, fault scarps are absent and the complexity of the topography is mainly the result of differential erosion acting on rocks of differential hardness, resulting from a much longer geological history of uplift and erosion of the South African dome, and rejuvenation and headward erosion of the Limpopo and Crocodile Rivers (Bailey et al. 2011: 268–269).

On the western Arabian escarpment, faulting associated with the active rift mainly takes place deep beneath sea level in the axial trough of the Red Sea. Fault scarps are rare on land and the main expression of tectonics is extensive volcanic lava flows that impede drainage to create fertile local environments and create physical barriers and constraints on animal movements. A further challenge in this region is the effect of Quaternary sea-level change, which has obscured large parts of the landscape likely to have been important to prehistoric hunters and the archaeological evidence of their presence (Bailey et al. 2015).

Tectonically informed research is ongoing in many of these regions, and we believe that the application of the concepts and methods described here will in due course enable reconstruction of more landscapes with the same degree of confidence, producing a richer dataset of comparative examples, more extensive tests of existing hypotheses, and most likely a wider variety of patterns of human land use and site function associated with different sorts of archaeological and fossil deposits, different regions and different stages in the hominin evolutionary trajectory.

Acknowledgements

The African research reported here was funded by the European Research Council through ERC Advance Grant 269586 DISPERSE, under the 'Ideas' Specific Programme of FP7, granted to GB and GCPK in 2011, with support from the National Museums of Kenya, the Kenya Agricultural and Livestock Organisation, the Kenyan National Commission for Science, Technology and Innovation, and the British Institute in Eastern Africa. This is DISPERSE contribution no. 47.

References

Alperson-Afil, N., Goren-Inbar, N. 2016. Acheulian hafting: proximal modification of small flint flakes at Gesher Benot Ya'aqov, Israel. *Quaternary International* 411. Part B: 34–43.

Armijo, R., Meyer, B., King, G.C.P., Rigo, A., Papanastassiou, D. 1996. Quaternary evolution of the Corinth Rift and its implication for the late Cenozoic evolution of the Aegean. *Geophysical Journal International* 126: 11–53.

Ashley G.M., Dominguez-Rodrigo, M., Bunn, H.T., Mabuulla A.Z.P., Baquedano, E. 2010. Sedimentary geology and human origins: a fresh look at Olduvai Gorge, Tanzania. *Journal of Sedimentary Research* 80: 703–709. DOI: 10.2110/jsr.2010.066.

Bailey, G.N. 2005. Site catchment analysis. In C. Renfrew, P. Bahn (eds) *Archaeology: the key concepts*: 230–235. London: Routledge.

Bailey, G.N., King, G.C.P. 2011. Dynamic landscapes and human dispersal patterns: tectonics, coastlines, and the reconstruction of human habitats. *Quaternary Science Reviews* 30 (11): 1533–1553.

Bailey, G.N., Reynolds S.C., King, G.C.P. 2011. Landscapes of human evolution: models and methods of tectonic geomorphology and the reconstruction of hominin landscapes. *Journal of Human Evolution* 60: 257–280. DOI:10.1016/jjhevol.2010.01.004

Bailey, G.N., Devès, M.H., Inglis, R.H., Meredith-Williams, M.G., Momber, G., Sakellariou, D., Sinclair, A.G.M., Rousakis, G., Al Ghamdi, S., Alsharekh, A.M. 2015. Blue Arabia: Palaeolithic and underwater survey in SW Saudi Arabia and the role of coasts in Pleistocene dispersal. *Quaternary International* 382: 42–57. http://dx.doi.org/10.1016/j.quaint.2015.01.002.

Behrensmeyer, A.K., Potts, R., Deino, A., Ditchfield, P. 2002. Olorgesaille, Kenya: a million years in the life of a rift basin. In: Renaut, R.W., Ashley, G.M. (eds), *Sedimentation in Continental Rifts*. SEPM Society for Sedimentary Geology Special Publications, vol. 73: 97–106.

Bergner, A., Strecker, M., Trauth, M., Deino, A., Gasse, F., Blisniuk., P., Dühnforth, M. 2009 Tectonic and climatic control on evolution of rift lakes in the Central Kenya Rift, East Africa. *Quaternary Science Reviews* 28: 2804–2816. DOI: http://dx.doi.org/10.1016/j.quascirev.2009.07.008

Blumenschine, R.J., Peters, C.R. 1998 Archaeological predictions for hominid land use in the paleo-Olduvai Basin, Tanzania, during lowermost Bed II times. *Journal of Human Evolution* 34: 565–607. DOI: http://dx.doi.org/10.1006/jhev.1998.0216

Blumenschine, R.J., Stanistreet, I.G., Njau, J.K., Bamford, M.K., Masao, F.T., Albert, R.M., Stollhofen, H., Andrews, P., Prassack, K.A., McHenry, L.J., Fernández-Jalvo, Y., Camilli, E.L., Ebert, J.I. 2012. Environments and hominin activities across the FLK Peninsula during Zinjanthropus times (1.84 Ma), Olduvai Gorge, Tanzania. *Journal of Human Evolution* 63: 364–383. DOI: http://dx.doi.org/10.1016/j.jhevol.2011.10.001

Bunge, H.-P., Glasmacher, U.A. 2017. Models and observations of vertical motion (MoveOn) associated with rifting to passive margins: preface. *Gondwana Research*. https://doi.org/10.1016/j.gr.2017.07.005

Bunn, H., Harris, J.W., Isaac, G., Kaufulu, Z., Kroll, E., Schick, K., Toth, N., Behrensmeyer, A.K. 1980. FxJj50: an early Pleistocene site in northern Kenya. *World Archaeology* 12: 109–136. DOI: http://dx.doi.org/10.1080/00438243.1980.9979787

Burke, K. 1996. The 24th Alex L. Du Toit Memorial Lecture: the African plate. *South African Journal of Geology* 99(4): 339–409.

Burke, K, Gunnell Y. 2008. *The African erosion surface: a continental-scale synthesis of geomorphology, tectonics, and environmental change over the past 180 million years*. Memoir 201. Boulder, Colorado, Memoir of the Geological Society of America, DOI: 10.1130/2008.1201

Davies, G.F. 1999. *Dynamic Earth: Plates, Plumes and Mantle Convection*. Cambridge: Cambridge University Press. DOI: 10.1017/CBO9780511605802.

Devès, M., King, G.C.P., Klinger, Y., Agnon, A. 2011. Localised and distributed deformation in the lithosphere: modelling the Dead Sea region in 3 dimensions. *Earth and Planetary Science Letters* 308: 172–184.

Devès M., Sturdy, D., Godet, N., King, G.C.P. 2014. Hominin reactions to herbivore distribution in the lower Palaeolithic of southern Levant. *Quaternary Science Reviews*. DOI: http://dx.doi.org/10.1016/j.quascirev.2014.04.017 171.

Devès, M.H., Reynolds, S., King, G.C.P., Kübler, S., Sturdy, D., Godet, N. 2015. Insights from earth sciences into human evolution studies: the examples of prehistoric landscape use in Africa and the Levant. *Comptes Rendus Geoscience* 347(4): 201–211.

Ellis, M., King, G.C.P. 1991. Structural control of flank volcanism in continental rifts. *Science* 254: 839–842.

Enzel, Y., Bar-Yosef, O. (eds) 2017. *Quaternary of the Levant: environments, climate change, and humans.* Cambridge: Cambridge University Press.

Friedrich, A.M., Bunge, H.-P., Rieger, S.M., Colli, L., Ghelichkhan, S., Nerlic, R. 2017. Stratigraphic framework for the plume mode of mantle convection and the analysis of interregional unconformities on geological maps. *Gondwana Research.* Doi: 10.1016/j.gr.2017.06.003

Gowlett, J.A.J. 1991 Kilombe - review of an Acheulean site complex. In Clark, J.D. (ed.) *Approaches to understanding early hominid life-ways in the African savanna:* 129–136. Römisch-Germanisches Zentralmuseum Forschungsinstitut für Vorund Frühgeschichte in Verbindung mit der UISSP, 11 Kongress, Mainz, 31 August - 5 September 1987, Monographien Band 19. Bonn: Dr Rudolf Habelt GMBH.

Gowlett, J.A.J. 1999. Lower and Middle Pleistocene archaeology of the Baringo Basin. In: Andrews, P. Banham, P. (eds) *Late Cenozoic environments and hominid evolution: a tribute to Bill Bishop:* 123–141. London: Geological Society.

Gowlett, J.A.J., Crompton, R.H. 1994. Kariandusi: Acheulean morphology and the question of allometry. *African Archaeological Review* 12: 3–42.

Gowlett, J.A.J., Brink, J.S., Herries, A.I.R., Hoare, S., Onjala, I, Rucina, S.M. 2015. At the heart of the African Acheulean: the physical, social and cognitive landscapes of Kilombe. In Coward, F., Hosfield, R., Wenban-Smith, F. (eds) *Settlement, society and cognition in human evolution: landscapes in mind:* 75-93. Cambridge: Cambridge University Press

Goren-Inbar, N., Feibel, C.S., Verosub, K.L., Melamed, Y., Kislev, M.E., Tchernov, E., Saragusti, I. 2000. Pleistocene milestones on the out-of-Africa corridor at Gesher Benot Ya'aqov, Israel. Science 289: 944–947.

Guillocheau F, Simon B, Baby G, Bessin P, Robin C, Dauteuil, O. 2017. Planation surfaces as a record of mantle dynamics: the case example of Africa. *Gondwana Research.* Doi: 10.1016/j.gr.2017.05.015.

Higgs, E.S., Vita-Finzi, C. 1972. Prehistoric economies: a territorial approach. In Higgs, E.S. (ed.) *Papers in Economic Prehistory:* 27–36. London: Cambridge University Press.

Hubert-Ferrari, A., King, G.C.P., Manighetti, I., Armijo, R., Meyer, B., Tapponnier, P. 2003. Long-term elasticity in the continental lithosphere; modelling the Aden Ridge propagation and the Anatolian extrusion process. *Geophysical Journal International* 153: 111–132.

Isaac, G.L. 1977. *Olorgesailie: archaeological studies of a Middle Pleistocene lake basin in Kenya.* Chicago: University of Chicago Press.

Isaac, G.L. 1981. Stone Age visiting cards: approaches to the study of early land use patterns. In Hodder, I., Isaac, G., Hammond, N. (eds.) *Pattern of the past:* 131–155. Cambridge: Cambridge University Press.

King, B.C., King, G.C.P. 1967. The world rift system and plate tectonics or 1971 and all that. *Nature* 232: 37.

King, G.C.P., Bailey, G.N. 2010. Dynamic landscapes and human evolution. Geological Society of America, Special Paper 471 doi: 10.1130/2010.2471(01).

Kingston, J.D. 2007 Shifting adaptive landscapes: progress and challenges in reconstructing early hominid environments. *Yearbook of Physical Anthropology* 50: 20. DOI: http://dx.doi.org/1 0.1002/ajpa.20733

Kroll, E.M. 1994. Behavioral implications of Plio-Pleistocene archaeological site structure. *Journal of Human Evolution* 27: 107–138. DOI: http://dx.doi.org/10.1006/jhev.1994.1038

Kübler, S., Owenga, P., Reynolds, S.J., Rucina, S.M., King, G.C.P. 2014. Animal movements in the Kenya Rift and evidence for the earliest ambush hunting by hominins. *Nature Scientific Reports* 5, 14011: 1–7. doi:10.1038/srep14011.

Kübler, S., Owenga, P., Reynolds S.J., Rucina, S.M., Bailey, G.N, King, G.C.P. 2016. Edaphic and topographic constraints on exploitation of the Central Kenya Rift by large mammals and early hominins. *Open Quaternary* 2: 5:1–18. DOI: http://dx.doi.org/10.5334/oq.21.

Leakey, L.S.B. 1931. *The Stone Age cultures of Kenya Colony.* Cambridge: Cambridge University Press.

Murray, M.G. 1995. Specific nutrient requirements and the migration of the wildebeest. In Sinclair, A.R.E., Arcese, P. (eds) *Serengeti II - dynamics, management and conservation of an ecosystem:* 231–256. Chicago: University of Chicago Press.

Okada, Y. 1982. Internal deformation due to shear and tensile fault in a half-space. *Bulletin of the Seismological Society of America* 82:1018–1040.

Potts, R. 1989. Olorgesailie: new excavations and findings in Early and Middle Pleistocene contexts, southern Kenya rift valley. *Journal of Human Evolution* 18: 477–484.

Potts, R., Behrensmeyer, A.K., Ditchfield, P. 1999. Paleolandscape variation and early Pleistocene hominid activities: Members 12 and 7, Olorgesailie Formation, Kenya. *Journal of Human Evolution* 37: 747–788. DOI: http://dx.doi.org/10.1006/jhev.1999.0344.

Potts, R., Behrensmeyer, A.K., Deino, A., Ditchfield, P., Clark, J. 2004. Small Mid-Pleistocene hominin associated with East African Acheulean technology. *Science* 305:75–78.

Rabinovich, R., Gaudzinski-Windheuser, S., Kindler, L., Goren-Inbar, N. 2012. *The Acheulian site of Gesher Benot Ya'aqov Volume III. Mammalian taphonomy. The assemblages of Layers V-5 and V-6.* Dordrecht: Springer.

Reynolds, S.C., Bailey, G.N., King, G.C.P. 2011. Landscapes and their relation to hominin habitats: case studies from *Australopithecus* sites in eastern and southern Africa. *Journal of Human Evolution* 60: 257–280. DOI:10.1016/j.jhevol.2010.10.001.

Shipton, C. 2011. Taphonomy and behaviour at the Acheulean site of Kariandusi, Kenya. *African Archaeological Review* 28:141–155. DOI: http://dx.doi.org/10.1007/s10437-011-9089-1.

Stern, N. 1994. The implications of time-averaging for reconstructing the land-use patterns of early tool-using hominids. *Journal of Human Evolution* 27: 89–105. DOI: http://dx.doi.org/10.1006/jhev.1994.1037.

Sturdy, D.A., Webley, D.P. 1998. Palaeolithic geography: or where are the deer? *World Archaeology* 19:262–280.

Sturdy, D.A., Webley, D.P., Bailey, G.N. 1997. The Palaeolithic geography of Epirus. In Bailey, G.N. (ed.) *Klithi: Palaeolithic settlement and Quaternary landscapes in Northwest Greece, Volume 2: Klithi in its local and regional setting*: 541–559. Cambridge: McDonald Institute for Archaeological Research, Cambridge.

Suess, E. 1885. *Das Antlitz der Erde*. Prag-Leipzig: Tempsky, Freytag.

Trauth, M.H., Maslin, M.A., Deino, A., Strecker, M.R. 2005. Late Cenozoic moisture history of East Africa. *Science* 309: 2051–2053.

Vita-Finzi, C., Higgs, E.S. 1970. Prehistoric economy in the Mount Carmel area of Palestine: site catchment analysis. *Proceedings of the Prehistoric Society* 36: 1–37.

Winder, I.C., King, G.C.P., Devès, M., Bailey, G.N. 2013. Complex topography and human evolution: the missing link. *Antiquity* 87: 333–349.

Winder, I.C., King, G.C.P., Devès, M.H., Bailey, G.N. 2014. Human bipedalism and the importance of terrestriality: a reply to Thorpe et al. (2014). *Antiquity* 88: 915–916.

Simon Kübler

Department of Earth and Environmental Sciences, Ludwig Maximilians University, Luisenstrasse 37, 80333 Munich, Germany
Email: s.kuebler@lmu.de

Geoff (Geoffrey N.) Bailey

University of York, Department of Archaeology, The King's Manor, York, YO1 7EP, UK
and
College of Humanities, Arts and Social Sciences, Flinders University, GPO Box 2100, Adelaide, SA 5001, Australia
Email: geoff.bailey@york.ac.uk

Stephen Rucina

Earth Science Department, National Museums of Kenya, P.O. Box 40658 00100, Nairobi, Kenya
Email: stephenrucina@yahoo.com

Maud H. Devès

Institut de Physique du Globe de Paris, 1 rue Jussieu, 75238 Paris cedex 05, France
and
Centre de Recherche Psychanalyse Médecine et Société, CNRS EA 3522 – Sorbonne, France
Email: maud.deves@gmail.com

Geoffrey C.P. King

Institut de Physique du Globe de Paris, 1 rue Jussieu, 75238 Paris cedex 05, France
Email: kinggcp@gmail.com

How Many Handaxes Make an Acheulean? A Case Study from the SHK-Annexe Site, Olduvai Gorge, Tanzania

Ignacio de la Torre and Rafael Mora

Introduction

For better or for worse, discussions on the transition between archaeological cultures still receive significant attention today. Despite the culture- history baggage implicit in enquiries of this nature, in a historical science such as Archaeology it is probably unavoidable that we continue interrogating ourselves about what elements characterise new archaeological techno-complexes and differentiate them from the earlier ones. Until we are able to figure out the dynamics that led to the emergence of the earliest human technology (Panger et al. 2002; Rogers and Semaw 2008; de la Torre 2010, 2011a; Tennie et al. 2017), it is fair to state that the first major archaeological transition –and one that has received much attention– is that from the Oldowan to the Acheulean.

Once Louis Leakey (1951) established that the Oldowan-Acheulean transition took place in East Africa, the seminal work by Mary Leakey (1971) at Olduvai Gorge built the paradigm for the chronological, technological and evolutionary grounds over which such transition took place (see reviews in de la Torre and Mora 2014; de la Torre 2016). In subsequent years, Leakey's (1971) model was the basis for most discussions on the Oldowan-Acheulean transition (e.g., Davis 1980; Stiles 1980), discussions in which the recipient of this festschrift played a pivotal role (e.g., Gowlett 1986, 1988).

In recent years, we have learned that older Acheulean sites exist elsewhere in East Africa (e.g., Lepre et al. 2011; Beyene et al. 2013), but the rich archaeological sequence excavated by Leakey (1971) at Olduvai has still been the focus of the greater part of the debate (e.g., de la Torre and Mora 2005, 2014; Semaw et al. 2009). The past decade has also witnessed the renewal of fieldwork in early Acheulean contexts at Olduvai (Diez-Martin et al. 2015; Dominguez-Rodrigo et al. 2014, 2017; de la Torre et al. 2018), which is contributing new data to the debate on the dynamics of the Oldowan-Acheulean transition.

This paper aims to contribute to such debate by presenting a first-hand re-study of the lithic assemblage excavated by Mary Leakey (1971) in SHK-Annexe. Leakey (1971) included SHK within the Developed Oldowan B, and therefore it is a relevant assemblage to characterize technological dynamics during the onset of the Acheulean at Olduvai Gorge.

The S. Howard Korongo (SHK) site was found during the 1935 expedition led by Louis Leakey, but first excavations took place in 1953, and then in 1955 and 1957 (Leakey 1971). Louis Leakey (1958) reported briefly on the fauna and a possible bone tool, Kleindienst (1959) studied the handaxes, and then Mary Leakey (1971) presented the excavations in SHK-Main and SHK-Annexe. Fieldwork at SHK-Main has resumed in recent years (Diez-Martin et al. 2014, 2017), but apart from a revision of the fossil assemblage (Egeland and Dominguez-Rodrigo 2008) and of the cores and handaxes from SHK-Main (de la Torre and Mora 2014), we are unaware of any other studies involving the Leakey SHK lithic collections. Thus, this paper introduces the first re-study of the SHK-Annexe stone tool assemblage since it was originally published by Mary Leakey (1971).

Materials and methods

The SHK outcrop is located in the Side Gorge of Olduvai (Figure 1). A clay unit can be traced from SHK-Annexe to SHK-Main, with archaeological material on top of the clay unit at SHK-Annexe, and mostly concentrated in a fluvial conglomerate eroding the clay at SHK-Main (Leakey 1971). Leakey (1971) emphasized the resemblance of the tuff overlaying the clay unit to Tuff IID, but she and then Hay (1976) positioned the site stratigraphically in the upper part of Middle Bed II. However, new tephro-geochemistry results indicate that the tuff capping the archaeological deposits is indeed Tuff IID, and therefore the site should be placed in Upper Bed II (McHenry et al. 2016), rather than Middle Bed II. Despite these recent results, renewed fieldwork at SHK has positioned the site below Tuff IIC (Diez-Martin et al. 2017), although they do not discuss why the tephro-chemistry analyses should be disregarded. Therefore, we will follow here the latest tephro-chronology model (McHenry et al. 2016), and consider the site as positioned in Upper Bed II, along with geographically close sites such as BK and SC (Figure 1).

The archaeological material at SHK-Annexe was clustered in an area of approximately 4.5 x 3 metres, and included stone tools and bone specimens. Sediments were not systematically screened, and part of the smaller sized remains were not retained (Leakey 1971).

We analysed the SHK-Annexe assemblage in 2010, when the Leakey collection was still housed at the National Museums of Kenya in Nairobi, and before it was repatriated to Tanzania. Curation of the assemblage was excellent in the museum of Nairobi, where the material was organized in drawers and bags according to Mary Leakey's classification, therefore enabling to contrast our identifications with hers.

The assemblage was studied macroscopically only. All artefacts were measured and weighed. The analysis of flaked and detached artefacts followed criteria established by de la Torre and Mora (2005), and updated by de la Torre (2011b) and de la Torre and Mora (2018a). The distinction between the chaînes opératoires of small debitage and Large Cutting Tool (LCT) production, and the technological categories in the latter, are based on the criteria discussed by de la Torre and Mora (2018b). The study of battered artefacts followed groups established by Mora and de la Torre (2005) and updated in de la Torre and Mora (2010), and Arroyo and de la Torre (2018).

Results

Assemblage composition

The collection studied totals 492 artefacts that weigh ~71 kg. Metamorphic rocks are the most abundant raw material (n = 439; 58.9 kg), followed by basalt/trachyte-trachyandesite (T-Ta) and phonolite (see Table 1). Quantitatively, detached artefacts dominate the assemblage (n = 271; 55.1%), although flaked pieces also show high frequencies (35.6%). Whilst the number of pounded tools is comparatively insignificant (n = 46), they comprise 20 kg of used raw material and therefore are more relevant than detached pieces (13.8 kg) in terms of weight contribution, being only exceeded by flaked tools (37.4 kg).

The dominance of quartzite both in terms of number of artefacts and weight contribution is consistent in all technological categories (see Figure 2A), although Figure 2B indicates that the total weight of lava cores is considerably higher than in any other technological group. The Chi-square comparison of technological categories and raw material groups indicates the existence of significant differences (X^2 (6) = 13.15, p < 0.0407), and the Pearson's residues and Fisher's test link such differences to an overrepresentation of lava cores and quartzite retouched tools.

A minimum of 283 artefacts (57.5% of the assemblage) can be attributed to the chaîne opératoire of small debitage. Most of the debitage that shows no reduction sequence-defining features (and therefore are listed as indeterminable in Table 2) probably correspond to the small debitage chaîne opératoire as well. Therefore, the very low frequencies of artefacts attributable to the LCT production sequence (n = 14; 2.8%) is probably an accurate reflection of the actual technological pattern at the site (Figure 2C). Although in terms of weight contribution the LCT chaîne opératoire yields higher values (3.5 kg; 5%), Table 3 and Figure 2C show that the entire assemblage is still overwhelmingly dominated by the small debitage reduction sequence (39.4 kg; 55.3%).

Table 4 shows that 91% of the material for which roundness was analysed is fresh, with negligible frequencies of abraded artefacts. Quartzite artefacts are particularly fresh, often in mint conditions, and the few slightly abraded pieces are mostly in lava (Table 4), with statistically significant differences between the two raw materials (X^2 (2) = 49.10, p < 0.0001).

The chaîne opératoire of small debitage

Detached artefacts. Even if we added the indeterminable detached pieces from Table 2 to the small debitage chaîne opératoire, the overall number of shatter and flake fragments is comparatively small when compared to the combined sample of whole flakes (n = 120). Whole flakes show two morphological patterns: some are quadrangular, wide and short flakes, which are probably linked to the reduction of radially flaked cores. Other flakes have more elongated shapes, and suggest knapping from unidirectional cores.

Considering only the sample of flakes clearly attributable to the chaîne opératoire of small debitage (n = 111), average length is ~5 cm. Although length variability is significant (see Table 5), most flakes (71%) range between 40 and 59 mm (Table 6 and Figure 3C). No clear disparity is observed in length size per raw material (Table 5 and Figure 3A). More obvious differences could exist in terms of weight, as lava flakes are in average over a gram heavier than quartzite flakes (Table 5 and Figure 3B), but this disparity might be influenced by the much smaller sample available for lava flakes.

Flake striking platforms are predominantly cortex-free (92.5%), although prepared butts (i.e., bifaced or multifaceted striking platforms) are rare (see values in Table 6). Thus, clear preponderance of unifaceted butts in lava (94.1%) and quartzite (84.4%) flakes (Figure 3D) suggest core knapping platforms that were cleared of cortex but which were not further prepared prior to flake removal. Dorsal patterns are predominantly non-cortical (72.2%), although some variability is observed between lava and quartzite flakes (Table 6), with cortex more abundant on the dorsal side of lava flakes. The proportion of Toth's (1982) types (Table 6 and Figure 3E) shows that fully cortical flakes (Type I) are negligible in the assemblage (1.9%), and most flakes were obtained from fully (Type VI: 65.4%) or considerably (Type V: 25%) decorticated cores.

Cores. The number of small debitage cores (n = 98) is substantial when compared to other technological categories (see Table 2) and, in terms of weight contribution (~32 kg), is the largest group within the assemblage (Table 3). In addition to complete cores, a considerable number of quartzite core fragments (n = 28) are documented. Some of these fragments correspond to cores of very small size that broke in half during flaking. A large number of cores (n = 37; 37.8%) bear battering marks (see details in Table 7), usually on natural/cortical areas opposite the knapping surface. In several examples, this is interpreted as hammerstones that were then recycled as cores. In other instances, however, it was observed that battering was posterior to flaking, as shown by pitted areas that break through flaking scars.

Mean core length is ~7cm and ~330 g, with lava cores being in average large than quartzite cores (see values in Table 5). Figure 4A shows no clear breaks in quartzite core dimensions, whereas size of lava cores is clustered. Clustering of lava core length is better observed in Figure 4C, which shows most cores are in the 80-99 mm range (see also Table 7). Lava cores are substantially heavier in average than quartzite cores (Table 5 and Figure 4B), particularly in the 401-800 g range (Table 7 and Figure 4D). The Shapiro-Wilk test indicates the normal distribution of dimensions and weight of lava cores (alpha = 0.05; p-value = 0.52 [length], 0.67 [width], 0.83 [thickness], 0.93 [weight]), as opposed to quartzite cores, which do not follow a normal distribution (alpha = 0.05; p-value < 0.0001 in all variables). The Mann-Whitney test confirms that lava and quartzite core do not have the same size distribution (alpha = 0.05, p-value < 0.0001 [length, width, thickness], 0.0003 [weight]).

The wide variability of quartzite core sizes when compared to the lava sample is linked to the blanks used for flaking. As shown in Table 7, nearly all lava cores are on cobble blanks (see also Figure 4E). Conversely, blanks for quartzite cores range from small flakes to fragments, cobbles and blocks; some of the cores on small flakes and fragments weigh less than 50 g, whereas a number of cores on blocks have large dimensions and some weigh over 2 kg (see Figure 5 #1-2).

Table 7 shows that only 27.5% of cores preserve large quantities of cortex. A relatively higher reduction intensity can be inferred for quartzite, where 40.6% of cores preserve no cortex, as opposed to lavas, where 56.3% of cores contain predominantly cortical surfaces (see also Figure 4F). Cores have an average of 5.7 scars, although lava cores have a higher mean (6.7 scars) than metamorphic cores (5.5 scars). Considering the entire sample together, 79.5% of cores have four or more scars, with predominance of cores with 4-6 scars (see Table 7), which is generally consistent with cortex results and thus indicates moderate reduction of core blanks.

The bipolar technique is observed in 7.4% of cores (all of them in quartzite), while the rest (n = 88) are attributed to freehand flaking methods. Table 7 and Figure 6A show the predominance of unidirectional abrupt (particularly UAU1) and radial (BP and BHC) flaking schemes. Clear differences are observed by raw material, with most 'chopper' cores (USP and BSP) made of lava (Figure 6B), despite overwhelming predominance of quartzite in the entire sample.

As with core size, flaking schemes seem to be conditioned, at least in part, by blank type. Thus, choppers are usually made on lava cobbles, which have natural shapes prone to produce core morphologies with unifacial or bifacial partially flaked edges opposite a cortical surface (e.g., Figure 5 #7). Likewise, most of UAU cores are on quartzite tabular blocks, in which knappers used natural planar surfaces to remove longitudinal flakes. Such unidirectional cores are not structured, and often show no more than one series of flake removals, despite the large size of core blanks and the potential for further reduction (see Figure 5 #1-2). Alongside substantially large cores such as that in Figure 5 #1-2, bipolar cores are small (often < 40 mm), although not necessarily heavily exploited; instead, their small dimensions are due to the reduced size of blanks (Figure 5 #4). Some radial cores are also small (see Figure 5 #5) and, in this case, it is uncertain to what extent size correlates with reduction intensity; most of radial cores correspond to BP schemes, where central volumes are not exploited and thus flaking sequences cannot be sustained for long. Similarly, BHC cores at SHK-Annexe did not undergo long reduction sequences and their small size may also be explained by blank selection, rather than flaking intensity.

Retouched tools. All but one of the 46 retouched pieces identified in SHK-Annexe are made of quartzite, and constitute an abundant category within the small debitage chaîne opératoire (Table 2). With an average length of ~40 mm and of ~22 g in weight (Table 5), the size distribution of retouched tools (Figure 7A) shows no clear clustering. Results of the Shapiro-Wilk test indicate that dimensions of retouched tools do not follow a normal distribution (alpha = 0.05, p-value =0.0036 [length], 0.0220 [width], <0.0001 [thickness, weight]).

Blanks used for retouched tools were flakes or flake fragments, but Table 5 shows that average dimensions of retouched pieces are consistently smaller than those of flakes (see also Figure 7B). The Mann-Whitney test (alpha = 0.05) confirms that the two groups do not share the same size distribution in all variables (p-value < 0.0001 for length and width, and 0.0490 for weight) apart from thickness (p-value = 0.7859). It is unclear, however, whether such differences are due to selection of smaller blanks for retouching, size reduction due to retouching intensity, or to the fact that some retouched pieces were made on flake fragments (instead of complete flakes).

As shown in Figure 8, there is no clear standardization of retouched shapes, although there is predominance of unifacial denticulates and notches, normally using the ventral face as striking platform to retouch the dorsal face (i.e., direct retouch). A number of these tools are retouched around their entire perimeter, which is interesting given the small size of some of the blanks.

The chaîne opératoire of Large Cutting Tool production

LCT production is attested in a very small sample of the SHK-Annexe assemblage (n = 14 and 3.5 kg; see Table 2 and Table 3). As shown in Table 8, small flakes associated to handaxe shaping (n = 5), and potential LCT blanks (n =5) are the best represented categories, while actual LCTs are very rare, with one complete specimen (Figure 9A) and two LCT fragments (e.g., Figure 9B). Interestingly, while both the complete LCT and all flakes are of quartzite, the two LCT fragments correspond to broken handaxes of lava.

The LCT in Figure 9A shows the usual features of knives from the early Acheulean of Olduvai Gorge (de la Torre and Mora, 2018b); a large blank (19 cm of length and over 1 kg of weight; see Table 8), probably a flake, was shaped unifacially on one edge, through denticulate retouch. This shaping is on the abrupt edge of the artefact (probably the butt of the flake), and is opposite a sharper, unmodified edge. The rest of the artefact remained mostly unmodified. Figure 9B shows a handaxe tip with bilateral and partially bifacial retouch, and thus shows more extensive shaping than the complete handaxe from Fig. 9A. Emphasis on the shaping of tips is also characteristic of most handaxe morphotypes in the Olduvai early Acheulean (de la Torre and Mora, 2018b).

Pounding tools

A minimum of 20 kg of raw materials were used in pounding tools (n = 46). Nonetheless, relevance of battering was probably higher, since many cores (n = 37) show also evidence of impact marks (Table 7). Admittedly, on occasions it is difficult to distinguish impact marks produced by missed blows on knapping platforms aimed at removing flakes, from battered areas produced during pounding. Nevertheless, at least in the case of those that show clustered battering on natural surfaces opposite a flaking surface (Figure 10 #3 and #4), it is clear that some cores were also involved in percussive activities.

Although we chose not to assign any of the pounding tools to either the small debitage or LCT production chaînes opératoires (see Table 2), Figure 7D shows that

two of the knapping hammerstones are large (with one of them weighing over 1.7 kg; Table 8), and therefore may have been associated with production of LCT blanks.

Anvils (n = 5) amount to 4.4 kg (Table 1; see dimensions in Table 8), and show battering over planar surfaces of quartzite blocks and chipping of the edges (e.g., Fig. 10 #6). Nonetheless, their role also as active hammers cannot be excluded, as evidenced by the clustered battering in Fig. 10#6. Only one artefact was categorised as a spheroid, but many of the pounding tools classified here as hammerstones with fracture angles (n = 25; 7.5 kg) show various stages of shape rounding that is linked to blunting of edges through battering.

With an average weight of >400 g, many pounding tools do not seem fit to flake some of the tiny cores present in the SHK-Annexe collection. Thus, both the large size (Table 8 and Figure 7D) and weight (Figure 7C) of most pounding tools may be indicative of additional activities to knapping at the site.

Discussion

Comparing earlier and current classifications of the SHK-Annexe collection

Leakey (1971) reported that most SHK sediments were not sieved, and that only a sample of flakes was retained. Since Leakey did not state the exact number of flakes she curated from SHK-Annexe, attempts to consider the completeness of the collection for the present study have to be loosely based on the materials Leakey (1971: 171) listed as tools (n= 185). This figure is only slightly lower than the non-detached component of our study (n = 175 flaked and 46 pounded tools; Table 1), and hence it is inferred here that we accessed most, if not all, of the assemblage originally curated by Mary Leakey.

Under this premise, divergences between Leakey's and our own technological attributions concern inter-analyst variability on the identification of retouched tools, subspheroids/ spheroids, bifaces and core tools. Most of Leakey's core tools (e.g., choppers, heavy-duty scrapers) are here attributed to a range of core flaking schemes (e.g. USP, UAP). Conversely, some of the pieces classified by Leakey as discoids are seen here as small retouched tools or irregular fragments, rather than as centripetally- flaked and biconically- reduced cores.

Leakey (1971) identified a substantial number of artefacts as spheroids and subspheroids. We agree that intensity of battering at SHK-Annexe is high, although our analysis raises some uncertainties. For example, some of the pieces originally listed as spheroids/ subspheroids bear battering over cortical surfaces –cortex is conspicuous in river cobbles, and in SHK-Annexe it is not rare to document quartzites that are clearly fluvial, rather than derived from the primary source at Naibor Soit (see discussion in McHenry and de la Torre 2018). Thus, blank rounding is natural –rather than produced by battering– in many pieces, which therefore questions their attribution to spheroidal shaping (see discussion in de la Torre and Mora 2005). On the other hand, uncertainties on the analysis of Olduvai subspheroids/ spheroids should be acknowledged (see Arroyo and de la Torre 2018), and therefore our attribution of many of Leakey's subspheroids to hammerstones with fracture angles should also be considered with caution.

We generally agree with Leakey's identification of a substantial number of artefacts as retouched tools. However, we did not recognise any of the alleged burins as such, and several of them are Siret burins (i.e., split flakes), rather than intentional burins –see de la Torre and Mora (2005) for a discussion of this misidentification in other Olduvai sites. Some of the retouched tools are substantially modified (at least when compared to other Olduvai Beds I and II assemblages), but they can all be considered as denticulate side scrapers and notches, with no clearly standardized shapes apart from some possible 'pointed' tools (e.g., Figure 11B, #10).

The most relevant divergence in our techno-typological attribution refers to artefacts originally classified as bifaces. Apart from the three specimens retained in this paper as LCT or handaxe fragments, Leakey (1971) identified six further artefacts. However, we are unconvinced that they should be considered within that category, with some best qualifying as small retouched tools or cores (see Figure 11C), rather than as part of the LCT chaîne opératoire.

Interpreting assemblage composition in SHK-Annexe

Any consideration of the SHK-Annexe assemblage as a whole should take into account the deficit of debitage due to collection bias. Nonetheless, the material is fresh (Table 4) and Leakey (1971) considered the site as a living floor on a clay surface covered by tuff, so it can be assumed that, apart from the smaller pieces, the studied collection resembles the original configuration of the assemblage.

To overcome post-depositional and collection biases, we can use the proportion of larger artefacts to run inter-assemblage comparisons. In this line, ratios of relevant tools in the Olduvai post-Tuff IIB sequence (de la Torre and Mora, 2018b) shed interesting results as far as SHK-Annexe is concerned. Thus, SHK-Annexe has the highest core:LCT ratio (value=98) in all of the twenty-seven sites considered between Beds II and Masek (average ratio= 6.1). This means that SHK-Annexe contains proportionally (when compared to

handaxes) the highest number of cores across the entire post-Tuff IIB sequence at Olduvai. Equally interesting is that, according to the calculations by de la Torre and Mora (2018b), SHK-Annexe also yields the lowest LCT:retouched tool ratio (value=0.02) of the entire Olduvai sequence (average ratio =1.9), which is again influenced by the near absence of handaxes at the site. It is only once LCTs are removed from comparisons that SHK stops being at one end of the sample; when, in addition to indices produced by de la Torre and Mora (2018b), we calculate a new ratio of retouched tools to cores (mean value =1.2), SHK-Annexe sits in a more intermediate position (ratio = 0.46).

Overall, the most salient features of the SHK-Annexe assemblage are the disproportionate abundance of cores, and the merely testimonial presence of handaxes. It is clear then that small debitage flaking was the most relevant technological activity at the site, while tasks involving production and/or use of handaxes were marginal, or else handaxes were removed from the site.

With over 20 kg of pounding tools, it is also evident that battering tasks played an important role at the site. SHK-Annexe not only includes knapping hammerstones, but also contains anvils and hammerstones with fracture edges (Figure 10) that were probably involved in other pounding tasks beyond flaking. A number of such tools are heavily battered, which suggests significant intensity of pounding activities. A recent comparison of pounded tool frequencies in Oldowan and Acheulean assemblages does not discern differences between the two periods (Arroyo and de la Torre 2018), but it is now manifest that percussive activities played an important role throughout the Olduvai sequence (Mora and de la Torre 2005). Although some have questioned the relevance of pounding tasks at Olduvai (Diez-Martin et al. 2009), they have subsequently acknowledged their abundance in the new excavations at SHK-Main (Sánchez-Yustos et al 2015). This is interesting given the geographic proximity between SHK-Main and SHK-Annexe and the pene-contemporaneity of the two sites (Leakey 1971: 166), and could indicate an emphasis on battering activities in some particular spots of the Olduvai paleo-landscape. Bones were found spatially associated to stone tools at SHK-Annexe (Leakey 1971), but recent reassessments of the Leakey fossil collection (Egeland and Dominguez-Rodrigo 2008) do not provenance materials to either the SHK-Main or SHK-Annexe, and therefore it is unfeasible to link the zooarchaeological and technological data within the latter assemblage.

What is the technology of SHK-Annexe?

The SHK-Annexe assemblage contains elements from two clearly different chaînes opératoires. Figure 11A illustrates such difference by comparing flakes from the small debitage reduction sequence (#1-5) to potential blanks for LCT shaping (#6-8). This is even clearer in Figure 11D, where flakes attributed to the LCT reduction sequence (#14, #17) are 3-4 times larger than the sources of flakes (that is, cores: #15-16) in the small debitage chaîne opératoire. The target of the chaîne opératoire of small debitage is to produce 4-5 cm flakes such as those in Figure 11A #1-5, which in occasions are retouched (Figure 11B #9-10). In contrast, large flakes produced in the LCT sequence, such as those in Figure 11A #6-8 and Figure 11D #14, #17), are potential blanks to configure LCTs such as that in Figure 9A.

Despite the obvious differences between the two chaînes opératoires, it is also clear that LCT production is very poorly represented at the site, which contains isolated elements of a fragmented reduction sequence only evident in a few large blanks, broken handaxes and a single LCT. The absence of LCT cores is not particularly surprising, as this is a pattern shared by most early Acheulean sites at Olduvai (de la Torre and Mora 2018b). However, the remarkably rare presence of handaxes in the collection –with only one undisputable and complete specimen– pushes the limits of its attribution to the Acheulean techno-complex; is it meaningful to categorize SHK-Annexe as an Acheulean assemblage when most of the assemblage show no signs of handaxe production?

Most specialists will agree that it would be best not to return to the –now superseded– term of Developed Oldowan B (Leakey 1971). Apart from the culture-history connotations of the term (see review in de la Torre and Mora 2014), the alleged techno-typological features of the Developed Oldowan B at SHK-Annexe presents the same problems as we have previously reviewed for BK, TK and FC West (de la Torre and Mora 2005). This includes the dubious character of the so-called diminutive bifaces (see Figure 11C), the similarity of 'true' handaxes (Figure 9A) to those of undisputed Acheulean assemblages such as EF-HR, and the existence of other elements of the LCT chaîne opératoire that are invisible in typological recounts of normative tools (e.g., handaxe shaping flakes, LCT blanks; Table 2 and Table 3).

We subscribe Gowlett's (1986) reasoning that, if the Acheulean is characterised by the presence of handaxes, then one should be enough to consider an assemblage as such. And as mentioned in the paragraph above, SHK-Annexe contains other elements apart from handaxes per se to typify the site as Acheulean; for instance, the ability to produce large flakes has long been claimed to be a defining feature of the Acheulean (Isaac 1969), and such ability is well attested at SHK-Annexe.

Nonetheless, even if most present-day specialists will concur in including SHK-Annexe within the Acheulean –on the basis that the Acheulean does not solely convey the presence of handaxes but refers to a techno-complex with shared biological, cognitive, technological and subsistence affinities (de la Torre 2016)–, this still does not satisfactorily explain why we encounter such a substantial inter-assemblage variability at Olduvai (de la Torre and Mora 2014), and elsewhere in East Africa (de la Torre 2011b).

Here, consideration of another defining feature of the Acheulean might provide an important clue; and that is the fragmented character of reduction sequences during this period, which again is well documented both at Olduvai (de la Torre and Mora 2018b) and at other East African Acheulean sequences (Gowlett 1982; de la Torre et al 2014). For instance, raw material source distance is key to explain variability in Kilombe (Gowlett 1993) and Olorgesailie (Potts et al. 1999), and site function and their paleoecological context at Olduvai have also been used to decipher the Acheulean/ Developed Oldowan B dichotomy (Isaac 1971; Hay 1976). In principle, this may be seen as a truism, as we all expect technological strategies will correlate with transport distance and site function. However, such patterns are not so obvious during the Oldowan; indeed, transport-decay and reduction intensity patterns are observed to correlate with raw material distance in pre-Acheulean contexts (e.g., Blumenschine et al. 2008; Toth 1982), but Oldowan toolkit composition does not seem to vary substantially. In contrast, inter-assemblage variability in the East African Acheulean is significantly higher (e.g., Isaac 1977) and is linked to a massive fragmentation of the chaînes opératoires (Gowlett 1982), which in turn may respond to a much more structured use of the landscape (de la Torre and Mora 2005; de la Torre et al. 2014).

What if there were not handaxes?

In response to the question posed in the title of this paper, we hope that reflections in the previous section are persuasive to conclude that one handaxe should usually be enough to include a site within the Acheulean techno-complex. But how about when we do not even have the one? There is consensus that assemblages with few or no handaxes in < 1-million-year-old sequences such as the Hope Fountain Industry (Posnasky 1959), Developed Oldowan C (Leakey and Roe 1994) and the Clactonian (White 2000) should still be accounted for as part of the technological variability within the Acheulean (see summary in de la Torre and Mora 2014). However, it gets more complicated for sites without handaxes in the chronological boundary between the Oldowan and the Acheulean, and SHK-Annexe may prove to be an excellent case study to contribute to the debate.

Firstly, because of its sample size. The SHK-Annexe collection studied here contains around 500 artefacts, and the original number sample was larger, as we know of the debitage discarded by Leakey; however, despite this relatively large number of artefacts, only one complete handaxe was recovered. Some of the handaxe-free assemblages in the Oldowan-Acheulean boundary such as Peninj-Type Section (de la Torre and Mora 2004), Nyabusosi (Texier 1995) and Chesowanja (Gowlett et al. 1981), yield similar or even smaller absolute frequencies. From this perspective, there is a possibility that an increase in the collection size of those sequences eventually led to a documentation of handaxes, even if in a small percentage as in SHK-Annexe.

In the absence of handaxes, the presence of structured flaking methods of small debitage has been proposed as an additional technological proxy to track Acheulean innovations (de la Torre 2009, 2011b). Relatively structured small debitage methods in East African sites at the Oldowan/ Acheulean boundary have been reported in Peninj (de la Torre 2009), Melka Kunture (Gallotti 2013), Nyabusosi (Texier 1995), and in TK and BK at Olduvai (de la Torre and Mora 2005; Sánchez-Yustos et al. 2017; Santonja et al. 2014), thus supporting yet another prescient observation by Gowlett (1986, 1990), who linked the appearance of handaxes with the development of structured centripetal methods. While handaxe-rich assemblages such as EF-HR exhibit expedient small debitage flaking techniques, the SHK-Annexe collection contains examples of centripetal cores (de la Torre and Mora 2014), some of them considerably small (e.g., Figure 5 #5), and similar to those of TK and BK (de la Torre and Mora, 2005). Thus, even though flaking methods in SHK-Annexe are generally simple (see Figure 6), more structured small debitage schemes appear consistently, and separate the assemblage from the flaking patterns typical in the Oldowan sequence at Olduvai (Proffitt, 2018; de la Torre and Mora 2005, 2018a).

The average size of stone tools in SHK-Annexe is also clearly different from dimensional patterns in Oldowan assemblages. It is unfeasible to embark here in a quantitative assessment of the Olduvai archaeological sequence, but a cursory comparison between Table 5 and the metric datasets for pre-Tuff B sites available in de la Torre and Mora (2005) will show that small debitage technological categories in SHK-Annexe are consistently larger than in any of the Olduvai Oldowan assemblages. This pattern stands despite the huge intra-assemblage size variability observed within SHK-Annexe (see examples of such disparities in Figure 5), and clearly points to the overall larger dimension of all stone tools –i.e., not only those associated to LCT production but also those from small debitage chaînes opératoires– in the Acheulean when compared to the Oldowan.

Conclusions

The publication of Mary Leakey's (1971) seminal work on Olduvai Beds I and II was the starting point for the modern debate on the origins of the Acheulean in East Africa. Five decades on, the debate is still very much alive today, and John Gowlett's ideas have been pivotal in shaping it. He was a pioneer in advocating the need to understand cognitive and technical abilities behind the Oldowan and Acheulean stone tools (Gowlett 1982, 1986, 1990). When the debate was still typological, Gowlett (1982) was already stressing the fragmentation of chaînes opératoires in the Acheulean sequences. While discussions were still based on comparison of tool frequencies, he was proposing that the proportion of handaxes is irrelevant and that it is the mental templates involved what matters (Gowlett 1986). When, more recently, debates on handaxe variability have turned purely quantitative, he has reminded us that there is a set of technical parameters intrinsic to them all (Gowlett 2006).

SHK-Annexe highlights the interest of conducting first-hand assessments of Acheulean assemblages, and that such studies should include the entire collections. Direct inspection allows to better understanding analysts' decisions in artefact attribution, which in the case of the so-called diminutive handaxes bear relevant implications on the character of handaxes during the Developed Oldowan B. Also, the focus on entire assemblages instead of on particular categories enables us to identify additional elements attributable to the handaxe reduction sequence (e.g., LCT shaping flakes, LCT blanks), and finding alternative proxies to handaxes (e.g. features of small debitage systems) to refine the features of the Acheulean technology as a whole.

As reassessment of classic collections and data from new excavations accrue, Gowlett's views on the Oldowan -Acheulean gradient have proved visionary. Our restudy of the SHK-Annexe assemblage is no exception, and it only proves right arguments that Gowlett (1986) had already put forward thirty years ago; one handaxe is as important as forty as far as the mental templates required for handaxe manufacture are concerned. A handaxe is a techno-unit that requires a hierarchical construction through an interval of manufacture, and entails manipulation of a set of instructions in a tri-dimensional state (Gowlett 2002). Such instructions are 'packaged' around a few concepts or 'imperatives' that impose a heavy cognitive load (Gowlett 2006), and are those that truly define the Acheulean character of the assemblage.

Acknowledgements

We thank the National Museum of Nairobi for permits to study the Leakey collection from SHK. Funding by the NSF (BCS-0852292) and the European Research Council-Starting Grants (283366) is gratefully acknowledged. We thank the editors of this festschrift for inviting us to participate. Our special thanks go to John Gowlett, for being a constant source of inspiration to all of us interested in early stone tool technologies. IT also wishes to thank John for his support and friendship along the years.

References

Arroyo, A., de la Torre, I., 2018. Pounding tools in HWK EE and EF-HR (Olduvai Gorge, Tanzania): percussive activities in the Oldowan - Acheulean transition. *Journal of Human Evolution* 120, 402-421.

Beyene, Y., Katoh, S., WoldeGabriel, G., Hart, W.K., Uto, K., Sudo, M., Kondo, M., Hyodo, M., Renne, P.R., Suwa, G., Asfaw, B., 2013. The characteristics and chronology of the earliest Acheulean at Konso, Ethiopia. *Proceedings of the National Academy of Sciences* 110, 1584-1591.

Blumenschine, R.J., Masao, F.T., Tactikos, J.C., Ebert, J.I., 2008. Effects of distance from stone source on landscape-scale variation in Oldowan artifact assemblages in the Paleo-Olduvai Basin, Tanzania. *Journal of Archaeological Science* 35, 76-86.

Davis, D.D., 1980. Further Consideration of the Developed Oldowan at Olduvai Gorge. *Current Anthropology* 21, 840-843.

de la Torre, I., 2009. Technological Strategies in the Lower Pleistocene at Peninj (West of Lake Natron, Tanzania), in: Schick, K., Toth, N. (Eds.), *The Cutting Edge: New Approaches to the Archaeology of Human Origins. Stone Age Institute Press*, Bloomington, pp. 93-113.

de la Torre, I., 2010. Insights on the Technical Competence of the Early Oldowan, in: Nowell, A., Davidson, I. (Eds), *Stone Tools and the Evolution of Human Cognition*. University Press of Colorado, Boulder, pp. 45-65.

de la Torre, I., 2011a. The origins of stone tool technology in Africa: a historical perspective. *Phil. Trans. R. Soc. B* 366, 1028-1037.

de la Torre, I., 2011b. The Early Stone Age lithic assemblages of Gadeb (Ethiopia) and the Developed Oldowan / early Acheulean in East Africa. *Journal of Human Evolution* 60, 768-812.

de la Torre, I., 2016. The origins of the Acheulean: past and present perspectives on a major transition in human evolution. *Philosophical Transactions of the Royal Society of London B: Biological Sciences* 371, 20150245.

de la Torre, I., McHenry, L.J., Njau, J. (Eds) 2018. From the Oldowan to the Acheulean at Olduvai Gorge (Tanzania). *Journal of Human Evolution* 120, 1-421.

de la Torre, I., Mora, R., 2004. *El Olduvayense de la Sección Tipo de Peninj (Lago Natron, Tanzania)*. CEPAP, vol. 1, Barcelona.

de la Torre, I., Mora, R., 2005. *Technological Strategies in the Lower Pleistocene at Olduvai Beds I & II*. ERAUL 112, Liege.

de la Torre, I., Mora, R., 2010. A technological analysis of non-flaked stone tools in Olduvai Beds I & II. Stressing the relevance of percussion activities in the African Lower Pleistocene, in: Mourre, V., Jarry, M. (Eds.), *Entre le marteau et l'enclume. La percussion directe au percuteur dur et la diversité de ses modalités d'application. Actes de la table ronde de Toulouse, 15-17 mars 2004, Paleo 2009-2010*, numéro spécial, pp. 13-34.

de la Torre, I., Mora, R., 2014. The Transition to the Acheulean in East Africa: an Assessment of Paradigms and Evidence from Olduvai Gorge (Tanzania). *Journal of Archaeological Method and Theory* 21, 781–823.

de la Torre, I., Mora, R. 2018a. Oldowan technological behaviour at HWK EE (Olduvai Gorge, Tanzania). *Journal of Human Evolution* 120, 236-273.

de la Torre, I., Mora, R. 2018b. Technological behaviour in the early Acheulean of EF-HR (Olduvai Gorge, Tanzania). *Journal of Human Evolution* 120, 329-377.

de la Torre, I., Mora, R., Arroyo, A., Benito-Calvo, A., 2014. Acheulean technological behaviour in the Middle Pleistocene landscape of Mieso (East-Central Ethiopia). *Journal of Human Evolution* 76, 1-25.

Diez-Martín, F., Fraile, C., Uribelarrea, D., Sánchez-Yustos, P., Domínguez-Rodrigo, M., Duque, J., Díaz, I., De Francisco, S., Yravedra, J., Mabulla, A., Baquedano, E., 2017. SHK Extension: a new archaeological window in the SHK fluvial landscape of Middle Bed II (Olduvai Gorge, Tanzania). *Boreas* 46, 831-859.

Diez-Martín, F., Sánchez, P., Domínguez-Rodrigo, M., Mabulla, A., Barba, R., 2009. Were Olduvai Hominins making butchering tools or battering tools? Analysis of a recently excavated lithic assemblage from BK (Bed II, Olduvai Gorge, Tanzania). *Journal of Anthropological Archaeology* 28, 274-289.

Diez-Martín, F., Sánchez Yustos, P., Uribelarrea, D., Baquedano, E., Mark, D.F., Mabulla, A., Fraile, C., Duque, J., Díaz, I., Pérez-González, A., Yravedra, J., Egeland, C.P., Organista, E., Domínguez-Rodrigo, M., 2015. The Origin of The Acheulean: The 1.7 Million-Year-Old Site of FLK West, Olduvai Gorge (Tanzania). *Scientific Reports* 5, 17839.

Diez-Martín, F., Sánchez-Yustos, P., Uribelarrea, D., Domínguez-Rodrigo, M., Fraile-Márquez, C., Obregón, R.-A., Díaz-Muñoz, I., Mabulla, A., Baquedano, E., Pérez-González, A., Bunn, H.T., 2014. New archaeological and geological research at SHK main site (Bed II, Olduvai Gorge, Tanzania). *Quaternary International* 322–323, 107-128.

Domínguez-Rodrigo, M., Baquedano, E., Mabulla, A., Diez-Martín, F., Egeland, C.P. (Eds.) 2017. Oldowan and Acheulian archaeology of Olduvai Gorge. *Boreas*, 46: 605–936.

Domínguez-Rodrigo, M., Diez-Martín, F., Mabulla, A., Baquedano, E., Bunn, H.T., Musiba, C., (Eds.). 2014. The evolution of hominin behavior during the Oldowan–Acheulean transition: Recent evidence from Olduvai Gorge and Peninj (Tanzania). *Quaternary International* 322–323, 1-6.

Egeland, C.P., Dominguez-Rodrigo, M., 2008. Taphonomic perspectives on hominid site use and foraging strategies during Bed II times at Olduvai Gorge, Tanzania. *Journal of Human Evolution* 55, 1031–1052.

Gallotti, R., 2013. An older origin for the Acheulean at Melka Kunture (Upper Awash, Ethiopia): Techno-economic behaviours at Garba IVD. *Journal of Human Evolution* 65, 594-620.

Gowlett, J.A.J., 1982. Procedure and form in a Lower Palaeolithic industry: Stoneworking at Kilombe, Kenya. *Studia Praehistoria Belgica* 2, 101-109.

Gowlett, J.A.J., 1986. Culture and conceptualisation: the Oldowan-Acheulian gradient, in: Bailey, G.N., Callow, P. (Eds.), *Stone Age Prehistory: studies in memory of Charles McBurney*. Cambridge University Press, Cambridge, pp. 243-260.

Gowlett, J.A.J., 1988. A case of Developed Oldowan in the Acheulean? *World Archaeology* 20, 13-26.

Gowlett, J.A.J., 1990. Technology, skill and the psychosocial sector in the long term of human evolution. *Archaeological Review from Cambridge* 9, 82-103.

Gowlett, J.A.J., 1993. Le site Acheuléen de Kilombe: stratigraphie, géochronologie, habitat et industrie lithique. *L'Anthropologie* 97, 69-84.

Gowlett, J.A.J., 2002. Apes, hominids and technology, in: Harcourt, C.S., Sherwood, B.R. (Eds.), *New Perspectives in Primate Evolution and Behaviour*. Linnean Society, London, pp. 147-171.

Gowlett, J.A.J., 2006. The elements of design form in Acheulian bifaces: modes, modalities, rules and language, in: Goren-Inbar, N., Sharon, G. (Eds.), *Axe Age. Acheulian Toolmaking from Quarry to Discard*. Equinox, London, pp. 203-221.

Gowlett, J.A.J., Harris, J.W.K., Walton, D., Wood, B.A., 1981. Early archaeological sites, hominid remains and traces of fire from Chesowanja, Kenya. *Nature* 294, 125-129.

Hay, R.L., 1976. *Geology of the Olduvai Gorge*. University of California Press, Berkeley.

Isaac, G.L., 1969. Studies of early culture in East Africa. *World Archaeology* I, 1-28.

Isaac, G.L., 1971. The diet of early man: aspects of archaeological evidence from lower and middle Pleistocene sites in Africa. *World Archaeology* 2, 278-299.

Isaac, G.L., 1977. *Olorgesailie. Archeological Studies of a Middle Pleistocene Lake Basin in Kenya*. University of Chicago Press, Chicago.

Jorayev, G., Wehr, K., Benito-Calvo, A., Njau, J., De la Torre, I., 2016. Imaging and photogrammetry models of Olduvai Gorge (Tanzania) by Unmanned Aerial Vehicles: A high-resolution digital database for research and conservation of Early Stone Age sites. *Journal of Archaeological Science* 75, 40–56.

Kleindienst, M. R. 1959. Composition and significance of a Late Acheulian assemblage based on an analysis of East African occupation sites. Chicago: Unpublished PhD, University of Chicago.

Leakey, L.S.B., 1951. *Olduvai Gorge. A Report On The Evolution Of The Hand-Axe Culture In Beds I-IV*. Cambridge University Press, Cambridge.

Leakey, L.S.B., 1958. Recent Discoveries at Olduvai Gorge, Tanganyika. *Nature* 181, 1099-1103.

Leakey, M.D., 1971. *Olduvai Gorge. Vol 3. Excavations in Beds I and II, 1960-1963*. Cambridge University Press, Cambridge.

Leakey, M.D., Roe, D.A., 1994. *Olduvai Gorge. Volume 5. Excavations in Beds III, IV and the Masek Beds, 1968-1971*. Cambridge University Press, Cambridge.

Lepre, C.J., Roche, H., Kent, D.V., Harmand, S., Quinn, R.L., Brugal, J.-P., Texier, P.-J., Lenoble, A., Feibel, C.S., 2011. An earlier origin for the Acheulian. *Nature* 477, 82-85.

Proffitt, T. 2018. Is there a Developed Oldowan A at Olduvai Gorge? A diachronic analysis of the Oldowan in Bed I and Lower-Middle Bed II at Olduvai Gorge, Tanzania. *Journal of Human Evolution* 120, 92-113.

McHenry, L.J., Njau, J.K., de la Torre, I., Pante, M.C., 2016. Geochemical 'fingerprints' for Olduvai Gorge Bed II tuffs and implications for the Oldowan–Acheulean transition. *Quaternary Research* 85, 147-158.

McHenry, L., de la Torre, I. 2018. Hominin raw material procurement in the Oldowan-Acheulean transition at Olduvai Gorge. *Journal of Human Evolution* 120, 378-401.

Mora, R., De la Torre, I., 2005. Percussion tools in Olduvai Beds I and II (Tanzania): Implications for early human activities. *Journal of Anthropological Archaeology* 24, 179-192.

Panger, M.A., Brooks, A.S., Richmond, B.G., Wood, B., 2002. Older Than the Oldowan? Rethinking the Emergence of Hominin Tool Use. *Evolutionary Anthropology* 11, 235-245.

Posnasky, M., 1959. A Hope Fountain Site at Olorgesailie, Kenya Colony. *The South African Archaeological Bulletin* XIV, 83-89.

Potts, R., Behrensmeyer, A.K., Ditchfield, P., 1999. Paleolandscape variation and Early Pleistocene hominid activities: Members 1 and 7, Olorgesailie Formation, Kenya. *Journal of Human Evolution* 37, 747-788.

Rogers, M.J., Semaw, S., 2009. From Nothing to Something: The Appearance and Context of the Earliest Archaeological Record, in: Camps, M., Chauhan, P. (Eds.), *Sourcebook of Paleolithic Transitions. Methods, Theories, and Interpretations.* Springer, New York, pp. 155-171.

Sánchez Yustos, P., Diez-Martín, F., Díaz, I.M., Duque, J., Fraile, C., Domínguez, M., 2015. Production and use of percussive stone tools in the Early Stone Age: Experimental approach to the lithic record of Olduvai Gorge, Tanzania. *Journal of Archaeological Science: Reports* 2, 367-383.

Sánchez-Yustos, P., Diez-Martín, F., Domínguez-Rodrigo, M., Duque, J., Fraile, C., Baquedano, E., Mabulla, A., 2017. Diversity and significance of core preparation in the Developed Oldowan technology: reconstructing the flaking processes at SHK and BK (Middle-Upper Bed II, Olduvai Gorge, Tanzania). *Boreas* 46, 874-893.

Santonja, M., Panera, J., Rubio-Jara, S., Pérez-González, A., Uribelarrea, D., Domínguez-Rodrigo, M., Mabulla, A.Z.P., Bunn, H.T., Baquedano, E., 2014. Technological strategies and the economy of raw materials in the TK (Thiongo Korongo) lower occupation, Bed II, Olduvai Gorge, Tanzania. *Quaternary International* 322–323, 181-208.

Semaw, S., Rogers, M.J., Stout, D., 2009. The Oldowan-Acheulian Transition: Is there a 'Developed Oldowan' Artifact Tradition?, in: Camps, M., Chauhan, P. (Eds.), *Sourcebook of Paleolithic Transitions. Methods, Theories, and Interpretations.* Springer, New York, pp. 173-193.

Stiles, D., 1980. Industrial Taxonomy in the Early Stone Age of Africa. *Anthropologie* XVIII, 189-207.

Tennie, C., Premo, L.S., Braun, D.R., McPherron, S.P., 2017. Early Stone Tools and Cultural Transmission: Resetting the Null Hypothesis. *Current Anthropology* 58, 652-672.

Texier, P.-J., 1995. The Oldowan assemblage from NY 18 site at Nyabusosi (Toro-Uganda). *C.R. Acad. Sc. Paris* 320, 647-653.

Toth, N., 1982. The Stone Technologies of Early Hominids at Koobi Fora, Kenya; An Experimental Approach, Unpublished Ph. D., Berkeley, University of California.

White, M.J., 2000. The Clactonian Question: On the Interpretation of Core-and-Flake Assemblages in the British Lower Paleolithic. *Journal of World Prehistory* 14, 1-63.

Ignacio de la Torre
Instituto de Historia- CSIC, Albasanz, 26-28, Madrid, 28037, Spain.
ignacio.delatorre@csic.es

Rafael Mora
Centre d'Estudis del Patrimoni Arqueologic de la Prehistoria, Facultat de Lletres, Universitat Autonoma de Barcelona, 08193 Bellaterra, Spain

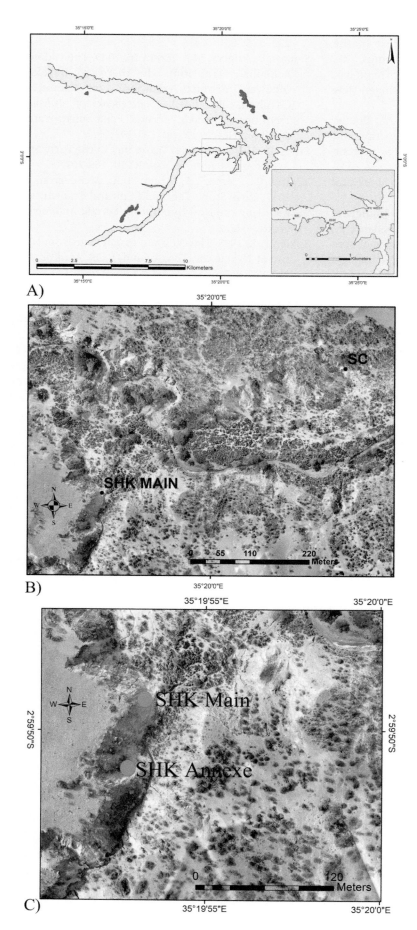

A)

B)

C)

Figure 1. A) Map of the Olduvai Gorge with the position of SHK and nearby localities in the Side Gorge. B) Orthomosaic of the area of SHK and the nearby SC site. C) Location of SHK Main and SHK-Annexe –the position of SHK-Annexe is approximate, and based on Leakey's (1971: 165) account of the site being at around 91 metres from SHK Main. Olduvai Gorge outline, aerial pictures and orthomosaics after Jorayev et al. (2016).

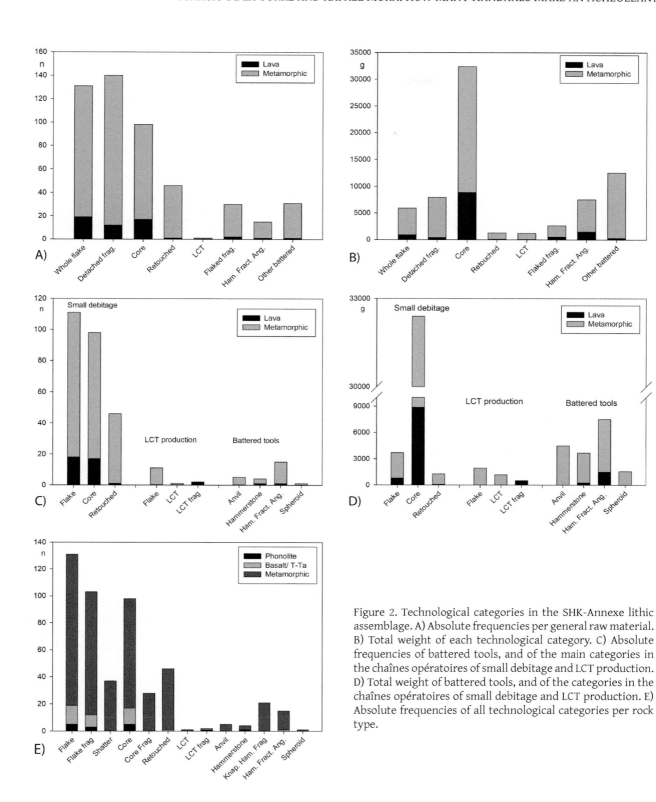

Figure 2. Technological categories in the SHK-Annexe lithic assemblage. A) Absolute frequencies per general raw material. B) Total weight of each technological category. C) Absolute frequencies of battered tools, and of the main categories in the chaînes opératoires of small debitage and LCT production. D) Total weight of battered tools, and of the categories in the chaînes opératoires of small debitage and LCT production. E) Absolute frequencies of all technological categories per rock type.

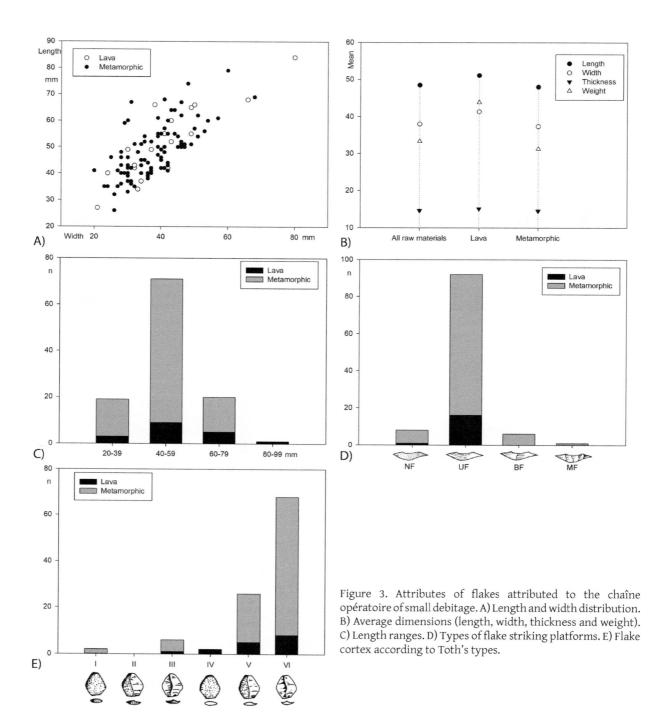

Figure 3. Attributes of flakes attributed to the chaîne opératoire of small debitage. A) Length and width distribution. B) Average dimensions (length, width, thickness and weight). C) Length ranges. D) Types of flake striking platforms. E) Flake cortex according to Toth's types.

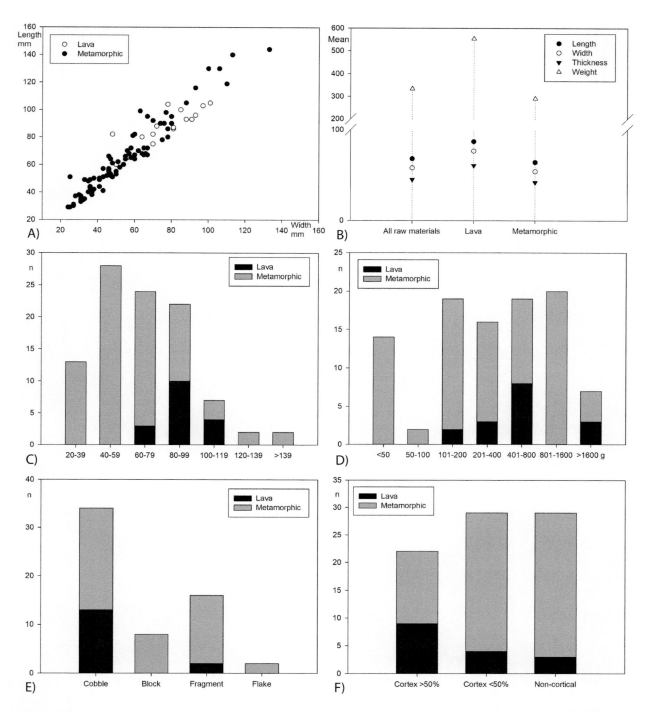

Figure 4. Attributes of small debitage cores. A) Length and width distribution. B) Average dimensions (length, width, thickness and weight). C) Length ranges. D) Weight ranges. E) Core blanks. F) Cortex percentage remaining in cores.

Figure 5. Small debitage cores in the SHK-Annexe collection. #1-3) Quartzite (#1-2) and gneiss UAU1 flaking schemes. #4) Quartzite bipolar core. #5-6) Quartzite BHC flaking schemes. #7) USP lava core.

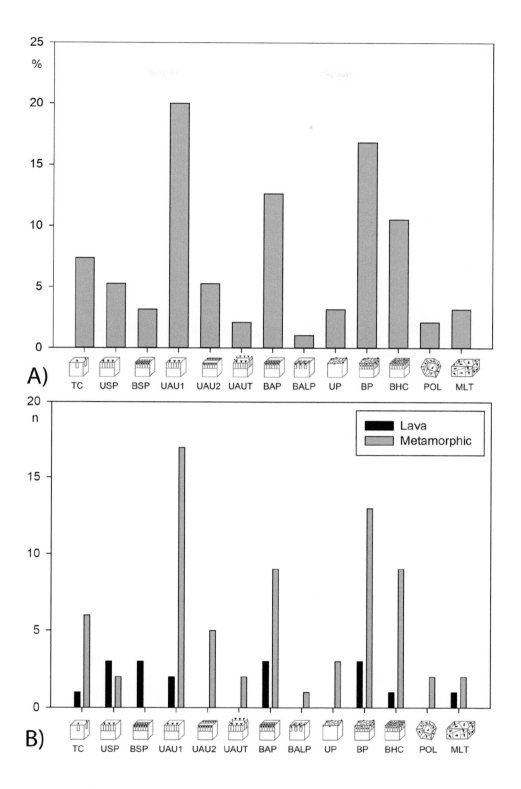

Figure 6. Freehand knapping schemes identified in the SHK-Annexe core assemblage. A) Entire freehand core assemblage. B) Freehand knapping schemes per raw materials. Abbreviations (extracted from de la Torre and Mora 2018b): TC: Test core. USP: Unifacial simple partial exploitation. USP2: Unifacial simple partial exploitation on two independent knapping surfaces. BSP: Bifacial simple partial. UAU1: Unidirectional abrupt unifacial exploitation on one knapping surface. UAU2: Unidirectional abrupt unifacial exploitation on two independent knapping surfaces. UAUT: Unifacial abrupt unidirectional total exploitation. UABI: Unifacial abrupt bidirectional. BAP: Bifacial abrupt partial. BALP: Bifacial alternating partial. BALT: Bifacial alternating total. UP: Unifacial peripheral. BP: Bifacial peripheral. UC: Unifacial centripetal. BHC: Bifacial hierarchical centripetal.

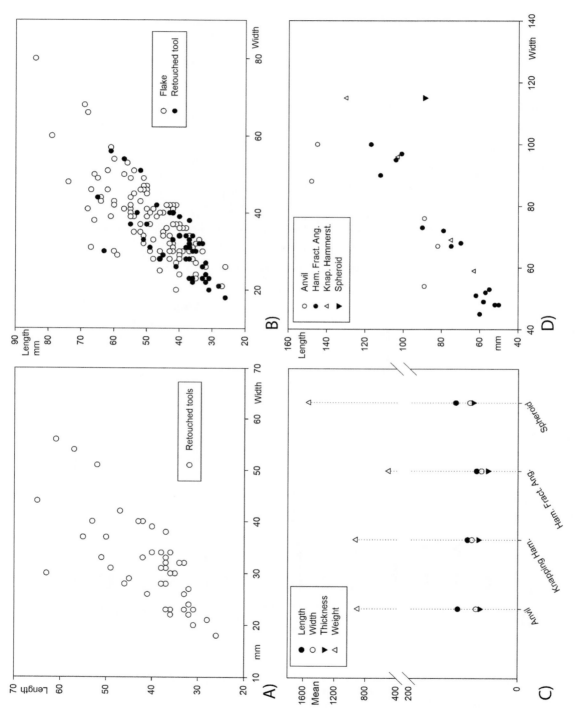

Figure 7. Dimensional features of relevant technological categories. A) Scatter plot of length and width of retouched tools attributed to the chaîne opératoire of small debitage. B) Length and width distribution of small debitage retouched pieces and whole flakes. C) Average dimensions of pounding categories. D) Length and width distribution of pounding categories.

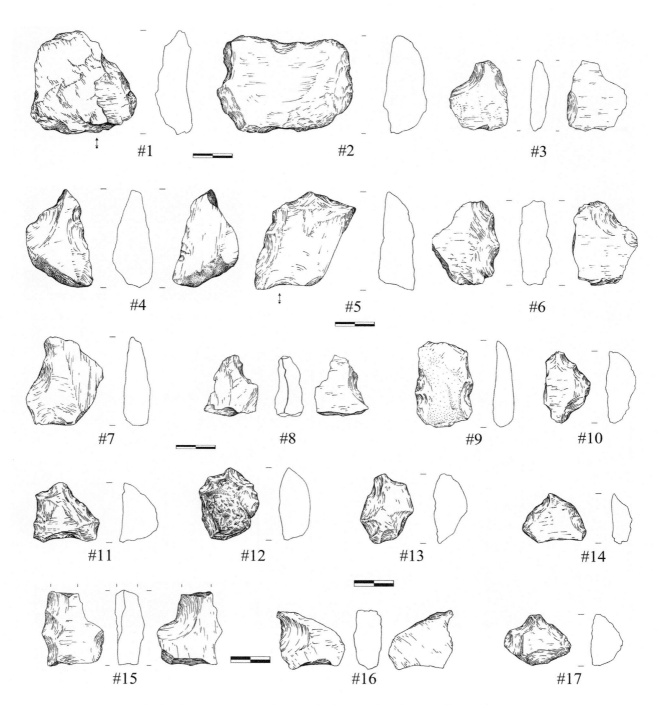

Figure 8. Retouched tools (denticulates and notches) of the small debitage chaîne opératoire at SHK-Annexe.

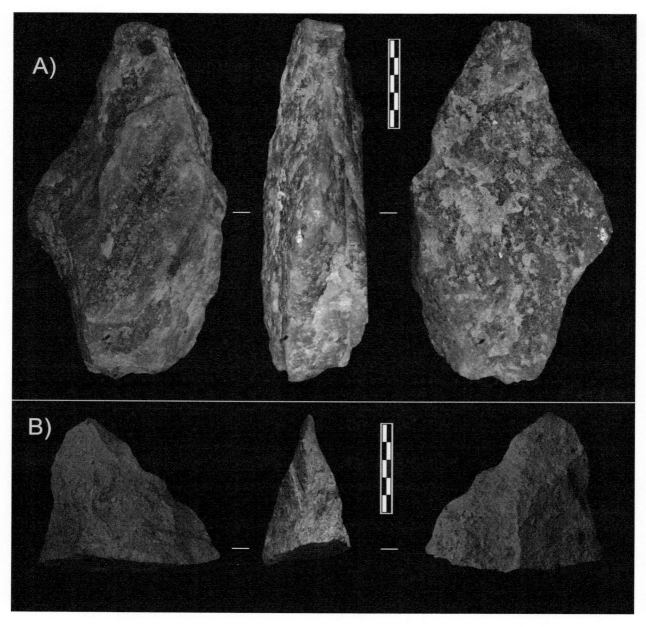

Figure 9. Handaxe evidence in SHK-Annexe. A) Complete quartzite knife. B) Tip of a broken LCT.

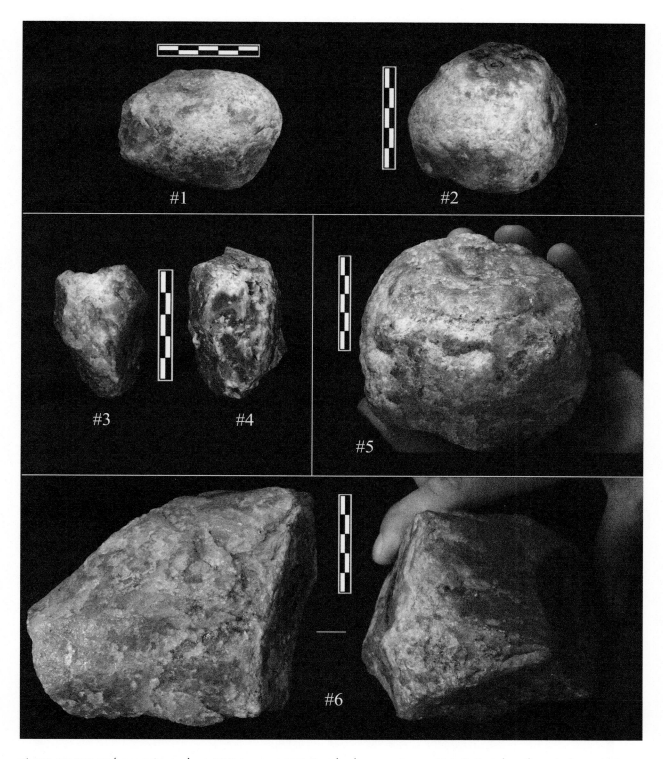

Figure 10: Battered quartzite tools at SHK-Annexe. #1-2: Regular hammerstones. #3-4: Battered surfaces in hammerstones recycled as cores. #5: Hammerstone with active edges. #6: Anvil with edge scarring (left) and clustered battering (right).

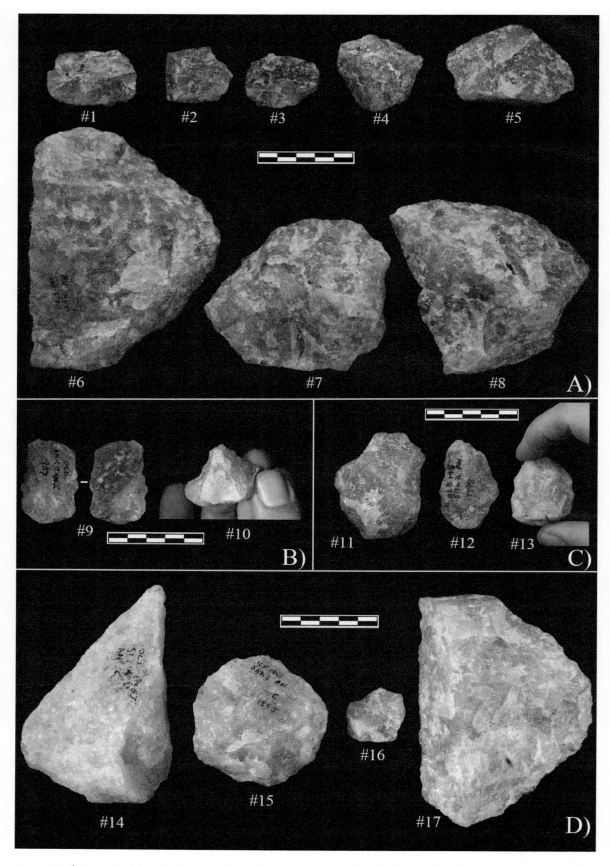

Figure 11. A) Quartzite flakes attributed to the reduction sequences of small debitage (#1-5) and of LCT production (#6-8). B) Small retouched tools of the small debitage chaîne opératoire. C) Examples of diminutive bifaces according to Leakey. D) LCT large flakes (#14, #17) compared to small debitage cores (#15, #16).

Table 1. General breakdown of technological categories in the SHK-Annexe assemblage.

		Phonolite				Basalt/T-Ta				Metamorphic*				Total			
		Frequency		Weight		Frequency		Weight		Frequency		Weight		Frequency		Weight	
		n	%	Sum	%	n	%	Sum	%	n	%	Sum	%	n	%	Sum	%
Detached	Flake	5	62.5	249	87.0	14	60.9	605	63.0	112	46.7	5048	40.0	131	48.3	5902	42.6
	Flake frag	3	37.5	37	13.0	9	39.1	356	37.0	91	37.9	4824	38.2	103	38.0	5217	37.6
	Shatter		0.0				0.0			37	15.4	2750	21.8	37	13.7	2750	19.8
	Total Detached	8	3.0	286	2.1	23	8.5	961	6.9	240	88.6	12623	91.0	271	55.1	13869	19.4
Flaked	Core	5	83.3	1891	90.2	12	85.7	6976	95.1	81	52.3	23528	84.1	98	56.0	32395	86.6
	Core Frag		0.0				0.0			28	18.1	2094	7.5	28	16.0	2094	5.6
	Retouched tool		0.0			1	7.1	62	0.8	45	29.0	1195	4.3	46	26.3	1257	3.4
	LCT		0.0				0.0			1	0.6	1160	4.1	1	0.6	1160	3.1
	LCT frag	1	16.7	206	9.8	1	7.1	301	4.1		0.0		0.0	2	1.1	507	1.4
	Total Flaked	6	3.4	2097	5.6	14	8.0	7339	19.6	155	88.6	27977	74.8	175	35.6	37412	52.5
Pounded	Anvil		0.0				0.0			5	11.4	4494	24.5	5	10.9	4494	22.4
	Hammerstone	1	100.0	258	100.0		0.0			3	6.8	3426	18.7	4	8.7	3684	18.4
	Knap. Ham. Frag		0.0				0.0			21	47.7	2823	15.4	21	45.7	2823	14.1
	Ham. Fract. Angles		0.0			1	100.0	1469	100.0	14	31.8	6045	33.0	15	32.6	7514	37.5
	Spheroid		0.0				0.0			1	2.3	1532	8.4	1	2.2	1532	7.6
	Total Pounded	1	2.2	258	1.3	1	2.2	1469	7.3	44	95.7	18320	91.4	46	9.3	20047	28.1
	Grand Total	**15**	**3.0**	**2641**	**3.7**	**38**	**7.7**	**9768**	**13.7**	**439**	**89.2**	**58919**	**82.6**	**492**	**100.0**	**71328**	**100.0**

*All quartzite except two gneiss artefacts (one piece of shatter and a core).

Table 2. Frequencies of artefacts in the chaînes opératoires of small debitage and LCT production.

| | Small debitage | | | | | | LCT production | | | | | | Indeterminable* | | | | | | Grand Total | |
| | Lava | | Metamorphic | | Total | | Lava | | Metamorphic | | Total | | Lava | | Metamorphic | | Total | | | |
	N	%	N	%	N	%	N	%	N	%	N	%	N	%	N	%	N	%	N	%
Flake	18	100.0	93	100.0	111	100.0		0.0	11	100.0	11	100.0	1	7.7	8	5.9	9	6.0	131	48.3
Flake frag		0.0				0.0		0.0		0.0		0.0	12	92.3	91	66.9	103	69.1	103	38.0
Shatter		0.0		0.0		0.0		0.0		0.0		0.0		0.0	37	27.2	37	24.8	37	13.7
Detached Total	18	6.6	93	34.3	111	41.0		0.0	11	4.1	11	4.1	13	4.8	136	50.2	149	55.0	271	100.0
Core	17	94.4	81	52.6	98	57.0		0.0		0.0		0.0		0.0		0.0		0.0	98	56.0
Core Frag		0.0	28	18.2	28	16.3		0.0		0.0		0.0		0.0		0.0		0.0	28	16.0
Retouched tool	1	5.6	45	29.2	46	26.7		0.0		0.0		0.0		0.0		0.0		0.0	46	26.3
LCT		0.0		0.0		0.0		0.0	1	100.0	1	33.3		0.0		0.0		0.0	1	0.6
LCT frag		0.0		0.0		0.0	2	100.0		0.0	2	66.7		0.0		0.0		0.0	2	1.1
Flaked Total	18	10.3	154	88.0	172	98.3	2	1.1	1	0.6	3	1.7		0.0		0.0		0.0	175	100.0
Anvil		0.0		0.0		0.0		0.0		0.0		0.0		0.0	5	11.4	5	10.9	5	10.9
Hammerstone		0.0		0.0		0.0		0.0		0.0		0.0	1	50.0	3	6.8	4	8.7	4	8.7
Knap. Ham. Frag		0.0		0.0		0.0		0.0		0.0		0.0		0.0	21	47.7	21	45.7	21	45.7
Ham. Fract. Angles		0.0		0.0		0.0		0.0		0.0		0.0	1	50.0	14	31.8	15	32.6	15	32.6
Spheroid		0.0		0.0		0.0		0.0		0.0		0.0		0.0	1	2.3	1	2.2	1	2.2
Pounded Total		0.0		0.0		0.0		0.0		0.0		0.0	2	4.3	44	95.7	46	100.0	46	100.0
Grand Total	36	7.3	247	50.2	283	57.5	2	0.4	12	2.4	14	2.8	15	3.0	180	36.6	195	39.6	492	100.0

*Stone tools that show no defining features attributable to the small debitage or LCT reduction sequences.

Table 3. Weight contribution of artefacts in the chaînes opératoires of small debitage and LCT production.

	Small debitage						LCT production						Indeterminable*						Grand Total	
	Lava		Metamorphic		Total		Lava		Metamorphic		Total		Lava		Metamorphic		Total			
	g	%	g	%	g	%	g	%	g	%	g	%	g	%	g	%	g	%	g	%
Flake	790	100.0	2908	100.0	3697	100.0		0.0	1922	100.0	1922	100.0	64	14.0	219	2.8	282	3.4	5902	42.6
Flake frag		0.0		0.0		0.0		0.0		0.0		0.0	393	86.0	4824	61.9	5217	63.2	5217	37.6
Shatter		0.0		0.0		0.0		0.0		0.0		0.0		0.0	2750	35.3	2750	33.3	2750	19.8
Detached Total	790	5.7	2908	21.0	3697	26.7		0.0	1922	13.9	1922	13.9	457	3.3	7793	56.2	8250	59.5	13869	100.0
Core	8867	99.3	23528	87.7	32395	90.6		0.0		0.0		0.0		0.0		0.0		0.0	32395	86.6
Core Frag		0.0	2094	7.8	2094	5.9		0.0		0.0		0.0		0.0		0.0		0.0	2094	5.6
Retouched tool	62	0.7	1195	4.5	1257	3.5		0.0		0.0		0.0		0.0		0.0		0.0	1257	3.4
LCT		0.0		0.0		0.0		0.0	1160	100.0	1160	69.6		0.0		0.0		0.0	1160	3.1
LCT frag		0.0		0.0		0.0	507	100.0		0.0	507	30.4		0.0		0.0		0.0	507	1.4
Flaked Total	8929	23.9	26817	71.7	35746	95.5	507	1.4	1160	3.1	1666	4.5		0.0		0.0		0.0	37412	100.0
Anvil		0.0		0.0		0.0		0.0		0.0		0.0		0.0	4494	24.5	4494	22.4	4494	22.4
Hammerstone		0.0		0.0		0.0		0.0		0.0		0.0	258	15.0	3426	18.7	3684	18.4	3684	18.4
Knap. Ham. Frag		0.0		0.0		0.0		0.0		0.0		0.0		0.0	2823	15.4	2823	14.1	2823	14.1
Ham. Fract. Angles		0.0		0.0		0.0		0.0		0.0		0.0	1469	85.0	6045	33.0	7514	37.5	7514	37.5
Spheroid		0.0		0.0		0.0		0.0		0.0		0.0		0.0	1532	8.4	1532	7.6	1532	7.6
Pounded Total		0.0		0.0		0.0		0.0		0.0		0.0	1727	8.6	18320	91.4	20047	100.0	20047	100.0
Grand Total	9719	13.6	29725	41.7	39443	55.3	507	0.7	3082	4.3	3588	5.0	2184	3.1	26113	36.6	28297	39.7	71328	100.0

*Stone tools that show no defining features attributable to the small debitage or LCT reduction sequences.

Table 4. Edge abrasion in a sample of the SHK-Annexe assemblage.

	Lava						Metamorphic						Grand Total	
	Detached		Flaked		Lava Total		Detached		Flaked		Metamorphic Total			
	n	%	n	%	n	%	n	%	n	%	n	%	n	%
Fresh	12	63.2	9	56.3	21	60.0	106	99.1	74	93.7	180	96.8	201	91.0
Slight	6	31.6	7	43.8	13	37.1	1	0.9	5	6.3	6	3.2	19	8.6
Medium	0	0.0	0	0.0	0	0.0	0	0.0	0	0.0	0	0.0	0	0.0
Severe	1	5.3		0.0	1	2.9		0.0		0.0		0.0	1	0.5
Grand Total	19	8.6	16	7.2	35	15.8	107	48.4	79	35.7	186	84.2	221	100.0

Table 5. Dimensions (in mm) and weight (in grams) of the main categories in the chaîne opératoire of small debitage.

			Minimum	Maximum	Mean	Std. Deviation
Small flake	Total (n=111)	Length	26.0	84.0	48.5	11.0
		Width	20.0	80.0	38.0	9.9
		Thickness	6.0	34.0	14.6	5.1
		Weight	5.4	160.3	33.3	27.3
	Lava (n=18)	Length	27.0	84.0	51.1	14.6
		Width	21.0	80.0	41.3	14.3
		Thickness	8.0	26.0	15.1	5.3
		Weight	5.4	160.3	43.9	43.8
	Metamorphic (n=93)	Length	26.0	79.0	48.0	10.2
		Width	20.0	68.0	37.3	8.8
		Thickness	6.0	34.0	14.6	5.1
		Weight	6.6	124.2	31.3	22.6
Core	Total (n=98)	Length	29.0	144.0	68.2	26.1
		Width	24.0	133.0	58.1	22.8
		Thickness	20.0	108.0	45.3	18.4
		Weight	15.3	2657.6	334.0	433.5
	Lava (n=17)	Length	60.0	105.0	87.0	13.0
		Width	48.0	101.0	76.7	15.7
		Thickness	35.0	83.0	60.7	13.4
		Weight	126.6	995.1	554.2	257.4
	Metamorphic (n=81)	Length	29.0	144.0	64.2	26.4
		Width	24.0	133.0	54.2	22.3
		Thickness	20.0	108.0	42.1	17.7
		Weight	15.3	2657.6	290.5	448.9
Retouched Tool	Total (n=46)	Length	26.0	65.0	40.6	9.5
		Width	18.0	56.0	31.6	8.7
		Thickness	7.0	37.0	15.4	5.4
		Weight	5.3	116.5	27.3	22.1
	Lava (n=1)	Length	57.0	57.0	57.0	.
		Width	54.0	54.0	54.0	.
		Thickness	17.0	17.0	17.0	.
		Weight	61.8	61.8	61.8	.
	Metamorphic (n=45)	Length	26.0	65.0	40.2	9.3
		Width	18.0	56.0	31.1	8.1
		Thickness	7.0	37.0	15.3	5.4
		Weight	5.3	116.5	26.6	21.7

Table 6. Main features of whole flakes attributed to the chaîne opératoire of small debitage.

	Small debitage						LCT production		Indet						All flakes					
	Lava		Metamorphic		Total		Metamorphic		Lava		Metamorphic		Total		Lava		Metamorphic		Total	
	n	%	n	%	n	%	n	%	n	%	n	%	n	%	n	%	n	%	n	%
20–39 mm	3	16.7	16	17.2	19	17.1		0.0		0.0	2	25.0	2	22.2	3	15.8	18	16.1	21	16.0
40–59 mm	9	50.0	62	66.7	71	64.0	1	9.1		0.0	5	62.5	5	55.6	9	47.4	68	60.7	77	58.8
60–79 mm	5	27.8	15	16.1	20	18.0	4	36.4	1	100.0	1	12.5	2	22.2	6	31.6	20	17.9	26	19.8
80–99 mm	1	5.6		0.0	1	0.9	3	27.3		0.0		0.0		0.0	1	5.3	3	2.7	4	3.1
100–119 mm		0.0		0.0		0.0	1	9.1		0.0		0.0		0.0		0.0	1	0.9	1	0.8
>119 mm		0.0		0.0		0.0	2	18.2		0.0		0.0		0.0		0.0	2	1.8	2	1.5
Total	18	13.7	93	71.0	111	84.7	11	8.4	1	0.8	8	6.1	9	6.9	19	14.5	112	85.5	131	100.0
Non-faceted	1	5.9	7	7.8	8	7.5		0.0	1	100.0	1	33.3	2	50.0	2	11.1	8	7.7	10	8.2
Unifaceted	16	94.1	76	84.4	92	86.0	11	100.0		0.0	2	66.7	2	50.0	16	88.9	89	85.6	105	86.1
Bifaceted		0.0	6	6.7	6	5.6		0.0		0.0		0.0		0.0		0.0	6	5.8	6	4.9
Multifaceted		0.0	1	1.1	1	0.9		0.0		0.0		0.0		0.0		0.0	1	1.0	1	0.8
Total	17	13.9	90	73.8	107	87.7	11	9.0	1	0.8	3	2.5	4	3.3	18	14.8	104	85.2	122	100.0
Cortical	4	23.5	14	15.4	18	16.7	3	27.3		0.0		0.0		0.0	4	22.2	17	16.3	21	17.2
Cortex >50%	1	5.9	7	7.7	8	7.4		0.0		0.0	1	50.0	1	33.3	1	5.6	8	7.7	9	7.4
Cortex <50%	2	11.8	2	2.2	4	3.7		0.0		0.0		0.0		0.0	2	11.1	2	1.9	4	3.3
Non-cortical	10	58.8	68	74.7	78	72.2	8	72.7	1	100.0	1	50.0	2	66.7	11	61.1	77	74.0	88	72.1
Total	17	13.9	91	74.6	108	88.5	11	9.0	1	0.8	2	1.6	3	2.5	18	14.8	104	85.2	122	100.0
I		0.0	2	2.3	2	1.9		0.0		0.0		0.0		0.0		0.0	2	2.0	2	1.7
II		0.0		0.0		0.0		0.0		0.0		0.0		0.0		0.0		0.0		0.0
III	1	6.3	5	5.7	6	5.8		0.0	1	100.0	1	50.0	2	66.7	2	11.8	6	5.9	8	6.8
IV	2	12.5		0.0	2	1.9		0.0		0.0		0.0		0.0	2	11.8		0.0	2	1.7
V	5	31.3	21	23.9	26	25.0	3	27.3		0.0	1	50.0	1	33.3	5	29.4	25	24.8	30	25.4
VI	8	50.0	60	68.2	68	65.4	8	72.7		0.0		0.0		0.0	8	47.1	68	67.3	76	64.4
Total	16	13.6	88	74.6	104	88.1	11	9.3	1	0.8	2	1.7	3	2.5	17	14.4	101	85.6	118	100.0

Table 7. Main attributes of small debitage cores.

		Lava		Metamorphic		Total	
		n	%	n	%	n	%
Length class	20-39 mm		0.0	13	16.0	13	13.3
	40-59 mm		0.0	28	34.6	28	28.6
	60-79 mm	3	17.6	21	25.9	24	24.5
	80-99 mm	10	58.8	12	14.8	22	22.4
	100-119 mm	4	23.5	3	3.7	7	7.1
	120-139 mm		0.0	2	2.5	2	2.0
	>139 mm		0.0	2	2.5	2	2.0
	Total	17	17.3	81	82.7	98	100.0
Weight class	<50 g		0.0	14	17.3	14	14.4
	50-100 g		0.0	2	2.5	2	2.1
	101-200 g	2	12.5	17	21.0	19	19.6
	201-400 g	3	18.8	13	16.0	16	16.5
	401-800 g	8	50.0	11	13.6	19	19.6
	801-1600 g		0.0	20	24.7	20	20.6
	> 1600 g	3	18.8	4	4.9	7	7.2
	Total	16	16.5	81	83.5	97	100.0
Cortex	Cortex >50%	9	56.3	13	20.3	22	27.5
	Cortex <50%	4	25.0	25	39.1	29	36.3
	Non-cortical	3	18.8	26	40.6	29	36.3
	Total	16	20.0	64	80.0	80	100.0
Blank	Cobble	13	86.7	21	46.7	34	56.7
	Block		0.0	8	17.8	8	13.3
	Fragment	2	13.3	14	31.1	16	26.7
	Flake		0.0	2	4.4	2	3.3
	Total	15	25.0	45	75.0	60	100.0
Battering	Esquillees		0.0	7	8.6	7	7.1
	Impacts	5	29.4	32	39.5	37	37.8
	Absent	12	70.6	42	51.9	54	55.1
	Total	17	17.3	81	82.7	98	100.0
Number of scars	1-3 scars	1	6.7	17	23.3	18	20.5
	4-6 scars	9	60.0	33	45.2	42	47.7
	7-9 scars	2	13.3	21	28.8	23	26.1
	> 9 scars	3	20.0	2	2.7	5	5.7
	Total	15	17.0	73	83.0	88	100.0
Flaking method	TC	1	5.9	6	7.7	7	7.4
	USP	3	17.6	2	2.6	5	5.3
	BSP	3	17.6		0.0	3	3.2
	UAU1	2	11.8	17	21.8	19	20.0
	UAU2		0.0	5	6.4	5	5.3
	UAUT		0.0	2	2.6	2	2.1
	BAP	3	17.6	9	11.5	12	12.6
	BALP		0.0	1	1.3	1	1.1
	UP		0.0	3	3.8	3	3.2
	BP	3	17.6	13	16.7	16	16.8
	BHC	1	5.9	9	11.5	10	10.5
	POL		0.0	2	2.6	2	2.1
	MLT	1	5.9	2	2.6	3	3.2
	BIPO		0.0	7	9.0	7	7.4
	Total	17	17.9	78	82.1	95	100.0

Table 8. Dimensions (in mm) and weight (in grams) of pounding tools and artefacts attributed to the chaîne opératoire of LCT production.

			Minimum	Maximum	Mean	Std. Deviation
LCT production	Small flake (n= 5)	Length	58.0	74.0	66.6	7.1
		Width	42.0	58.0	52.4	6.5
		Thickness	11.0	25.0	19.0	5.6
		Weight	28.4	102.6	68.0	32.4
	Intermediate flake (n= 1)	Length	83.0	83.0	83.0	.
		Width	60.0	60.0	60.0	.
		Thickness	28.0	28.0	28.0	.
		Weight	158.9	158.9	158.9	.
	LCT blank (n= 5)	Length	92.0	130.0	109.0	17.6
		Width	50.0	120.0	89.8	25.9
		Thickness	25.0	41.0	34.0	6.0
		Weight	152.3	417.2	298.3	98.1
	LCT (n=1)	Length	190.0	190.0	190.0	.
		Width	117.0	117.0	117.0	.
		Thickness	56.0	56.0	56.0	.
		Weight	1159.6	1159.6	1159.6	.
Pounding tools	Anvil (n= 5)	Length	82.0	148.0	110.6	32.9
		Width	54.0	100.0	77.0	17.9
		Thickness	62.0	75.0	70.0	5.4
		Weight	426.5	1541.9	898.8	475.4
	Hammerstone (n= 4)	Length	63.0	130.0	92.8	30.0
		Width	59.0	115.0	84.8	25.5
		Thickness	50.0	106.0	72.8	24.4
		Weight	258.2	1766.0	921.0	697.1
	Ham. Fract. Angles (n= 15)	Length	50.0	117.0	76.1	23.1
		Width	45.0	100.0	67.2	19.9
		Thickness	12.0	95.0	55.5	21.7
		Weight	134.3	1468.8	500.9	446.1
	Spheroid (n=1)	Length	115.0	115.0	115.0	.
		Width	89.0	89.0	89.0	.
		Thickness	83.0	83.0	83.0	.
		Weight	1531.5	1531.5	1531.5	.

An Acheulian Balancing Act:
A Multivariate Examination of Size and Shape in Handaxes from Amanzi Springs, Eastern Cape, South Africa

Matthew V. Caruana and Andy I. R. Herries

Introduction

For the past fifty years, analysing the shape of handaxes has defined approaches towards investigating the significance of Acheulian large cutting tools (LCTs) (Ashton and McNabb, 1995; Ashton and White, 2003; Bordes, 1961; Brink et al., 2012; Crompton and Gowlett, 1993; Gowlett, 2013, 2006; Gowlett and Crompton, 1994; Graham and Roe, 1970; Hodgson, 2015; Iovita and McPherron, 2011; Isaac, 1977; Lycett and Gowlett, 2008; McNabb, 2009; McNabb et al., 2004; McNabb and Cole, 2015; McPherron, 2003, 2000, 1999, Roe, 1976, 1964; Shipton, 2013; Shipton and Clarkson, 2015; White, 1995; Wynn, 1979; Wynn and Teirson, 1990). It is generally accepted that the continuity of handaxe shape across Acheulian assemblages is a product of imposed form by early hominins, which represents a critical transition towards tool-shaping in the evolution of lithic technology (e.g. Ambrose, 2001; Klein, 2009). From this perspective, the study of shape can provide insight into the manufacturing processes that governed this consistency, as well as underlying cognitive and behavioural capacities (Ashton and McNabb, 1995; Ashton and White, 2003; Brink et al., 2012; Iovita and McPherron, 2011; McPherron, 2006, 2003, 2000, 1999; White, 1995; Wynn, 1995, 1979; Wynn and Teirson, 1990). The seminal work of Bordes (1961) and Roe (1968, 1964) established methods for quantifying handaxe shape through sets of metric measurements, which have since been adapted and applied to Acheulian assemblages from Africa, Europe and Western Asia (Crompton and Gowlett, 1993; Gowlett, 2009; Gowlett and Crompton, 1994; Grosman et al., 2008; Isaac, 1977; Li et al., 2018; Sharon, 2007). An important aspect of this research is how handaxe shape varies at the intra- and inter-assemblage levels, which has been a central topic for debates focused on cognitive and behavioural capacities underlying handaxe production (Ashton and McNabb, 1995; Ashton and White, 2003; Iovita and McPherron, 2011; McPherron, 2006, 2003, 2000, 1999; Nowell et al., 2003; Park et al., 2003; White, 1995; Wynn, 1995, 1979; Wynn and Teirson, 1990).

While shape has played an important role in advancing our understanding of handaxes, size-based variation is an equally significant factor for examining the complexity of their forms. In fact, size proportions and shape are interconnected as the combination of length, breadth and thickness determines the resulting form, i.e. pointed vs. ovate handaxes. While size-based variation remains an under-explored aspect of lithic technology, Gowlett and colleagues (Crompton and Gowlett, 1993; Gowlett, 2013, 2011, 2009, 2006; Gowlett and Crompton, 1994) have pioneered methods for investigating how size and shape co-vary in handaxes across African Acheulian sites. This has highlighted the multivariate nature of these tools with specific focus on how variability in handaxes shape directly relates to variation in geometric size.

The focal point of Gowlett's work has described an allometric relationship between size and shape in handaxes, finding that metric proportions (length, breadth, thickness and mass) vary disproportionally in relation to geometric size. This has established that the maintenance of metric proportions is an important constraint on the consistency of shape, e.g. the ratio of length to breadth. In fact, Gowlett (2011) has found that the L/B ratio is relatively uniform throughout Acheulian assemblages at ~0.61. Gowlett (Crompton and Gowlett, 1993; Gowlett and Crompton, 1994) has suggested that such proportional relationships represents manufacturing 'rule-sets'. For instance, the L/B ratio demonstrates that as handaxe length is extended, breadth decreases, resulting in a narrow plan-view shape (Crompton and Gowlett, 1993). He concluded that this is likely a functional relationship in controlling the overall mass of the artefact, where if length and breadth grew at an isometric rate (linearly), weight would increase exponentially. In turn, this would affect the functionality of handaxes as object mass is a critical concern for manual manipulation in tool use (Bril et al., 2009; Visalberghi et al., 2009).

Building upon this insight, Gowlett (2006) has argued that the trends highlighted by multivariate covariation and proportional consistency in handaxes further reveals 'imperatives' relating to their production. These are the essential features of these tools that underlie their consistency in size and shape. One of the most significant examples of this is the principal of elongation, where handaxes increase in length relative to breadth exponentially, although this can vary considerably on an individual artefact basis (Gowlett,

2013). This likely relates to the extension of useable cutting edge in these tools, but also relates to other factors including the positioning of centre of mass and cutting edge angle. In this sense, the factors that are critical for the use-life of these tools are 'true variables' that can be investigated through Gowlett's multivariate approach.

Shape continues to be a focal point for Acheulian research on LCTs, increasingly through the application of geometric morphometric methods (Archer and Braun, 2010; Iovita and McPherron, 2011; Lycett and von Cramon-Taubadel, 2008). Yet 'size-free' methods have specifically been developed in lithic analyses to trace the influence of shape (Buchanan, 2006; Lycett et al., 2006; Shott et al., 2007). While these avenues of research are undoubtedly valuable, the question remains does removing size effects in examining shape variability in lithic artefacts limit (to some degree) an important aspect of their production? With respect to handaxes, Gowlett and Crompton (Crompton and Gowlett, 1993; Gowlett and Crompton, 1994) have described ubiquitous allometric trends in size and shape covariation across Acheulian sites. This suggests that understanding changes in shape requires a parallel insight into changes in size. Gowlett's interpretations of rule-sets based on investigating this covariation supports the notion that 'size-reduction' in handaxe analysis can potentially diminish a critical aspect of shape variability.

While Gowlett has thoroughly examined multivariate allometry in East African handaxes to understand shape variation, the application of these techniques in South African assemblages remains preliminary (Gowlett and Crompton, 1994; Brink et al., 2012). Here we present an initial investigation into the effects of allometric variation within South African handaxe assemblages with a focus on material from Hilary Deacon's (1970) Area 1 excavations of Amanzi Springs, Eastern Cape, South Africa (Figure 1). This site is one of the few Acheulian localities in southern Africa that preserves layered stratigraphy representing primary deposition of Acheulian technology. It is also unique in being a spring deposit rather than a more common secondary alluvial (e.g. Vaal River) or cave setting (e.g. Sterkfontein) (Herries, 2011). Although Amanzi Springs has yet to be dated using reliable, modern techniques, Deacon (1970: 111) described the material as a Late Acheulian assemblage. If this is accurate, then it should compare with other assemblages from this period, such as Cave of Hearths (<780 ka), Montagu Cave, and potentially also Wonderwerk Cave whose oldest Acheulian deposits have been dated to either side of the Brunhes-Matuyama Reversal at 780 ka based on uranium-lead dating (Herries, 2011; Herries and Latham, 2009; Kuman, 2007; Pickering, 2015; Stammers

et al., 2018). Other sites such as Duinefontein 2 (~1.1 to <~0.3 Ma) and Elandsfontein (Cutting 10; 1.1-0.6 Ma) could also fall within this time range or are just slightly younger than the Acheulian site of Cornelia-Uitzoek at 1.07-1.01 Ma (dates recalculated based on Singer [2014]) (see Braun et al., 2013; Brink et al., 2012; Herries, 2011).

To analyse allometric variation, the Amanzi Springs handaxes are compared with a recently published dataset of measurements from Cave of Hearths (<780 ka; Late Acheulian predominantly on quartzite; Herries and Latham, 2009) and Rietputs 15 (sometime between ~1.5 and ~1.1 Ma; Early Acheulian; on hornfels & andersite; Herries, 2011; Kuman & Gibbon, 2018) by Li et al. (2018). As such, due to its younger age and the similarity of raw material (quartzite) it is expected that the size and shape of handaxes at Amanzi Springs will compare closely to Cave of Hearths. This would support trends in 'refinement' as Late Acheulian handaxes are thought to be more standardized in shape and thinner than cruder forms from earlier Acheulian periods (Hodgson, 2015; Kuman, 2007; Shipton, 2013; Wynn, 1979; although see Li et al., 2018). The aim of this study is twofold: 1) to characterize the handaxes of Amanzi Springs and identify any differences in size and shape with Cave of Hearths and Rietputs 15; and 2) to characterize what 'rule-sets' guided the production of South African handaxes, which will test Gowlett's hypothesis that allometric trends are consistent across Africa.

Multivariate allometry

The multivariate methods developed by Gowlett and Crompton (Crompton and Gowlett, 1993; Gowlett and Crompton, 1994) provide a rigorous agenda for understanding allometric effects on size adjustments in relation to handaxe shape. They employed principal component analysis (PCA) and discriminant analysis (DA) to understand the relationship between size and shape, and how these variables distinguish handaxe assemblages There are several important assumptions that must be considered when implementing this approach, one of the most significant being that size and shape are covariates. As briefly mentioned above, increases in length, breadth and thickness in handaxes are managed through proportional adjustments to maintain specific shape parameters. As such, multivariate tests can be used to identify what variables drive variation in handaxe forms. This is a valuable approach for understanding what aspects of handaxes vary most on chronological and/or geographical scales. Secondly, measuring 'size' cannot be interpreted from a single metric measurement, such as length or breadth. Although, PCA is based on multidimensional scaling of variables that can be used to condense multiple measurements into a single 'size component'. Gowlett and Crompton (Crompton and Gowlett, 1993;

Gowlett and Crompton, 1994) used PCA to this effect in their analyses of Kilombe and Kariandusi handaxes. Analysing multiple localities at these sites and others, they subjected metric measurements, based on Bordes (1961) and Roe (1968, 1964) systems, to independent PCAs using a covariance matrix. The first principal component (PC1) then condensed size effects into a single variable (Buchanan, 2006; Buchanan and Collard, 2007; Shott *et al.*, 2007).

PC1 was then used to calculate allometric coefficients (ACs) for handaxes from East African sites that represented patterns of growth on positive, neutral (isometric) and negative allometric scales (Crompton and Gowlett, 1993; Gowlett and Crompton, 1994). They then compared shape changes in handaxes of different sizes and raw materials within and between these sites to characterize patterns of dimensional relationships. Herein lies the crux of Gowlett's multivariate approach, which is to calculate what variables demonstrate statistically significant relationships throughout the adjustment of dimensional proportions. Gowlett and Crompton (Crompton and Gowlett, 1993; Gowlett and Crompton, 1994) used purpose-written software to calculate ACs, which was based on the angle of PC1 coordinates. The principal behind this is based on the allometric equation, $y= bx^a$, which has been used in biological sciences to model shape changes in response to size growth in animal species (Jolicoeur, 1963). Essentially, ACs are calculated by rescaling the loading scores of PC1 to a mean of 1.0, where AC scores greater than 1.0 indicate positive allometric growth, those equal to 1.0 indicate isometry (neutral growth) and scores less than 1.0 indicate negative allometric growth (Diniz-Filho *et al.*, 1994; Strauss, 1985).

Gowlett and Crompton (*ibid*) used this as a means of comparing trends in growth between measurement variables. In turn, this highlighted features of handaxes that are important for understanding their consistency in form. For instance, the PCA for Kilombe found that the first 3 PC scores described 90% of variation in handaxes, of that PC1 (i.e. geometric size) accounted for approximately 60%. PCs 2 and 3 were associated with thickness and breadth in planform, which equally accounted for the remaining 30% of variation. These relationships between variables were used to identify what dimensions play a key role in defining shape changes in handaxes. The authors then described rule-sets for their production that predict proportional relationships between size and shape.

Lastly, they used DA to examine the statistical power of size and shape variables to discriminate *a priori* handaxe groups (defined by geographic location). The assumption here is that the consistency in handaxe form should result in considerable overlap between groups, particularly handaxes derived from a close geographical proximity, i.e. a region or individual site (Crompton and Gowlett, 1993). They tested this in grouping Kilombe handaxes according to excavation locality (Z, EH, AC, AH & AD), where locality Z handaxes exhibited some different trends in allometric growth. Results showed that 54% of handaxes were correctly assigned to their respective groups with the highest proportion of correct attributions belonging to locality Z handaxes at 79%. Thickness and breadth measurements were the most reliable discriminators, which demonstrated that multivariate analysis of allometry in handaxes can distinguish important size variables for investigating handaxe shape variability.

Amanzi Springs

Reviewing Gowlett's approach towards multivariate allometry demonstrates its statistical power to highlight correlations in dimensional proportions relating to shape. The focus of this research is to apply these methods for investigating similarities and differences in handaxes from Amanzi Springs. This assemblage is important for characterizing the South African Acheulian, although it is rarely discussed in archaeological research despite being a stratified site and also having an apparent association with wood and botanical remains (Deacon, 1970). This is perhaps in part because it is undated. Deacon (1970) confidently assigned the material to the Late Acheulian, albeit recognized that some elements of the assemblage seemed to demonstrate an unstandardized appearance (see below). A brief review of Amanzi Springs is presented below to discuss some of the points of comparison with Cave of Hearths and Rietputs 15 and the significance of using a multivariate approach to examine how these assemblage relate.

Site description and excavation history

Amanzi Springs (~10 km NE of Uitenhage, Eastern Cape, South Africa) is a large thermal spring mound located on a hill which borders the Coega River Valley to the southwest and contains at least 11 spring eyes (Figure 1). The site was first described by Ray Inskeep in 1963 who took note of the abundance of Acheulian lithic material and wood eroding out of the rim of one of the eyes (Inskeep, 1965). He described the spring eyes as circular craters largely comprised of clay sediments overlain by an ironstone crust. They would originally have been horseshoe shaped, with the spring flowing out of the open end, although our initial resurvey of the site has indicated that many of the springs have been extensively altered by historic use for irrigating some of the earliest citrus farms in South Africa. This includes furrowing as noted by Deacon (1970) as well

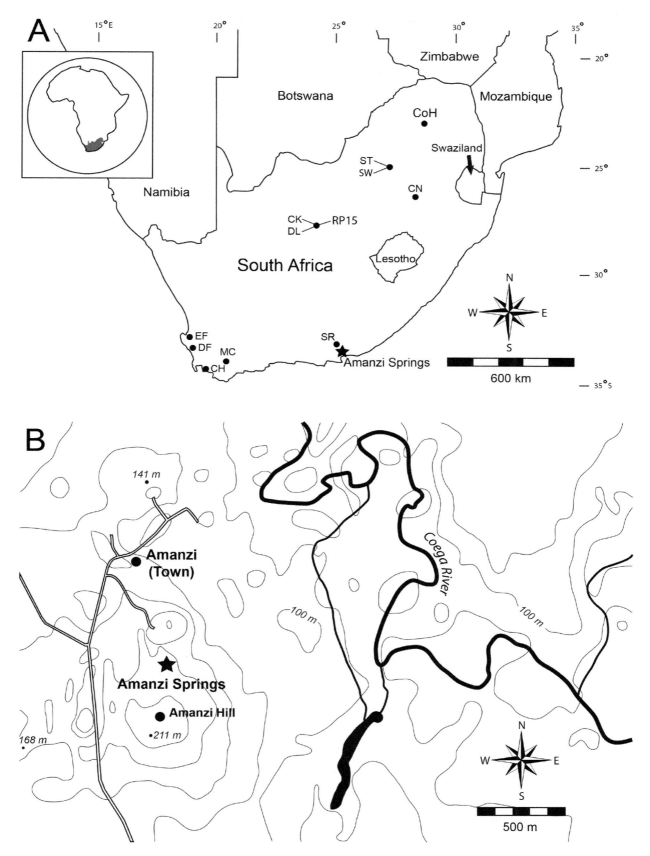

Figure 1. A. Geographical map of Amanzi Springs. A. Geographic positioning of Amanzi Springs within South Africa in relation to other important Acheulian sites. CK= Canteen Kopje, CH= Cape Hangklip, CoH= Cave of Hearths, CN= Cornelia-Uitzoek, DL= Doornlaagte, DF= Duinefontein 2, EF= Elandsfontein, MC= Montagu Cave, RP15 = Rietputs 15, SR= Sunday's River, ST= Sterkfontein Cave, SW= Swartkrans Cave. B. Topographic positioning of Amanzi Springs in relation to the Coega River valley.

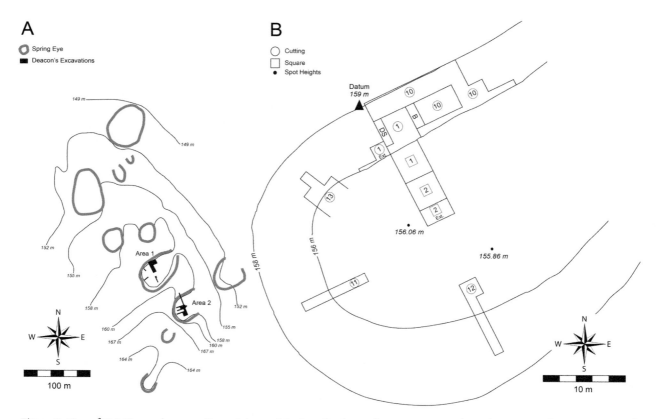

Figure 2. Map of H. J. Deacon's excavations at Amanzi Springs (Redrawn from Deacon 1970). A. The topographic positioning of the known spring eyes recorded by H. J. Deacon. B. Area 1 excavation map.

as the building of dams across the originally open ends of the horseshoe shaped springs. The sediment for the building of these dams seems to have come from the furrows or the more general scraping out of the centre of the spring eyes. This created large heaps of Acheulian stone tools that are now ex-situ. Inskeep excavated (Cutting 1) into in-situ deposits on the northern edge of one of the largest of these spring eyes towards the southern end of the site (defined as Area 1 by Deacon), which yielded an assemblage of Acheulian artefacts (N=1109) (Figure 2).

Deacon took over excavations at the site in 1964 through 1966 for his Master's thesis (Deacon, 1966), additionally opening 13 cuttings, three squares and two 'deep soundings' across Area 1 and a second spring eye (Area 2) at the southernmost extent of the site (Figure 2). After briefly extending Inskeep's cutting in Area 1 (Square 2), Deacon's unpublished records (as well as Deacon, 1966; 1970) show that much of his first season was spent excavating Area 2 (Cuttings 4, 5, 6, 7, 8 and 9 and Square 3). In the following season Deacon returned to Area 1 further extending Cutting 1 and opening Cutting 10 and Square 1. His final phase of excavation then explored other parts of Area 1 (Cuttings 11, 12 and 13), which was aimed at exposing an expanded stratigraphic sequence. Cutting 10, which is our main focus here is essentially a large scale extension of Cutting 1 and is around 9 – 15 m long E-W and 5 – 6 m wide N-S.

Local geology and stratigraphy

Amanzi Springs is an outlet for the Uitenhage Artesian Basin (UAB) system, which is the largest of its kind in South Africa (Mclear, 2001). The springs are part of the Coega Ridge Aquifer formed within quartzites of the Table Mountain Group and the basal sandstone and conglomerate layers of the Enon Formation. This is overlain by impermeable mudstones and siltstones of the Uitenhage Group which forms an aquiclude (Mclear, 2001). The aquifer is restricted to relatively narrow, well-defined zones of intense fracturing and stretches from immediately west of Amanzi Springs, eastward along Coega Ridge to the coast. Major changes to the UAB were caused by borehole drilling at Amanzi Springs between 1908 and 1916, which impeded spring discharge (Maclear, 2001).

The UAB was an open basin flanked in the South and West by mountains of quarzitic Table Mountain Sandstone (TMS) during the earlier Jurassic period (~204-146 Ma; Mclear, 2001). Due to further borehole drilling the springs stopped flowing altogether in 2018. The TMS forms the bedrock in the area within which the Coega Ridge Aquifer developed, as well as the core of Amanzi hill on which the springs are situated. Pebble to boulder alluvial deposits were washed from these mountains under a high energy environment and accumulated along the western margin of the basin, forming the Enon Formation conglomerate during the late Jurassic to early Cretaceous (~146 Ma) (Mclear, 2001). Clays were then deposited uncomfortably on the Enon Formation to form the mudstones and siltstones of the early Cretaceous Kirkwood Formation (~146-100 Ma) (Mclear, 2001). These deposits are the basal sediments within and around the spring eyes as shown by Deacon (1970) in his Area 2 Cutting 5 excavation (termed 'basal clays' or 'variegated marl'). Younger deposits also occur in the form of a silcrete cap at the top of Amanzi hill (likely early Cenozoic: from ~65 Ma; Mclear, 2001). Silcrete artefacts of Middle Stone Age character occur in some of the spring eyes at the site. Silcrete does not appear to have been used for Acheulian artefacts at least in the Area 1 deposits, as it was at Elandsfontein (Braun *et al.* 2013).

In Areas 1 and 2, Deacon (1970) and Butzer (1973) defined three stratigraphic members (Figure 3) across the two areas excavated at Amanzi Springs primarily comprised of sediments that welled up from deep subsurface strata (Butzer, 1973), presented from oldest to youngest:

Enqhura: Consists of 'basal clays' overlain by white sands and then marginal clays. Although this member was only minimally excavated by Deacon, Acheulian artefacts were recovered from its surface (white sands) in Square 1 of Area 1 and from three surfaces within the white sands in Area 2 (Cutting 6 and Deep Sounding). Butzer (1973) described the sands as well-laminated and coarse in Area 1, but lacking silty-clay inclusions as seen in similar deposits within Area 2. The sand represents periods of high spring volume, while the underlying and overlying clays represent periods of low flow.

Rietheuvel: In Area 1, the Rietheuvel Member is described as a brown herbaceous sand (BHS; with wood preservation) grading S to N to a Greenish Clayey Sand (GCS) at the base, and overlain by Grey Black Silts (GBS) in the northern sector of Cutting 1 and 10. In Area 2, it is described as an upper unit of grey-black silts and lower brown humic sands (BHS; aka 'Brown Sands' in Butzer, 1973). However, throughout Deacon's excavation notes, this member is also described as white sandy silt grading S to N to a yellow clayey silts overlain by blue-black material (clay) in the western area of Cutting 10 and in Cutting 1. Butzer (1973) further described the deposits as grey sands grading to pale yellow silt loam. This phase ends with a major truncation and disconformity. The majority of the Acheulian artefacts were said to occur in the top of this member in Cutting 1 based on Deacon's observations. This is in the base of GBS and top of the BHS-GCS. Deacon's (1970) artefact samples 1 – 3 come from this zone. Sample 3 is stratigraphically the lowest and was recovered entirely within GCS, although at the eastern end of Cutting 10 this same unit

is also defined as yellow clayey sands in Deacon's notes. In this same area, the GCS (called Blue Black material in Deacon's notes) covers the entire N-S width of Cutting 10 and is not restricted to the northern area as in the western part of Cutting 10 and Area 1. This sample was recovered from five trenches spaced across the northern part of Cutting 10. Deacon's (1970) sample 1 is described in his notebooks as stratigraphically the next set of lithics coming from the surface of and within the yellow clayey sands (thus BHS-GCS). This was recovered from an easterly extension of Cutting 10. The uppermost sample 2 was recovered entirely within GBS along the northern wall of Cutting 10. While all of samples 1 and 3 were recovered from spits within the various units, some of Sample 2 were piece plotted. Contrary to Inskeep (1965), Deacon stated that the wood-bearing zones in the stratigraphic sequence at Amanzi Springs were likely naturally accumulated and that no positive relationship with artefact accumulations could be identified.

Balmoral: Consists of poorly sorted sands with some well-stratified facies and channel fill deposits that form after the disconformity (Deacon, 1970). This unit generally grades upwards from loamy sands to sandy loam (Butzer, 1973). It is marked by the occurrence of ironstone deposits and layers cemented by iron that cap the older members and have stopped their erosion. The iron originates from the spring water, which in turn is derived from pyrite-rich sandstones from the basement TMS (Butzer, 1973). Deacon (1970) noted that Acheulian artefacts were excavated from this unit (sample 4), although he never analysed this material. While his published plans suggest Sample 4 artefacts only come from the central area of Cutting 10 (all of the artefacts from this sample were piece plotted), they are generally more concentrated in the southern part of this area and similar pothole fill artefacts were also excavated from this Member in Cutting 1. Whether this 'pothole fill' (Deacon, 1970) represents accumulation of artefacts within the time period represented by the younger Balmoral Member or a deflation, lag surface from the erosion of Acheulian artefacts out of the Rietheuvel Member during the major phase of erosion represented by the disconformity, is not certian, but our preliminary analysis of the stratigraphy suggests this is likley the case. The base of this Member was only excavated in Deacon's (1970) 'Deep Sounding' within Cutting 1 of Area 1 and was archaeologically sterile. Due to the question over the relationship of this material to the Rietheuvel Member Acheulian it has not been included in our analysis?

Assemblage characteristics

The Amanzi Springs assemblages were initially described by Deacon (1970: 98), who highlighted the 'heavy and unstandardized' nature of this material.

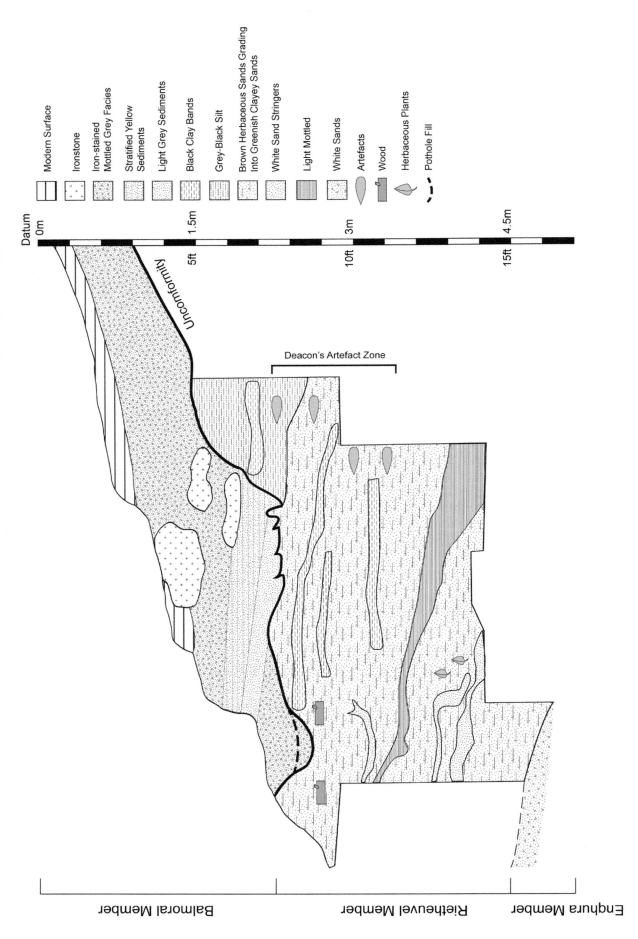

Figure 3. The Stratigraphy of Area 1 at Amanzi Springs (Redrawn from Deacon [1970]). This shows the complex relationship between the three members of the site: Balmoral, Rietheuvel and Enqhura.

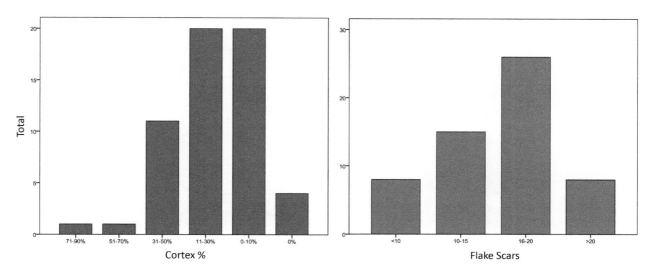

Figure 4. Bar graphs of handaxe attributes from Amanzi Springs (Area 1). A. Cortex percentage. B. Flake scar count.

Figure 4 shows ordinal categories relating to cortex percentages and flake scar counts for handaxes, which demonstrate that while they preserve an average of 10 – 30% cortical surfaces, the amount of flaking is fairly extensive (averaging 16 scars per specimen). However, flaking patterns show that the majority of reduction is restricted to primary shaping phases as these tools lack thinning and retouch (Figure 5). Large flaking blanks (>10cm) were used to produce handaxes are prevalent within the assemblage, including end- and side-struck and 'cobble opening' (eclát entame) methods (Inizan et al., 1999; Sharon, 2011, 2009). There is some evidence for tip preparation within the handaxes and rough-out forms. Deacon (1970) also noted this, referring to a 'five flake pattern' of tip-shaping restricted to cobble-reduced tools, which suggests a trend towards the production of pointed forms versus ovates (Figure 6).

In terms of raw materials, our recent review of the lithic assemblage found that the majority of lithics are made on Enon quartzites and all handaxes analysed here are composed on this material type. Our survey of the sites suggests that Enon quartzite clasts are scattered all over the Amanzi Springs hillside and appear to form the main raw material source for the Acheulian artefacts. These materials are also located within the poorly-sorted terrace gravels of the Coega River, within 2 km of the site (see Figure 2). This suggests a local raw material sourcing and transport strategy, which can be observed at some of the more well-known Acheulian sites in South Africa including Cave of Hearths (McNabb, 2009), Sterkfontein Member 5 West (Acheulian Infill) (Kuman, 1994; Kuman and Gibbon, 2018), Doornlaagte (Mason, 1988), Canteen Kopje (Mcnabb and Beaumont, 2011) and the Rietputs Formation (Gibbon et al., 2009; Kuman and Gibbon, 2018). In general, quartzite exploitation plays an important role in the South African Acheulian tradition, which becomes a widely utilized material

type during this period. However, the structural and fracture properties of quartzites vary greatly on a regional scale in South Africa.

Deacon (1970: 98) noted that the Amanzi Springs handaxes varied considerably in their size and shape, which pointed towards an unstandardized assemblage. Although, this did not exclude it from an 'advanced' phase of the Acheulian industry. Moreover, Deacon's analysis of Amanzi Springs LCTs led him to conclude that 'refinement' qualities in handaxes (i.e. symmetry in plan-view and/or thinness) cannot be used as a definitive chronological marker for Acheulian assemblages (although see Kuman, 2007; Shipton 2013). In fact, recent debate on handaxe shapes have found that examining assemblages as a collection of 'finished' artefacts downplays potential influences of raw materials and continuous reduction and resharpening (Ashton and McNabb, 1995; Ashton and White, 2003; Iovita and McPherron, 2011; McPherron, 2006, 2003, 2000, 1999; White, 1995). From this perspective, every discarded handaxe recovered from the archaeological record has not reached some sort of end-point on a production scale, especially when some show evidence of being re-worked over time or constrained by raw material properties. For this reason, the theoretical assumptions held here are that handaxes, like all other lithic artefacts, represent various stages of production and any variation in size and shape related back to practical concerns such as functionality and constraints on reduction.

Materials

The focus of this analysis is on the handaxes of Amanzi Springs and understanding how they compare in terms of size and shape to other South African Acheulian assemblages. A sample of 57 handaxes from the Area

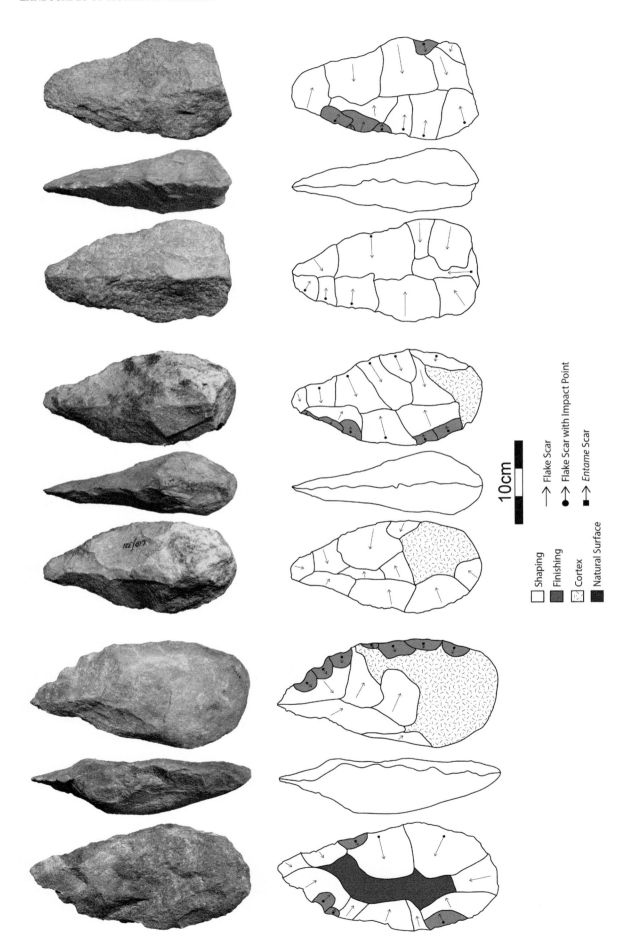

Figure 5. Handaxes from Amanzi Springs. Photographs (above) and schematic drawings (below) showing flaking patterns.

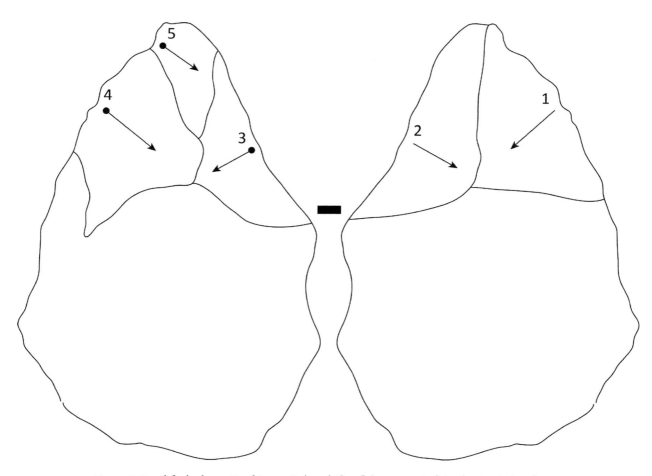

Figure 6. Simplified schematic of Deacon's (1970) 'five flake pattern' of tip-shaping in handaxes.

1 artefact collection were selected for comparison with the Cave of Hearths and Rietputs 15 assemblages. Concerning the latter two sites, measurements were taken from a useful dataset published in Li, et al. (2018), which compares handaxes from these sites to understand changes in production techniques through the Early versus Late Acheulian periods. The contexts of these assemblages have been reviewed in numerous publications, although brief descriptions are necessary for understanding their comparability to Amanzi Springs.

Cave of Hearths is located within the Makapan Valley system near the town of Mokopane, Limpopo Province (see Figure 1). It was originally discovered by van Riet Lowe (1938) in 1937, although it was first systematically excavated by Mason (1988) between 1953-4. This exposed an expansive cultural sequence from the Later, Middle and Earlier Stone Ages preserved in 11 stratigraphic layers (Beds), overlain by an Iron Age deposit (Bed 12) (Mason, 1988; McNabb and Sinclair, 2009). Beds 1 – 3 have yielded a large, Late Acheulian assemblage, which has since been dated to a maximum age of 780 ka based on palaeomagnetism (Herries and Latham, 2009). The artefacts are largely made on locally-sourced quartzite materials and show increased amounts of thinning and

retouch when compared to Amanzi Springs. While this site has often been referenced as a 'type assemblage' for the Late Acheulian in South Africa, McNabb's (2009) most recent assessment of the Beds 1 – 3 materials found that the LCTs were also unstandardized. This was based on the variability of shape within their tips, which further questions the notion of 'refinement' in the Late Acheulian period of South Africa (also see McNabb et al., 2004). However, the unstandardized nature of this assemblage and the consistency of quartzite use should find some parallels to the Amanzi Springs materials. Given this material comes from a cave that has no immediate quartzite outcrops surrounding it, the artefacts likely represent curated materials and are extensively reduced. (McNabb, 2009). As such, the site is broadly comparable to Amanzi Springs with regards its raw material, if not its depositional setting.

Rietputs 15 is located on the Vaal River near the town of Windsorton (Northern Cape Province) (see Figure 1), where artefacts were identified and collected within alluvial gravels during a mining operation (Gibbon et al., 2009; Kuman and Gibbon, 2018). The Vaal River gravel sequence has been used as a benchmark for dating Acheulian sites in South Africa since the 1940's (see van Riet Lowe, 1952), which has been divided into sequential terraces ('Older' and 'Younger') that preserve artefacts from all Stone Age techno-complexes in South Africa (Butzer *et al.*, 1973). Rietputs 15 fits into the 'younger' gravel sequence (Butzer *et al.*, 1073; Gibbon *et al.*, 2009; Helgren, 1978) and the Pit 5 artefacts (analysed in Li *et al.*, 2018) have been dated to sometime between ~1.5 and ~1.1 Ma (1.31 ± 0.21 to 1.27 ± 0.20 Ma) based on cosmogenic burial methods (Herries, 2011; Kuman and Gibbon, 2018). The assemblage is almost exclusively made on Ventersdorp lava, which include andesite and other forms of diabase rocks. Handaxes are not extensively flaked and vary considerably in shape (Kuman and Gibbon, 2018). Handaxes from Pit 5 at Rietputs 15 are included here to test how the extent of size and shape variation in this assemblage compares to Amanzi Springs. As stated above, Deacon (1970) described Amanzi Springs' lithics as unstandardized, which may show some affinities to an Early Acheulian assemblage in terms of handaxe shapes. Although given the possible age restriction on the formation of Amanzi Springs to a younger Pleistocene period, it is expected that the assemblage will compare more closely to the Late Acheulian and thus Cave of Hearths,. Some variation in these assemblages based on the fracture properties and hardness of medium grained igneous rocks versus quartzite is expected. However, the increased use of quartzite materials throughout the Acheulian industry in South Africa should reflect a mastery over any raw material constraints on handaxe production.

Methods

The dataset used here combined published measurements of handaxes from Rietputs 15 and Cave of Hearths handaxes by Li, et al. (2018) with similar measurements from Amanzi Springs. Gowlett included breadth-at-midpoint (BM) and thickness-at-midpoint (TM) measurements, which were not used in the Li, et al. (2018) dataset and thus excluded in this analysis for comparability. Gowlett and Crompton (1994: 30) stated that TM is largely 'a redundant variable if T [thickness] is available,' which was assumed the same for BM. Similarly, mass was also excluded from these analyses as the variables used here account for geometric size and thus volumetric measurements (i.e. weight) are redundant (Crompton and Gowlett, 1993). Non-parametric tests were chosen here for comparison as approximations to the normal distribution in archaeological data cannot be assumed. The initial exploration of the data involved a basic analysis comparing length, breadth, thickness and mass (g), as well as elongation (L/B) and refinement (B/Th). These data were explored through boxplots and Kruskal-Wallis tests to accommodate more than two samples. Results will reveal potential differences in metric proportions between assemblages. In following Gowlett's assumptions, length, breadth, thickness and mass should demonstrate a power relationship consistent with allometric growth patterns. To test

this, regression analyses were used to plot length, breadth and thickness against mass. The reasoning for using mass as a dependent variable is that it is most comparable to the total volume (and thus the size) of artefacts.

Next, multivariate techniques employed by Gowlett and Crompton (Crompton and Gowlett, 1993; Gowlett and Crompton, 1994), discussed above, were used to explore allometric patterns in the combined dataset. The first step consisted of log-transforming all variables for PCAs analysing individual sites using a covariance matrix. The PC1 scores were then used to calculate ACs for each measurement variable through rescaling loadings to a mean of 1.0 (Diniz-Filho *et al.*, 1994; Strauss, 1985). ACs were used to compare allometric relationships between handaxe shapes at South African sites. Coefficients of variation for measurement variables were also calculated for comparison with ACs to understand how they vary on an allometric scale (Crompton and Gowlett, 1993). A second PCA was then run on the complete dataset to explore the dispersion of variation on component scores. This was used to highlight what variables play significant roles in handaxe shape variability between these South African assemblages.

It is expected that most of the variation will be explained by PC1 with similar trends in PC2 (thickness) and PC3 (breadth in planform) as seen in the Kilombe and Kariandusi analyses (Crompton and Gowlett, 1993; Gowlett and Crompton, 1994). This will be used to test Gowlett's hypothesis that 'rule-sets' in handaxe production are consistent across Africa. A set of PCAs were then run on individual sites to verify these results and test the consistency of PC loading trends in the PCA run on the combined dataset to identify allometric trends. Finally, a DA was used to test the ability of the variables to discriminate between sites. It should be expected that Cave of Hearths and Amanzi Springs would exhibit a fair amount of cross-classification due to their assignment to the Late Acheulian and their more similar raw material (e.g. quartzite). This will test the ability of handaxe size and shape to distinguishing between assemblages and test the notion of refinement that Deacon (1970) questioned when reviewing the Amanzi Springs lithic materials.

Results

Exploratory results

The combined dataset was subjected to an exploratory analysis of basic dimensions (L, B, Th & M) between sites, which yielded unexpected results. Figure 7 demonstrates that Amanzi Springs is considerably larger in overall geometric size when compare to both Rietputs 15 and Cave of Hearths. A Kruskal-Wallis test found significant differences for all variables between groups (L: x^2= 40.57, p< 0.001; B: x^2= 27.57, p< 0.001; Th: x^2= 39.25, p< 0.001; M: x^2= 53.98, p< 0.001). Another test was then run to compare Cave of Hearths and Amanzi Springs as Late Acheulian assemblages, which also confirmed significant differences (L: x^2= 14.65, p< 0.001; B: x^2= 16.87, p< 0.001; Th: x^2= 31.45, p< 0.001; M: x^2= 43.97, p< 0.001). Elongation (L/B) and refinement (B/Th) showed mixed patterns, where Amanzi Springs demonstrates that handaxes are more elongated, yet are the least refined (Figure 8). Kurskal-Wallis results confirmed significant differences between these variables as well (El: x^2= 20.46, p< 0.001; Rf: x^2= 13.58, p= 0.001). A separate test was again run between Cave of Hearths and Amanzi Springs for elongation and refinement, which only returned a significant difference for refinement (El: x^2= 1.43, p= 0.231; Rf: x^2= 12.28, p< 0.001). This is unexpected for two Late Acheulian assemblages, especially considering that Rietputs 15 shows displays more refinement that Amanzi Springs.

Examining the regression results, a power relationship was observed between all linear measurements and mass as previously predicted by Gowlett (Crompton and Gowlett, 1993; Gowlett and Crompton, 1994) (Figure 9). An interesting trend to note is that r^2 values were higher for Amanzi Springs in length and slightly for width, although not for thickness. This suggests the

presence of an allometric relationship between basic dimensions (L, B & Th) when compared to volumetric size (mass), which generally correlates highest in the Amanzi Springs handaxes. Another significant feature of the regression analyses is the overall clustering patterns of plot scores for Amanzi Springs when compared to the other sites, which consistently group towards the higher spectrum of values. This shows that similar to Gowlett's conclusions about allometric effects on size, shape and mass, handaxes from Amanzi Springs approximately 15cm in length will weigh around 500 g, while an increase in length to 20cm doubles the weight (~1000 g; see Crompton and Gowlett, 1993).

Next, Table 1 compares mean, standard deviation and coefficient of variation (CV) values for all measurements used in the multivariate analysis, also including mass, elongation and refinement. An interesting pattern emerges demonstrating that the Late Acheulian assemblages (Cave of Hearths and Amanzi Springs) are considerably more variable than Rietputs 15. This result was also found in Li et al. (2018), who suggested that variability is perhaps a more characteristic factor of Late Acheulian handaxes in South Africa than refinement. CV values for variables are graphically displayed in Figure 10, which shows some distinct patterns between sites. Rietputs 15 is consistently less variable aside from thickness dimensions, which may

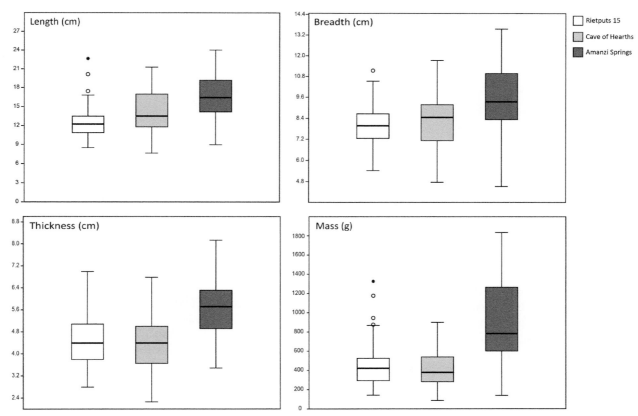

Figure 7. Comparison of basic metric measurements showing that Amanzi Springs handaxes are larger in size when compared to Cave of Hearths and Rietputs 15.

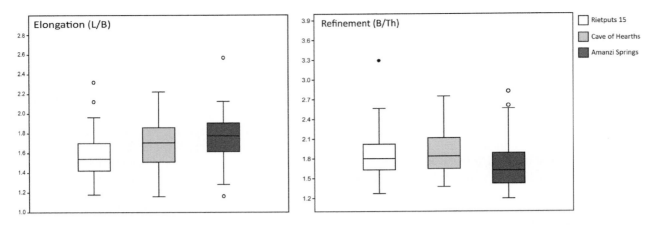

Figure 8. Comparison of elongation and refinement indices demonstrating that handaxes from Amanzi Springs are comparatively more elongated yet less refined.

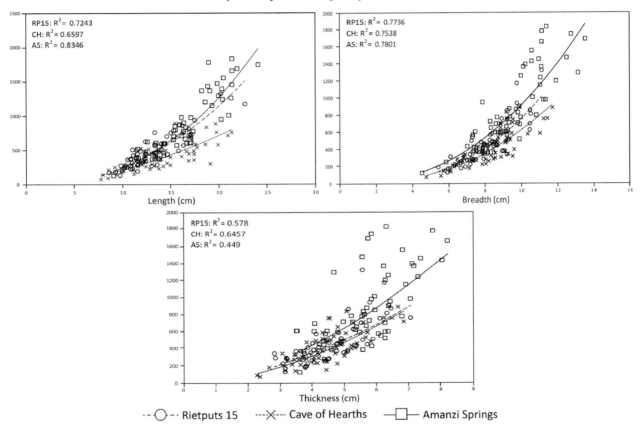

Figure 9. Regression of metric variables against mass (g) highlighting allometric trends.

be a consequence of the package size and shapes of Ventersdorp lavas available in the Vaal River gravels. Cave of Hearths is more variable in L, B and Th variables when compared to Amanzi Springs, which are only similar in PMB. Although when CV values for mass are considered, Amanzi Springs is nearly double the volumetric size of the other sites. This pattern suggests that while handaxes size for Amanzi Springs were considerably larger, they were more restricted in shape variation than Cave of Hearths, and in thickness aspects compared to Rietputs 15.

Multivariate allometry results

Allometric coefficients were then calculated according to Gowlett's multivariate procedures (Crompton and Gowlett, 1993). Table 2 displays both PC1 and AC scores by site. Figure 11 displays the graphic results for AC scores, which highlights allometric trends for shape variables at these South African sites. Upon a visual comparison to results for Kilombe and Kariandusi, the South African sample displays similar patterns to East African assemblages (see Figure 20 in Gowlett and

Table 1. Handaxe measurements (cm) for South African Acheulian assemblages.

Rietputs 15 (N= 57)	L	B	BA	BB	PMB	T	TA	TB	M	L/B	B/Th
Mean	12.71	8.09	4.60	6.70	4.73	4.51	2.39	3.78	458.23	1.57	1.84
S.D.	2.59	1.25	1.07	1.13	1.32	0.97	0.53	0.88	239.41	0.21	0.34
C.V.	20.38	15.39	23.32	16.86	27.82	21.52	22.29	23.33	52.25	13.47	18.23
Cave of Hearths (N= 64)											
Mean	14.10	8.29	4.82	7.20	4.99	4.44	2.19	3.63	418.97	1.70	1.91
S.D.	3.36	1.46	1.37	1.34	1.72	1.01	0.65	0.90	200.99	0.24	0.33
C.V.	23.82	17.59	28.41	18.64	34.48	22.71	29.60	24.69	47.97	14.35	17.14
Amanzi Springs (N= 57)											
Mean	16.63	9.55	7.37	8.74	6.52	5.71	3.69	4.37	887.07	1.76	1.71
S.D.	3.23	1.76	1.63	1.70	2.27	1.13	0.81	0.85	431.50	0.24	0.38
C.V.	19.40	18.47	22.05	19.44	34.76	19.80	22.05	19.44	48.64	13.40	21.91

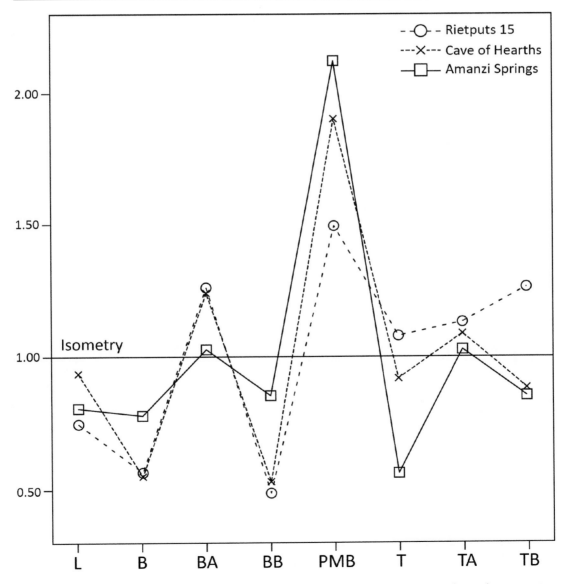

Figure 10. Coefficient of variation scores for variables used in the multivariate analysis, demonstrating differences in dimensional variation. L= Length, B= Breadth, BA= Breadth of Tip, BB= Breadth of Base, PMB= Point of Maximal Breadth, T= Thickness, TA= Tip Thickness, TB= Breadth of Base.

Crompton, 1994). As such, PMB shows the highest, positive allometric score, while BB displays the opposite pattern. B and BB show greater trends towards negative allometry in the South African sample than Kilombe and by comparison, allometric trends in South African handaxes plot closely with Kapthurin and Kariandusi (Gowlett and Crompton, 1994).

Further, AC and CV scores were compared in a bivariate plot according to mean values, which displays the general relationship between these factors for the measured variables. Figure 12 displays these results for allometric and variation patterns, which is comparable to results reported for

Table 2. PC1 loadings and allometric coefficient scores by site.

	Rietputs		Cave of Hearths		Amanzi Springs	
	PC1	AC	PC1	AC	PC1	AC
L	0.12	0.74	0.14	0.93	0.12	0.80
B	0.09	0.56	0.08	0.55	0.11	0.77
BA	0.20	1.26	0.19	1.24	0.15	1.02
BB	0.08	0.48	0.08	0.52	0.12	0.85
PMB	0.24	1.49	0.29	1.90	0.31	2.13
T	0.17	1.08	0.14	0.91	0.08	0.56
TA	0.18	1.13	0.16	1.08	0.15	1.02
TB	0.20	1.26	0.13	0.88	0.12	0.85

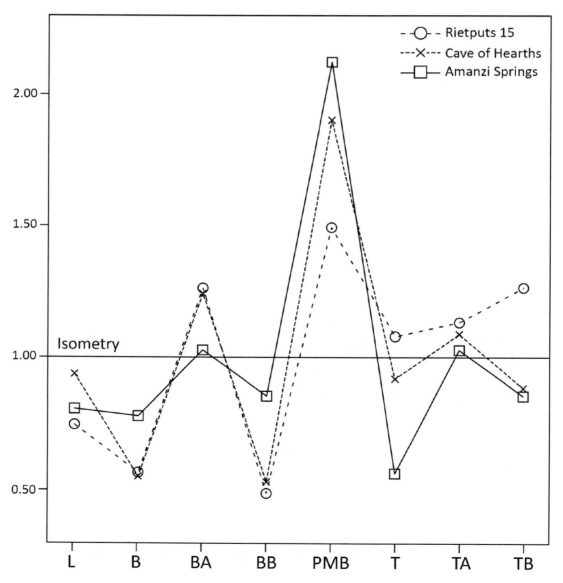

Figure 11. Allometric coefficients for variables used in the multivariate analysis demonstrating which variables are positively and negatively allometric. L= Length, B= Breadth, BA= Breadth of Tip, BB= Breadth of Base, PMB= Point of Maximal Breadth, T= Thickness, TA= Tip Thickness, TB= Breadth of Base.

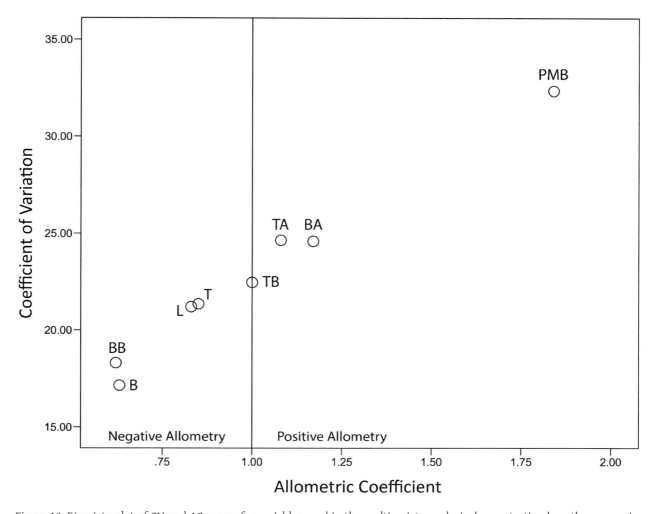

Figure 12. Bivariate plot of CV and AC scores for variables used in the multivariate analysis demonstrating how they group in terms of negative and positive allometric patterns. L= Length, B= Breadth, BA= Breadth of Tip, BB= Breadth of Base, PMB= Point of Maximal Breadth, T= Thickness, TA= Tip Thickness, TB= Breadth of Base.

Kilombe (see Figure 8 in Crompton and Gowlett, 1993). However, there are some disparities in these patterns that differentiate the East and South African handaxes to a degree. Crompton and Gowlett (1993) grouped variables according to trends in negative-to-positive allometry and high-to-low variation, identifying a linear relationship between allometric and variation trends. They found that variables fell into three basic patterns: 1) basic planform and butt variables (negative allometry/low variation); 2) thickness variables (negative allometry/average variation); and 3) pointedness variables (positive allometry/high variation).

A similar linear correlation was detected for the variables used to compare South African handaxes (r^2= 0.98, p< 0.001), where PMB (highest allometric coefficient) displays the highest amount of variation and B and BB show the opposite trend. However, the grouping of variables demonstrates differences in

correlative patterns. For the South African sample, four basic groups are found: 1) general plan-view and profile shape (negative allometry/low variation); 2) butt thickness (isometry/low variation); 3) tip shape (positive allometry/average variation); and 4) point of maximum breadth (positive allometry/high variation). The needs to create four groups was motivated by the isometric trend in butt thickness, which is closely related to tip shape variables, albeit should be separated on its trend of 'geometric' growth (i.e. 1:1 increase with size).

Assessing these groups, some general patterns emerge. Primarily, there is a relationship between the overall shape of handaxes and geometric growth. This shows that size increases in handaxes negatively correlate with maximum dimensions in plan and profile views. This suggests that knappers consistently constrain their overall proportions to maintain shape. The second pattern relates to butt thickness growing at an isometric rate with size, which suggests that this variable is relatively stable throughout production. This likely correlates to the location of the centre of mass in handaxes, which is a critical variable in their use (Gowlett, 2006; Grosman et al., 2008; Park et al., 2003). The third pattern includes both tip shape and PMB because they show similar positive allometric trends, albeit the latter variable shows maximum variation. Breaking this third pattern down, tip breadth and thickness grow at an increasingly faster rate to geometric increases in size. This likely correlates to maintaining elongation in handaxes as the widest and thickest aspects of these tools tend to be located towards the butt end. As such, accelerating growth in the tip then counteracts trends towards ovate shapes and maintains a prominent pointed tip. Lastly, the positioning of PMB is also critical towards maintaining the centre of mass in handaxes. As tool length increases, the need to position the PMB towards the mid-point of overall length is important for avoiding butt-heavy products. As Gowlett (2011, 2013) found, smaller handaxes tend to have wider plan-view shapes (increased breadth relative to length), typically concentrated towards the butt-end, which is manifested as an average ratio of 0.75. Whereas larger handaxes tend to be comparatively thinner in plan-view (decreased breadth relative to length), manifested as an L/B ratio of 0.50 (Gowlett, 2011).

These patterns are useful to understand some principals underlying the size/shape relationship of handaxes, which are assumed to reflect the knapper's control over their form. A second PCA on the entire dataset was then used to see if AC/CV patterns could be condensed into more meaningful 'rule-sets' (Crompton and Gowlett, 1993; Gowlett and Crompton, 1994). A KMO-Bartlett's test was used to test the strength of the PCA (KMO= 0.848; Bartlett's: x^2= 1508.53, df= 28, p< 0.001), which demonstrated that results were suitable for multivariate analysis. Figure 13 shows the results comparing the first 3 principal components, which represented over 93% of the variance. When the loading

scores for the first 3 PCs are compared (Figure 14), some patterns are different to Crompton and Gowlett's (1993) findings for Kilombe handaxes. For instance, length accounted for the most variation on the PC1 (79%) axis, with all other variables loading relatively weaker. Given that PC1 represents geometric size, length seems to correlate strongest with this factor in these handaxes. Yet as Crompton and Gowlett (1993) noted, this correlation doesn't account for most allometric variables discussed above. PC2 (8%) was most strongly correlated with both breadth (B, BA & BB) and thickness (T, TA & TB) variables. This suggests that positive allometric trends in tip shape (TA & BA) co-vary with plan-view (B & BB) and profile (T & TB) shape below the point of maximum breadth and thickness. In this sense, as handaxes increase in geometric size, tip shape increases while the lower portions of these tools decrease, which

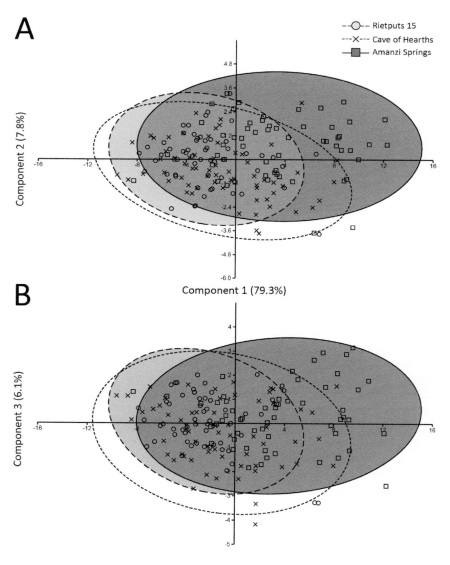

Figure 13. Principal Component Graphs for Amanzi Springs, Cave of Hearths and Rietputs 15. A. PC1 and PC2. B. PC1 and PC3. Amanzi Springs scores load strongest onto PC1, which correlated to the overall size of handaxes (length).

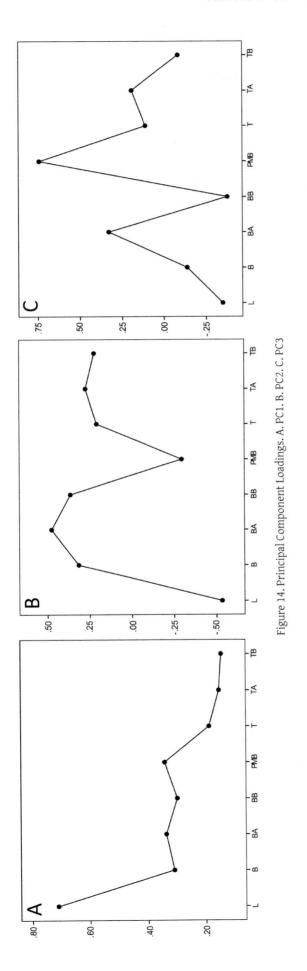

Figure 14. Principal Component Loadings. A. PC1. B. PC2. C. PC3.

likely relate to managing the positioning of the centre of mass. Finally, PC3 (6%) represents variation in PMB, which correlates to the trends highlighted above in Figure 12. These results substantiate the AC/CV patterns in allometric growth in handaxes, which are discussed in more detail below.

To verify these results, three PCAs for individual sites were run separately to test the representative nature of trends highlighted above. Component loadings for individual sites were then compared to test the findings displayed in Figures 13 and 14. This confirmed that PC loadings for variables were broadly similar between Amanzi Springs, Cave of Hearths and Rietputs 15 (data not presented here). Interestingly, Amanzi Springs and Rietputs 15 were similar in loading scores across PCs 1 – 3, which parallel those found in the PCA for the complete dataset discussed above. Cave of Hearths displayed slight deviation in component loadings for PMB, which loaded onto PC 2 more positively and similarly for BB onto PC3. This suggests that perhaps PMB plays more of a role in variation for the Cave of Hearths handaxes, possibly correlating to variability in raw material package size and the extent of flaking. Li, et al. (2018) have shown that Cave of Hearths handaxes are more extensively flaked when compared to Rietputs 15. Coupled along with CV scores in Table 1, Cave of Hearths shows the most variability in length, suggesting a correlation between this dimension and the positioning of PMB. Regardless of this difference, the PCAs for individual sites corroborate that PCs 2 and 3 reflect allometric patterns in how tip and butt shape variables co-vary during increases in geometric growth.

Finally, the results of the DA show that Amanzi Springs handaxes are distinct from the Rietputs 15 and the Cave of Hearths assemblages (Figure 15). In fact, the latter two assemblages show the most overlap, which is unexpected given one of these assemblages is meant to represent the Early Acheulian and the other the Late Acheulian. The greater correlation between Cave of Hearths and Rietputs 15 when comparing the former site to Amanzi Springs is even more unexpected given they are both classified as Late Acheulian. The first function captures 88% of the variance, in which length and tip breadth are the best discriminators. This suggests that geometric size is again correlated with length, which is distinguishing factor for Amanzi Springs handaxes. Thickness and breadth variables then load strongest onto the second function, which reiterates the results of the PCA above. Table 3 displays the cross-validated classification results for specimens assigned to groups by site. It is clear that the DA validates the *a priori* groups with 71.5% of specimens assigned correctly. However, this high percentage is mostly driven by 87.9% rate of Amanzi Springs handaxes correctly discriminated. Approximately 30% of Rietputs 15 and Cave of Hearths handaxes, respectively were incorrectly assigned to one another, exhibiting the most significant amount of overalp between groups. The

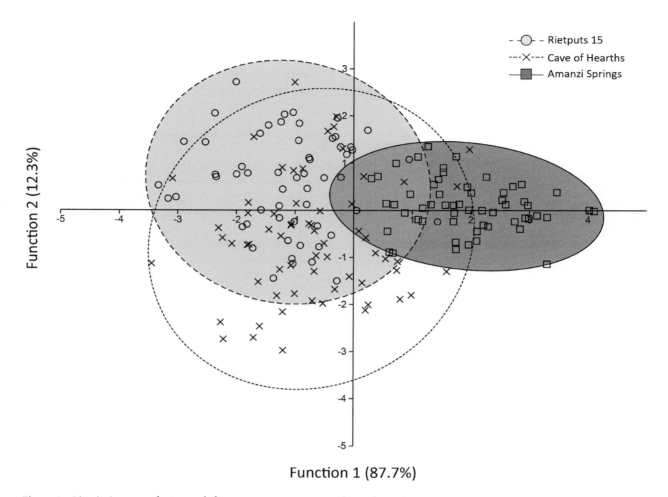

Figure 15. Discriminant Analysis Graph for Amanzi Springs, Cave of Hearths and Rietputs 15. Amanzi Springs scores load strongest onto Function 1, which correlated to overall size (length and breadth).

fact that Amanzi Springs remains undated needs to be considered when interpreting these results, yet if this assemblage is Late Acheulian, its unstandardized nature complicates the typological principal that 'refinement' is a characteristic of handaxes from this period of the Acheulian industry (Deacon, 1970; Li *et al.*, 2018).

Discussion

As Crompton and Gowlett (1993) discussed over 25 years ago, there was likely no strict set of rules governing the production of handaxes in Acheulian times. There are a number of factors including raw materials, function, style and the individual knapper that resulted in wide-ranging variability when handaxe size and shape are

Table 3. Cross-validated classification table for sites (75.1% correctly classified).

	RP15	CH	AS
RP15	66.7%	29.8%	3.5%
CH	29.7%	60.9%	9.4%
AS	3.4%	8.6%	87.9%

investigated in the modern era. When the South African sites compared here are examined, similarities suggest basic guidelines for understanding the relationship between size and shape. This can then be used to interpret rule-sets underlying handaxe production, albeit these are not rigidly applicable to every specimen (Crompton and Gowlett, 1993). Nonetheless, the results presented here demonstrate that allometry is a common factor of handaxe shape across the African continent and that rule-sets are indeed consistent (on a general level) between eastern and southern regions.

Allometric trends in South African handaxes

When allometric patterns are compared between the AC/CV and PCA results, two principal 'rule-sets' are detected relating to shape variables in these South African handaxes, which corroborate Gowlett's (Crompton and Gowlett, 1993; Gowlett and Crompton, 1994) previous findings for East African assemblages. The primary factor is that the basic dimensions (L, B, and Th) show negative allometric patterns, suggesting constrains on general proportions as handaxes increase in size (Figures 11 and 12). As such, shape is an important concern in the overall production process of these tools. Maintaining proportions of basic dimensions in this sense restricts the shape of handaxes to predetermined forms. However, variance in allometric patterns at individual sites and artefact levels supports the notion that production processes were likely only guided by a shared concept of form rather than a strict operational construct (McNabb et al., 2004). It is more probable that the resulting variability in handaxe shapes is a product of practical concerns including raw materials, functionality and reduction intensity (Ashton and McNabb, 1995; Ashton and White, 2003; McNabb and Cole, 2015; McPherron, 2006, 2003, 2000, 1999; White, 1995).

Secondly, the balance between plan-view and profile shape is a critical factor, albeit this relationship is inversely correlated. This pertains to the upper and lower halves of handaxes, above and below the point of maximum breadth and thickness. The upper portion relates to adjustments tip shape (i.e. tip breadth and thickness) that correlate positively with size, while the lower portions relates butt shape, which is isometric-to-negatively correlated with size. Combined with variability in the point of maximum breadth, this balance between tip and butt shape likely relates to the length-to-breadth ratio in handaxes observed by Gowlett (2011), which showed that increases in length correlated with 'skinnier' plan-view shapes. The critical variable operating here is likely the positioning of the centre of mass (Gowlett, 2006; Grosman et al., 2008; Saragusti et al., 2005). As stated above, Gowlett (2006) argued that specific 'imperatives' related to the proportions of handaxe shape. The centre of mass is critical for the use of handaxes in terms of the balance of the tool, reiterating the 'balance' of tip and butt shapes found the multivariate results above. In this respect, the motor-perceptual capacities of hominin tool-use (like all other primates) were linked to the overall shape and weight of the implement, which in terms effects manual dexterity (Bril et al., 2010, 2009; Visalberghi et al., 2009). The balance of handaxes likely impacted their manual manipulation in food-processing activities, thus tools too heavily weighted towards their tip or butt ends were likely not as efficient, as gripping and dexterity would have to compensate for these issues (Crompton and Gowlett, 1993).

Lastly, another remarkable point demonstrated in the PCA analyses is the stability of trends reflected in the South African handaxes across temporal, geographical and raw material boundaries. As stated above, PC loading trends in PCAs run for individual sites were broadly similar, even though Rietputs 15 is comprised on Ventersdorp lavas while quartzite dominates the other assemblages. In fact, the only deviation in PC loading patterns was seen in Cave of Hearths, which is expected to plot closely with Amanzi Springs due to similarities in raw materials and industrial affiliation.

This could reflect some differences in raw material package size and shape, function or skill, which McNabb et al. (2004) suggested to be a controlling factor in handaxe production. Nonetheless, the stability of patterns corroborate findings for allometric trends common in South African handaxe production, where the covariation between tip and butt shape determine the balance of tool mass.

The Amanzi Springs handaxes

When considering Amanzi Springs in the context of South African handaxe assemblages, a trend towards increased geometric size coupled with low variation in shape separates it from Cave of Hearths and Rietputs 15 (compare Figures 10 and 15). The results of the exploratory analysis show that Amanzi Springs handaxes are relatively larger and yet the variation is quite restricted across most length, breadth and thickness variables (see Table 1 and Figure 10). For a purported Late Acheulian assemblage, this is an unexpected result in terms of contradicting the general pattern of refinement (see Figure 8). On this point, the trend towards refinement from Early to Late Acheulian has been argued at length in Earlier Stone Age research (Hodgson, 2015; Klein, 2009; Kuman, 2007; Wynn, 1995, 1979; Wynn and Teirson, 1990). However, a number of studies have found that this principal is not ubiquitous of Late Acheulian assemblages, some of which show increased variability in handaxe forms (Li et al., 2018; McNabb, 2009; McNabb et al., 2004; McNabb and Cole, 2015). Possibly the clearest example of this is found in the southern African region. McNabb (2009; McNabb et al. 2004) found that handaxes from Cave of Hearths were highly variable in their overall shape. Li, et al. (2018) have also recently found that Cave of Hearths handaxes are more variable in shape than Rietputs 15 and that the refinement (B/Th) index does not discriminate these assemblages. They suggested that the overall coverage of flake scars, relating to primary and secondary flaking patterns to remove cortex and shape handaxes, are more abundant in the Cave of Hearths assemblage. While handaxes clearly show a general trend towards shape consistency, variability arises because of idiosyncratic circumstances on the individual knapper level, i.e. negotiating raw materials, package size and shapes, functional needs and general constraints on reduction intensity (McNabb et al. 2004; McNabb, 2009; Li et al., 2018).

When comparing the variability in handaxe size and shape between Amanzi Springs, Cave of Hearths and Rietputs 15, there are several hypotheses that should be considered when interpreting results: 1) Amanzi Springs is possibly an Early Acheulian assemblage; and 2) these patterns represent an adaptation to the raw material availability and overall functional needs of the tool-makers. Deacon (1970) assigned Amanzi Springs to the Late Acheulian partly on his C[14] dates of 60,600 ± 1,100 BP (GrN-4407) for the Riethuevel Member, which can be considered as an infinite age for the site, as well as the technological elements. Currently, there is no reason to doubt Deacon's hypothesis that Amanzi Springs is a Late Acheulian site, however reliable dating of the site is needed to understand where the assemblage fits into the Acheulian chronology.

Nonetheless, the second scenario presents a stronger case when viewing handaxe production from a practical perspective. This would argue that variability in size and shape are products of specific manufacturing processes including raw material selectivity and blank production, as well as discard patterns. It is difficult at present to evaluate the nature of deposition of the Amanzi Springs Acheulian artefacts. They occur in a distinct band within the Rietheuvel Member and then within the Pothole Fill of the Balmoral Member, although perhaps derived from the erosion of Rietheuvel. Deacon (1970) documented stone tools of all size grades, suggesting the Rietheuvel material does not represent size sorted, secondary deposition where smaller material has been winnowed away to leave larger handaxes. Given the springs are thermal, it also seems unlikely that the hominins who made them visited the site for water, especially given there are other freshwater springs ~7 km to the West. It was likely that the availability of Enon quartzite cobbles and boulders were in some part the attraction of the springs.

In terms of these materials, Deacon (1970) noted that the mechanical properties of Enon quartzites inhibit conchoidal fracture when compared to local silcrete materials that might have been available in Acheulian times. He described the material as comprised of large quartz grains with interstitial silica and a coarse, granular texture that caused irregular fracture patterns. Further, this material is very dense and requires increased percussive force to knap when compared to more isotropic materials such as chert or flint (pers. observ. MVC). When the overall size differences are considered from this perspective, it is possible that Acheulian hominins were selecting for larger blank materials to account for material constraints. Table 1 shows that the average handaxe length is over 16cm and 9cm in breadth, which suggests that boulder-sized, material packages were being exploited. Further, a variety of large flake-blank production techniques are present in the Amanzi Springs collections as stated above, which supports the notion that the Amanzi Springs tool-makers were well-adapted to negotiating the constraints of Enon quartzites. However, if this is the case then why would the Amanzi Springs collections exhibit a 'large and unstandardized' appearance if the hominins were

adept at manipulating difficult raw materials? One possibility is that the Acheulian hominins were not concerned about the overall shape of handaxes and that the production of one or more useable cutting edges were acceptable given the material constraints. However, it seems likely that if tool-makers were not concerned about the extent of reduction in handaxes, the cortex percentages and flaks scar counts would be more skewed towards higher amounts of the former and lesser of the latter. Another possible answer is that this is a production site preserving handaxes that either were rejected and discarded due to production flaws while the more well-made tools were transported off Amanzi hill. In this sense, the overall larger size of Amanzi Springs handaxes reflects primary stages of production where large flakes and large clasts blanks were initially worked and discarded because of production complications. This may better account for the larger and unstandardized appearance of Amanzi Springs handaxes, which could relate to manufacturing failures and the timing of discard. However, conclusions on this matter remain tentative and future work with these collections aims to investigate this issue.

Conclusion

Handaxe 'imperatives' in the South African Acheulian

The results discussed for three South African sites have found similar outcomes compared to Gowlett and Crompton's (1994; Crompton and Gowlett, 1993) multivariate analyses of Kilombe and Kariandusi. However, differences were detected that show some variation in allometric trends across the assemblages. The PCA results for Kilombe summarized four basic trends: 1) PC1 accounts for size variation (60%) and does not represent allometric trends therein; 2) PC2 (15%) represented thickness variation linked with allometry; 3) PC3 (15%) represented planform variation linked with allometry; and 4) PC4 (10%) represented tip width variation linked with allometry (Crompton and Gowlett, 1993). The PCA results here deviated from this pattern slightly in that PC1 accounted for a larger percentage of variation (79%) with PC2 (7%) and PC3 (6%) capturing smaller proportions when compared to the assessments of Kilombe and Kariandusi (Crompton and Gowlett, 1993; Gowlett and Crompton, 1994). Further, length was strongly correlated with PC1 variance, which is not highlighted in Gowlett and Crompton's work. Nonetheless, the effects of size on the PCA used in this study could be skewed to some degree by the overall differences between Amanzi Springs and Cave of Hearths and Rietputs 15. In this sense, the larger proportions of Amanzi Springs handaxes drives size variation in multivatiate ordination tests, which is reflected in the DA as well.

Nontheless, PCs 2 and 3 display some interesting patterns in allometric trends that are different from Gowlett and Crompton's (1994; Crompton and Gowlett, 1993) results. The two major trends suggest that tip shape and butt thickness relative to the positioning of the point of maximum breadth are the most critical factors in maintaining size and shape proportions in South African handaxes. Here, butt thickness and tip shape are closely related, suggesting a tight balance between these variables as discussed above. This demonstrates that South African hominins were adept at controlling proportions of handaxes relating to the weight of these tools towards their upper and lower halves. Increasing or decreasing these variables at an uneven rate will cause exponential increases in mass towards one half or the other and significantly impact the balance of the tool.

While this trend is perhaps unique to South African handaxes, its end goal is similar across African Acheulian assemblages, which is to maintain the center of mass (CM) (Gowlett, 2006). The focus on breadth and thickness in the tip and butt ends of handaxes suggest a balancing of two ends, which directly correlates to the positioning of the CM. Recent studies have shown that shape variation and the positioning of CM is an important relationship in understanding the uniformity of handaxe forms (Grosman *et al.*, 2008; Park *et al.*, 2003; Saragusti *et al.*, 2005). In fact, Grosman, et al. (2008) found that the position of CM relative to volume was consistent across Late Acheulian assemblages from the Western Asia This implies that CM is imperative for understanding covariation between handaxe size and shape and while tool-makers from different regions may have used diffent rule-sets, the results were similar.

The significance of this likely correlates to the use of these tools as cutting implements. The overall positioning of weight in tools is a critical aspect of their balance and efficiency. As such, maintaining tip and butt proportions in handaxes likely improved their efficiency. In fact, Key and Lycett (2016) found that mass and cutting efficiency correlated in a regression analysis, which showed that handaxes over a particular threshold of mass are more efficient in cutting activities than lighter (smaller) counterparts. They noted that handaxes ranging above 10cm correlated with cutting efficiency, which fit within the handaxe size range for assemblages analysed here and in Gowlett's work. In this sense, the positioning of mass within these tools likely aids in functionality and as such, Gowlett's (2006, 2013) imperatives argument for elongation and CM as 'true variables' that were important for handaxe production in Africa remains accurate.

In this context, Amanzi Springs represents an important assemblage for understanding the South African Acheulian in terms of morhological variation.

As the analysis above demonstrates, handaxes form this site are comparatively larger in geometric size, yet tightly constrained in length, breadth and thickness dimensions (see Table 1 and Figure 10). However, Key and Lycett (2016) also found that elongation in handaxe size does not necessarily correlate to increasing efficiency, which then begs the question of why the Amanzi Springs handaxes are so large when compared to Cave of Hearths and Rietputs 15. As mentioned above, one possibility for differences in size range may relate to the nature of these sites, where Amanzi Springs may preserve tools that were not shaped to their intended extent and rather are discarded pieces that failed during production. In this sense, Amanzi Springs may represent a production site, where the majority of these large and unstandardized tools are simply the 'handaxes that were left behind.' Further work is required to comfirm this tentative conclusion, although if this were the case, Amanzi Springs would be a rare site where production habits may provide further insight into both the varaition and constraints on handaxe size and shape.

Acknowledgments

This work was funded by Australian Research Council Discovery Grant DP170101139 to AIRH. MVC was supported by the DST-NRF Center of Excellence in Palaeosciences through the University of the Witwatersrand, South Africa. We thank Celeste Booth of the Albany Museum in Grahamstown for allowing us access to the collection and Clyde Niven, the landowner of Amanzi Farm for permission to work at the site. We thank Sello Mokhanya and the Eastern Cape Provincial Heritage Resources Authority (ECPRHA) for providing a permit to excavate Amanzi Springs.

References

Ambrose, S.H., 2001. Paleolithic technology and human evolution. Science 291, 1748–53.

Archer, W., Braun, D.R., 2010. Variability in bifacial technology at Elandsfontein, Western Cape, South Africa: a geometric morphometric approach. *Journal of Archaeological Science* 37: 201–209.

Ashton, N., McNabb, J., 1995. Bifaces in Perspective, in: Ashton, N., David, A. (Eds.), Stories in Stone. Lithic Studies Society, Oxford, U.K., pp. 182–191.

Ashton, N., White, M., 2003. Bifaces and Raw Materials, in: Soressi, M., Dibble, H.L. (Eds.), *Multiple Approaches to the Study of Bifacial Technologies.* University of Pennsylvania Museum of Archaeology and Anthropology, Philadelphia, PA, pp. 109–123.

Bordes, F., 1961. Typologie du Paléolithique Ancien et Moyen. CNRS, Paris, France.

Braun, D.R., Levin, N.E., Stynder, D., Herries, A.I.R., Archer, W., Forrest, F., Roberts, D.L., Bishop, L.C., Matthews, T., Lehmann, S.B., Pickering, R.,

Fitzsimmons, K.E., 2013. Mid-Pleistocene Hominin occupation at Elandsfontein, Western Cape, South Africa. *Quaternary Science Reviews* 82: 145–166.

Bril, B., Dietrich, G., Foucart, J., Fuwa, K., Hirata, S., 2009. Tool use as a way to assess cognition: how do captive chimpanzees handle the weight of the hammer when cracking a nut? *Animal Cognition* 12: 217–35.

Bril, B., Rein, R., Nonaka, T., Wenban-Smith, F., Dietrich, G., 2010. The role of expertise in tool use: skill differences in functional action adaptations to task constraints. *Journal of Experimental Psychology* 36: 825–39.

Brink, J., Herries, A.I.R, Moggi-Cecchi,, J., Gowlett, J., Adams, J.W., Hancox, J., Bousman, C.B., Grün, R., Eisenmann, V., Adams, J.W., Rossouw, L. 2012. First hominine remains from ~1 Ma bone bed at Cornelia-Uitzoek, Free State Province, South Africa. *Journal of Human Evolution.* 63: 527-535.

Buchanan, B., 2006. An analysis of Folsom projectile point resharpening using quantitative comparisons of form and allometry. *Journal of Archaeological Science* 33: 185–199.

Buchanan, B., Collard, M., 2007. Investigating the peopling of North America through cladistic analyses of Early Paleoindian projectile points. *Journal of Anthropological Archaeology* 26: 366–393.

Butzer, K.W., 1973. Spring sediments from the Acheulian site of Amanzi (Uitenhage District, South Africa). *Quaternaria* 17: 299–319.

Butzer, K.W., Helgren, D.M., Fock, G.J., Stuckenrath, R., 1973. Alluvial Terraces of the Lower Vaal River, South Africa: A Reappraisal and Reinvestigation. *The Journal of Geology* 81: 341–362.

Crompton, R.H., Gowlett, J.A.J., 1993. Allometry and multidimensional form in Acheulian bifaces from Kilombe, Kenya. *Journal of Human Evolution.* 25: 175-199

Deacon, H.J., 1966. The Early Stone Age Occupation at Amanzi Springs, Uitenhage District, Cape Province. Vol. 1. *Unpublished Master's Thesis.* University of Cape Town.

Deacon, H.J., 1970. The Acheulian Occupation at Amanzi Springs, Uitenhage District, Cape Province. Annals of the Cape Province Museum Vol. 8, Grahamstown, South Africa.

Diniz-Filho, J.A.F., Von Zuben, C.J., Fowler, H.G., Schlindwein, M.N., Bueno, O.C., 1994. Multivariate morphometrics and allometry in a polymorphic ant. *Insectes Sociaux* 41: 153–163.

Gibbon, R.J., Granger, D.E., Kuman, K., Partridge, T.C., 2009. Early Acheulian technology in the Rietputs Formation , South Africa, dated with cosmogenic nuclides. *Journal of Human Evolution* 56: 152–160.

Gowlett, J.A.J., 2006. The Elements of Design Form in Acheulian Bifaces: Modes, Modalities, rules and language, in: Axe Age: *Acheulian Tool-Making from Quarry to Discard.* pp. 203–221.

Gowlett, J.A.J., 2009. The Longest Transition or Multiple Revolutions?, in: Camps, M., Chauhan, P. (Eds.), *Sourcebook of Paleolithic Transitions.* Springer, New York, NY.

Gowlett, J.A.J., 2011. The Vital Sense of Proportion: Transformation, Golden Section, and 1:2 Preference in Acheulian Bifaces. *Paleoanthropology* 174–187.

Gowlett, J.A.J., 2013. Elongation as a factor in artefacts of humans and other animals: an Acheulian example in comparative context. *Philosophical transactions of the Royal Society of London. Series B, Biological Sciences* 368: 1–11.

Gowlett, J. a J., Crompton, R.H., 1994. Kariandusi: Acheulian morphology and the question of allometry. *The African Archaeological Review* 12: 3–42.

Graham, J.M., Roe, D., 1970. Discrimination of British Lower and Middle Palaeolithic handaxe groups using canonical variates. *World Archaeology* 1: 321–337.

Grosman, L., Smikt, O., Smilansky, U., 2008. On the application of 3-D scanning technology for the documentation and typology of lithic artifacts. *Journal of Archaeological Science* 35: 3101–3110.

Helgren, D.M., 1978. Acheulian settlement along the lower Vaal River, South Africa. *Journal of Archaeological Science* 5(1): 39-60.

Herries, A., Latham, A., 2009. Archaeomegnetic Studies at the Cave of Hearths, in: McNabb, J., Sinclair, A. (Eds.), The Cave of Hearths: Makapan Middle Pleistocene Research Project. BAR International Series, Oxford, U.K., pp. 59–64.

Herries, A.I.R., 2011. A chronological perspective on the Acheulian and its transition to the Middle Stone Age in Southern Africa: the question of the Fauresmith. *International Journal of Evolutionary Biology* 2011: 1-25.

Hodgson, D., 2015. The symmetry of Acheulian handaxes and cognitive evolution. *Journal of Archaeological Science: Reports* 2: 204–208.

Inizan, M.-L., Ballinger, M., Roche, H., Tixier, J., 1999. *Technology of Knapped Stone.* In: Pre´histoire de la Pierre Taille´ e, vol. 5. CREP, Nanterre.

Inskeep, R.R., 1965. Earlier stone age occupation at Amanzi: A preliminary investigation. *South African Journal of Science* 61: 229–242.

Iovita, R., McPherron, S.P., 2011. The handaxe reloaded: a morphometric reassessment of Acheulian and Middle Paleolithic handaxes. *Journal of Human Evolution* 61: 61–74.

Isaac, G.L., 1977. Olorgesailie: Archeological Studies of a Middle Pleistocene Lake Basin in Kenya. University of Chicago Press, Chicago, IL.

Jolicoeur, P., 1963. The multivariate generalization of the allometry equation. *Biometrics* 19: 497–499.

Key, A.J.M., Lycett, S.J., 2016. Influence of Handaxe Size and Shape on Cutting Efficiency: A Large-Scale Experiment and Morphometric Analysis. *Journal of Archaeological Method and Theory* 24: 514–541.

Klein, R.G., 2009. *The Human Career.* University of Chicago Press, Chicago, IL.

Kuman, K., 1994. The archaeology of Sterkfontein—past and present. *Journal of Human Evolution* 27: 471–495.

Kuman, K., 2007. The Earlier Stone Age in South Africa: Site Context and the Influence of Cave Studies, in: Pickering, T.R., Schick, K., Toth, N. (Eds.), *Breathing Life Into Fossils: Taphonomic Studies in Honor of C.K. (Bob) Brain.* Stone Age Institute Press, Gosport, IN, pp. 181–198.

Kuman, K., Gibbon, R.J., 2018. The Rietputs 15 site and Early Acheulian in South Africa. *Quaternary International* 480: 4-15.

Li, H., Kuman, K., Leader, G.M., Couzens, R., 2018. Handaxes in South Africa: Two Case Studies in the Early and Later Acheulian. *Quaternary International* 480: 29–42.

Lycett, S.J., von Cramon-Taubadel, N., Foley, R.A., 2006. A crossbeam co-ordinate caliper for the morphometric analysis of lithic nuclei: a description, test and empirical examples of application. *Journal of Archaeological Science* 33: 847e861

Lycett, S.J., Gowlett, J. a. J., 2008. On questions surrounding the Acheulian 'tradition.' *World Archaeology* 40: 295–315.

Lycett, S.J., von Cramon-Taubadel, N., 2008. Acheulian variability and hominin dispersals: a model-bound approach. *Journal of Archaeological Science* 35: 553–562.

Mason, R.J., 1988. The Cave of Hearths, Makapansgat, Transvaal. University of the Witwatersrand Press, Johannesburg.

Mclear, L.G.A., 2001. The hydrogeology of the Uitenhage Artesian Basin with reference to the Table Mountain Group aquifer. *Water SA* 27: 499–506.

McNabb, J., 2009. The ESA Stone Tool Assemblage from the Cave of Hearths, Beds 1-3, in: McNabb, J., Sinclair, A. (Eds.), The Cave of Hearths: Makapan Middle Pleistocene Research Project. *BAR Internal Series S1940. University of Southampton Series in Archaeology* 1, Southampton, U.K., pp. 75–104.

McNabb, J., Beaumont, P., 2011. Excavations in the Acheulian Levels at the Earlier Stone Age Site of Canteen Koppie, Northern Province, South Africa. *Proceedings of the Prehistoric Society* 78: 51–71.

McNabb, J., Binyon, F., Hazelwood, L., 2004. The Large Cutting Tools from the South African Acheulian and the Question of Social Traditions. *Current Anthropology* 45: 653–677.

McNabb, J., Cole, J., 2015. The mirror cracked: Symmetry and refinement in the Acheulian handaxe. *Journal of Archaeological Science: Reports* 3: 100–111.

McNabb, J., Sinclair, A. (Eds.), 2009. The Cave of Hearths, *BAR Internal Series S1940. University of Southampton Series in Archaeology* 1, Southampton, U.K.

McPherron, S.P., 1999. Ovate and pointed handaxe assemblages: Two points make a line. *Préhistoire Européenne* 14: 9–32.

McPherron, S.P., 2000. Handaxes as a Measure of the Mental Capabilities of Early Hominids. *Journal of Archaeological Science* 27: 655–663.

McPherron, S.P., 2003. Technological and typological variability in the bifaces from Tabun Cave, Israel, in: Soressi, M., Dibble, H. (Eds.), *Multiple Approaches to the Study of Bifacial Technologies.* University of Pennsylvania Museum of Archaeology and Anthropology, Philadelphia, PA, pp. 55–76.

McPherron, S.P., 2006. What Can Typology Tell Us About Handaxes Production, in: *Axe Age: Acheulian Tool-Making from Quarry to Discard.* Equinox Publishing Ltd., London, U. K., pp. 267–285.

Nowell, A., Park, K., Metaxas, D., Park, J., 2003. Deformable Modeling, in: Soressi, M., Dibble, H.L. (Eds.), Multiple. University of Pennsylvania Museum of Archaeology and Anthropology, Philadelphia, PA, pp. 193–208.

Park, K., Nowell, A., Metaxas, D., 2003. Deformable Model Based Shape Analysis Stone Tool Application. 2003 Conference on Computer Vision and Pattern Recognition Workshop 1: 1–6.

Pickering, R., 2015. U-Pb Dating Small Buried Stalagmites from Wonderwerk Cave, South Africa: A New Chronometer for Earlier Stone Age Cave Deposits. *African Archaeological Review* 32: 645–668.

Roe, D.A., 1964. The British Lower and Middle Palaeolithic: Some Problems, Methods of Study and Preliminary Results. *Proceedings of the Prehistoric Society* 30: 245–267.

Roe, D.A., 1968. British Lower and Middle Palaeolithic Handaxe Groups. *Proceedings of the Prehistoric Society* 34: 1–82.

Roe, D.A., 1976. Typology and the Trouble with Hand-Axes, in: Sieveking, G. de G., Longworth, I.H., Wilson, K.E. (Eds.), *Problems in Economic and Social Anthropology.* Duckworth, London, U. K., pp. 61–70.

Saragusti, I., Karasik, A., Sharon, I., Smilansky, U., 2005. Quantitative analysis of shape attributes based on contours and section profiles in artifact analysis. *Journal of Archaeological Science* 32: 841–853.

Sharon, G., 2007. Acheulian large flake industries: technology, chronology, and significance. *BAR International Series*, S1701 Oxford, U.K.

Sharon, G., 2009. Acheulian Giant-Core Technology. *Current Anthropology* 50(3): 335-367.

Sharon, G., 2011. Flakes Crossing the Straits? Entame Flakes and Northern Africa–Iberia Contact During the Acheulean. *African Archaeological Review* 28: 125-140.

Shipton, C., 2013. A Million Years of Hominin Sociality and Cognition. *BAR International Series*, S2468. Oxford, U.K.

Shipton, C., Clarkson, C., 2015. Handaxe reduction and its influence on shape: An experimental test and archaeological case study. *Journal of Archaeological Science*: Reports 3: 408–419.

Shott, M.J., Hunzicker, D.A., Patten, B., 2007. Pattern and allometric measurement of reduction in experimental Folsom bifaces. *Lithic Technology* 32: 203–217.

Singer, B.S., 2014. A Quaternary Geomagnetic Instability Time Scale. *Quaternary Geochronology* 21: 29-52.

Stammers, R.C., Caruana, M.V., Herries, A.I.R. 2018. The first bone tools from Kromdraai and stone tools from Drimolen, and the place of bone tools in the South African Earlier Stone Age. *Quaternary International.* doi.org/10.1016/j.quaint.2018.04.026

Strauss, R.E., 1985. Evolutionary allometry and variation in body form in the South American catfish genus Corydoras (Callichtydae). *Systematic Zoology* 34: 381–396.

van Riet Lowe, C., 1938. The Makapan Caves. *South African Journal of Science* 35: 971–381.

van Riet Lowe, C., 1952. The Vaal River Chronology: An Up-to-Date Summary. *The South African Archaeological Bulletin* 7: 135–149.

Visalberghi, E., Addessi, E., Truppa, V., Spagnoletti, N., Ottoni, E., Izar, P., Fragaszy, D., 2009. Selection of effective stone tools by wild bearded capuchin monkeys. *Current Biology* 19: 213–7.

White, M.J., 1995. Raw Materials and Biface Variability in Southern Britain. *Lithics* 15: 1–21.

Wynn, T., 1979. The Intelligence of Later Acheulian Hominids. Man, New Series 14, 371–391.

Wynn, T., 1995. Handaxe Enigmas. *World Archaeology* 27: 10–24.

Wynn, T., Teirson, F., 1990. Regional comparison of shapes of later Acheulian handaxes. *American Anthropologist* 92: 73–84.

Matthew V. Caruana
Palaeo-Research Institute, University of Johannesburg, Auckland Park, 2195, Johannesburg, South Africa

Andy I. R. Herries
Palaeoscience Labs, Department of Archaeology and History, La Trobe University, Melbourne Campus, Bundoora, 3086, VIC, Australia
and
Palaeo-Research Institute, University of Johannesburg, Auckland Park, 2195, Johannesburg, South Africa
a.herries@latrobe.edu.au

Reflections on Possible Zoomorphic Acheulean Bifaces from Southwestern Algeria

Thomas Wynn, Mohamed Sahnouni, Tony Berlant and Claude Douce

Abstract

The chapter presents three zoomorphic and one phallic biface from a surface collection made at Bentadjine in Southwestern Algeria. The techniques of manufacture, associated artifacts, and condition of the artifacts all suggest that these bifaces are Late Acheulean in age, making them the oldest zoomorphic artifacts yet recognized. The recent reassignment of the Jebel Irhoud fossils to early anatomically modern humans enhances the significance of these artifacts.

Zoomorphic Acheulean bifaces, Algeria

Occasionally in Palaeolithic studies serendipity presents researchers with artifacts that have less-than-certain provenience, but whose potential significance for understanding human evolution warrants their description and publication. Here we describe four such artifacts, and weigh their evidential pros and cons. Given the redating of the Jebel Irhoud *Homo sapiens* fossils to ca. 300k (Hublin et al., 2017), the estimated date range of 380 – 290,000 years for these probable zoomorphic productions is especially provocative.

In the course of examining museum collections for examples to be used in the exhibition *First Sculpture: Handaxe to Figure Stone* (Berlant & Wynn, 2018), two of us (TW and TB) visited the Musée de Préhistoires, Chateau de Sauveboeuf, Aubas, France. Among the collections in storage is a surface collection made by a French army officer in 1979 and 1980 on or near Bentadjine in Southwestern Algeria. The artifacts include a range of handaxes and cleavers of Late Acheulean technique and typology (M.-H. Alimen, 1978), and also this unusual biface:

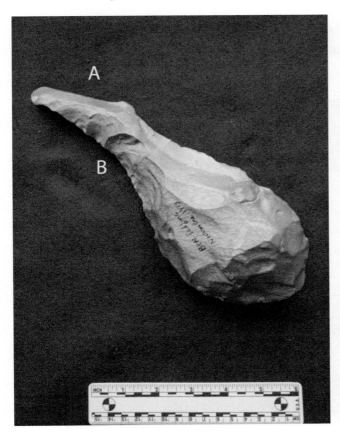

Figure 1a. Side A

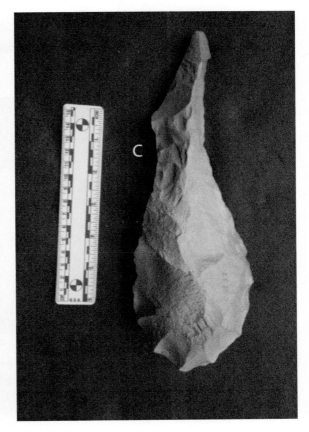

Figure 1b. Side B

The typological and technical features of the artifact correspond to no standard Palaeolithic artifact type. The blank on which artifact was made was a large flake from a prepared core (see following section on authenticity). There is extensive bifacial trimming around most of the perimeter. One short section of edge was left untrimmed (A). The position of the secondary, abrupt trimming (B and C) has transformed one end of the original blank into the profile of the head and neck of an animal, arguably an equid of some kind. The positioning of the trimming clearly demarcates the neck, an ear, and the nose/lips. The top of the snout is unmodified. The raw material is a fine-grained metamorphic rock.

Also in the collection are two other examples that are less clearly figurative:

Figure 2.

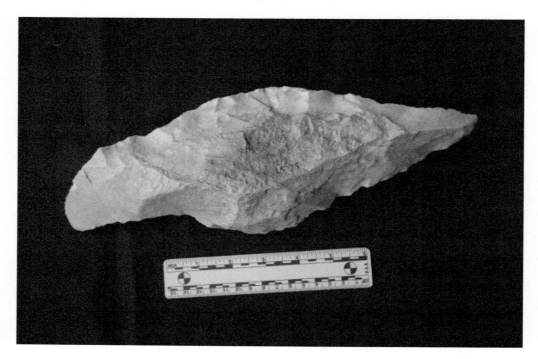

Figure 3.

Fig. 2 appears to be a less well-executed equid; Fig. 3 an unidentifiable creature (body, neck, and head). The raw material of both is quartzite. Without the clearly zoomorphic first example, we might not attribute figuration to the third, or even the second, despite their unusual asymmetric shapes. Their presence in the collection does, however, lend greater credence to the prime example.

Of particular importance in support of the authenticity of the collection is the inclusion of cleavers made on Kombewa and Levallois flakes, knapping techniques that appeared late in the Acheulean of the Western Sahara (Figures 4 and 5). Finally, though not technically a zoomorph, the collection includes the phallic biface manufactured on a quartzite cobble shown in Figure 6.

Figure 4a. Lava cleaver on Kombewa flake

Figure 4b.

Figure 5a. Quartzite cleaver on large Levallois flake

Figure 5b.

Figure 6. Phallic biface on quartzite cobble

Authenticity

Details surrounding the collection of these artifacts necessitate a tempering of enthusiasm. The artifacts are components of a surface collection. Without decent archaeological context it is impossible to eliminate completely the possibility that the collection includes artifacts of very different ages (e.g., the cleavers are Acheulean, but the zoomorphs much later). In addition, the artifacts bear two sets of collection dates, Novembre 1979, and Mars 1980, indicating two separate collecting trips, and therefore the possibility of two separate find spots. The collection has an uncertain chain of provenience. Although the artifacts themselves have labels marking the year and location, it is unclear when the labels were applied (the numbers on the artifacts are accession numbers of Musée de Préhistoires). Finally, the circumstances of the original collection are unknown and little is known about the experience or qualifications of the collector; he may simply have selected large, dramatic pieces.

Despite these discouraging factors, a number of other facts about the artifacts themselves, Saharan prehistory, the late Acheulean in the Tabelbala region in particular, and the condition of the artifacts suggests that the zoomorphs are authentic artifacts from the Late Acheulean:

1. Attributes of the equid biface itself point to a late Acheulean origin. The original blank appears to have been a large flake from a prepared core of some kind. The top of the 'snout' (letter A in Fig. 1, side A) is an unretouched flake edge. The knapping sequence is difficult to disentangle from the pattern of trimming, but the Side B view (Fig. 7) of the head and neck suggests a complex, multistep manufacture: first, preparation of the production surface of a core (Boeda, 1994) (including some flake truncations), followed by an initial removal of a large flake blank (truncating the arêtes on the snout), followed the removal of the blank for this artifact, followed by abrupt unifacial trimming on each face to produce the top and bottom of the neck and isolate the ear.

2. The large negative scar that constitutes the top of the neck on side A also suggests core preparation. Most of the perimeter of the biface has been trimmed, but the knapper used very few invasive flakes to reduce the thickness of the body of the biface (there is one invasive flake on side B, attesting to the knapper's ability to remove such flakes had he or she chosen to). The trimming flakes that delineate features of the neck and head are all small and often unifacial; their goal was not thinning the artifact or the edge, but producing a particular plan shape. The exact core preparation technique has been obscured by the subsequent trimming, but it could have been one of several typical of the late Acheulean of the Tabelbala region (we think that the complex pattern of arêtes and implied steps points to Tachenghit technique). The second and third zoomorphs also appear to have been knapped using a similar approach, i.e., modification of a large flake from a prepared core of some kind. In sum, attributes internal to

Figure 7. Detail of equid head and neck

the artifact itself argue for a late Acheulean age. Associated artifacts and condition of the finds support this assessment.

3. The collection includes cleavers on Kombewa flakes and other prepared cored techniques typical of the late Acheulean of the Tabelbala region.

4. The prime example is not unique; there are three zoomorphs, and the phallic biface.

5. The raw materials out of which the three zoomorphs are made (quartzites and another fine-grained metamorphic) are typical of the Late Acheulean of the area, but rare in later industries.

6. Artifacts in the collection have a similar degree of weathering.

7. There are no later prehistoric artifacts in the collection.

8. Bentadjine is extremely remote and an unlikely spot to deposit forgeries.

9. TW and TB encountered the artifacts quite by accident. No one brought the zoomorphs to our attention.

Alternative explanations

We believe that the preponderance of evidence supports a late Acheulean age for the zoomorphs. What other explanations are possible?

I. The zoomorphs are Neolithic in age. This is arguably the most serious alternative. The Western Sahara was occupied in Neolithic times during a moister climatic phase. Several Neolithic sites have been reported in the Tabelbala and Beni Abbes areas, including Hai el Hamaïda (Nougier, 1933), Kheneg et Tlaia (Chavaillon & Chavaillon, 1957), Hamama (Chavaillon, 1964), Erg er Raoui (Mateu, 1964), Tanezrouft (Mateu & Favergeat, 1965), Foum Seïada, Beni Abbes (Mateu, 1970), Erg Atchane (Savary, 1965b). These sites have yielded diagnostic Neolithic remains such as blades, bladelets, geometric microliths, pottery, ground stones, ornaments made of ostrich eggshell, and a sculpture made of stone. The latter is a sculpted broken piece of sandstone-quartzite (10.3 x 10.3 x 3.1 cm; 440g) bearing geometric ornament motifs on one face (Savary, 1965a). Further south in the Central Sahara, Neolithic sites yielded several sculptures made of stones called 'zoomorphic grinding stones,' most of them depicting bovids (Le Quellec, 2008; Savary, 1965b). However, no zoomorphic bifaces similar to the Bentadjine examples have ever been found in Saharan Neolithic contexts, or in the Neolithic of the adjacent Maghreb. The core preparation techniques used

for the equid are unknown for Neolithic stone knapping. Moreover, the Bentadjine collection itself includes no obviously Neolithic artifacts (pottery, projectile points, or ground stone).

II. The zoomorphs are Middle Stone Age (MSA), Later Stone Age (LSA), or Epipalaeolithic in age. As with the Neolithic, local MSA (Aterian), LSA (Iberomaurusian), and Epipalaeolithic (Typical Capsian and Upper Capsian) assemblages include no such artifacts. For instance, at the Iberomaurusian site of Tamar on the eastern Mediterranean coast of Algeria a ceramic figurine was discovered at a level dated to 20,200 B.P. (Saxon, Close, Cluzel, Morese, & Shackelton, 1974), and the excavation at the nearby cave site of Afalou yielded two zoomorphic statuettes in baked clay that dated to 11,000 B.P. and appear to depict heads of bovids of some kind (Hachi, 1996, 2006). In the Capsian, evidence of sculpture is limited to shaped and engraved pebbles, and limestone slabs engraved with representational scenes and abstract designs (Lubell, 2001). Nothing from these contexts resembles the Bentadjine zoomorphs. Moreover, the late Pleistocene stone knappers used flint and quartz almost exclusively. Large flakes from prepared cores were not the basis of these lithic technologies, and thus their knapping traditions were quite different from those used to make the zoomorphs (Foley, Maillo-Fernandez, & Lahr, 2013).

III. The zoomorphs are modern forgeries. We find this alternative to be extremely unlikely. The zoomorphs are weathered in way that is difficult to mimic. Moreover, Bentadjine is a very unlikely locale for someone to deposit fakes.

Conclusion

The zoomorphs of Bentadjine appear to be authentic, late Acheulean artifacts. The techniques of manufacture, the associated artifacts, and the condition of the artifacts all support this conclusion. Alternative interpretations simply require too many exceptions to known patterns to be credible.

The Late Acheulean of Southwestern Algeria

Bentadjine is situated about 100 km southwest of the oasis and town of Tabelbala. This oasis and its surrounding region were the focus of a classic study of the Acheulean tradition in the Western Sahara (M.-H. Alimen, 1978; Champault, 1966; Chavaillon, 1964). Relying on surface collections and test excavation at numerous localities, Alimen was able to document a sequence of technological developments comprising several phases, which she numbered I to VII. Phases

VI and VII characterized the latest periods of the Acheulean and included techniques and tool types similar to those in the Bentadjine collection. The late Acheulean in Northwestern Sahara witnessed significant technological developments including the use of soft hammers to produce thinner and more finely shaped bifaces, an increase in use of the Kombewa technique of flaking, and the innovation of the Tachenghit method for the manufacture of cleavers. The Kombewa technique involves manufacturing flakes with dual ventral faces, providing the Acheulean toolmakers with the advantage of shaping cleavers with a convex edge. The technique was also used to make large scrapers (Balout, 1967; Dauvois, 1981). The Tachenghit method is of particular interest as it was a novel technique for core preparation and cleaver detachment. As described by Tixier (Tixier, 1957), the method consisted of preforming the shape of the cleaver before knocking off the flake. The co-occurrence of Kombewa, Tachenghit, and Levallois techniques, along with finely finished bifaces makes the final stages of the Acheulean in the Tabelbala arguably the most elaborate of all known Acheulean industries, and marks the zenith of this tradition of stone knapping. Finding zoomorphic artifacts in this context serves to confirm the Acheulean of the Western Sahara as the most technically sophisticated Acheulean ever recognized.

Age estimate

There are few chronometric dates for the late Acheulean of the Western Sahara. Alimen's study included no chronometric dates of any kind because of the absence of datable volcanic and faunal materials. However, she positioned with precision the Acheulean deposits in the Quaternary stratigraphy of the Northwestern Sahara. The Acheulean artifacts from Tabelbala-Tachenghit were encased in deposits consisting of eolian white sand capped by a sandy whitish lake limestone called *Tachenghit* (H. Alimen, 1968). These deposits constitute the final Ougartian formation (Ougartien VI and VII), which is dated to the late Middle Pleistocene. Chavaillon (Chavaillon, 1964) correlates the final Ougartian with the Tensiftian continental climatic cycle of Morocco. According to Lefèvre and Raynal (Lefevre & Raynal, 2002), the Tensiftien corresponds to the MIS stage 10 of the Casablanca long Pleistocene sequence of Atlantic Morocco, which is dated to 367 ± 34 ka.

The climatic sequence in the adjacent Central Sahara corroborates this date. The hominins at Bentadjine must have lived in the area during a period of wetter climate when savanna habitats spread into portions of the Sahara, with attendant animal species, including equids of some kind. The chronology of these wet periods in the Central Sahara for the Middle Pleistocene is not firmly established. Evidence from the Fezzan region of southwestern Libya indicates four major periods of lake formation in the Middle and Late Pleistocene (Geyh & Thiedig, 2008), including one between 380 and 290,000 years ago, and one between 260 and 205,000 years ago, correlated to MIS stages 10-9 and 8-7 respectively. The earlier of the two periods corresponds well to the date inferred from the Moroccan sequence. However, the climate dynamics governing 'Green Sahara Periods' were complex, and Larrasoaña and colleagues (2013) argue that there were multiple short duration Green Periods, each of only a few thousand years duration, extending all the way back to the Miocene, making precise chronological assignment of the Bentadjine collection arguably impossible on climatic grounds alone.

Significance

Documented manufacture of zoomorphic artifacts in the Late Acheulean is significant for several reasons:

a. These are the oldest known zoomorphic artifacts (in the sense of manufactured objects) yet recognized, even assuming a conservative age estimate of 350,000 years.

b. The zoomorphs reinforce the implications of the Berekhat Ram figurine, which at 230,000 years is slightly more recent in age (d'Errico & Nowell, 2000; Goren-Inbar & Peltz, 1995; Marshack, 1997). Late Acheulean hominins were apparently capable of producing iconic representations of people and animals. This is congruent with the extensive evidence for pigment use (e.g., at Twin Rivers in Zambia (Barham, 2002)), which suggests body painting.

c. Hublin and colleagues have reassessed and re-dated the *Homo sapiens* fossils found at Jebel Irhoud in Morocco, concluding that the fossils were of early anatomically modern humans dating to at least 300,000 BP (Hublin et al. 2017). This reassessment dramatically increases the significance of the Bentadjine zoomorphs, which would now appear to be early examples of iconic artifacts produced by modern humans.

d. The zoomorphs confirm a long-held suspicion among some specialists that handaxes were more than just mechanical tools (Gowlett, 2011; Hodgson, 2015; Le Tensorer, 2006, 2009; Machin, 2009; Machin, Hosfield, & Mithen, 2006; T Wynn, 1995; Thomas Wynn & Gowlett, 2018). During the late Acheulean, at least, hominins (now, presumably, *Homo sapiens*) used handaxes for other purposes as well. The handaxes need not have been symbolic in the sense of carrying meaning beyond what they depict (they are obviously iconic), but knappers clearly made the zoomorphs to display something other than functionality.

e. Late Acheulean assemblages often contain unusual bifaces, including finely finished examples, giant examples, asymmetrical examples, and examples with inclusions (e.g., the West Tofts handaxe). In this company, figurative examples are perhaps not surprising. More provocative is the possibility that bifaces were the medium through which hominins constructed an understanding of iconic representation. Malafouris (2013) and Cole (2014, 2015) have explored the possible role of biface manufacture in the construction of concepts of self. As a medium for externalizing identity bifaces might also have ultimately enabled their makers to externalize and represent the identity of animals as well.

Precisely how scholars choose to weave this evidence into accounts of recent human evolution will be determined largely by the theoretical frameworks they bring to the task, which are likely to be varied and even contradictory. Our intent here is not to resolve any academic debate. It is simply to add an unusual set of artifacts to the list of puzzles that makes the study of human evolution an enduring challenge.

Acknowledgments

The research has been funded by the Nasher Sculpture Center of Dallas, Texas. We thank Jeremy Strick and the staff for their continued support and enthusiasm. We also thank the staff of Musée de Préhistoires, Chateau de Sauveboeuf, Aubas, France, for their assistance.

References

Alimen, H. (1968). Chronologie des formations contenant les industries acheuléennes dites de 'Tabelbala-Tachenghit' (Sahara nord-occidental). . *Comptes Rendus Académie des Sciences, Série D, 267*, 839-842.

Alimen, M.-H. (1978). *L'Evolution de l'Acheuleen au Sahara Nord-Occidental (Saoura - Ougarta - Tabelbala)*. Meudon: Centre National de la Recherche Scientifique.

Balout, L. (1967). Procédé d'analyse et questions de terminologie dans l'étude des ensembles industriels du Paléolithique inférieur en Afrique du Nord. In W. a. C. Bishop, J.D. (Ed.), *Background to Evolution in Africa* (pp. 707-736). Chicago: The University Press of Chicago.

Barham, L. (2002). Systematic pigment use in the Middle Pleistocene of South-Central Africa. *Current Anthropology, 43*(1), 181-190.

Berlant, T., & Wynn, T. (2018). *First Sculpture: Handaxe to Figure Stone*. Chicago: Studio Blue.

Boeda, E. (1994). *Le Concept Levallois: Variabilite des Methodes*. Paris: CNRS Editions.

Champault, B. (1966). *L'Acheuléen évolué du Sahara occidental. Notes sur l'Homme au paléolithique ancien.* . (Doctorat), Université de Paris, Paris.

Chavaillon, J. (1964). *Les formations quaterniares du Sahara Nord-Occidental.* . Paris: Centre National de la Recherche Scientifique.

Chavaillon, J., & Chavaillon, N. (1957). Présence d'industrie acheuléenne du Kheneg et Tlaia (Sahara Nord-Occidental). . *Bulletin de la Société Préhistorique Française, 54*(10), 636-644.

Cole, J. (2014). The identity model: A theory to access visual display and hominin cognition within the Palaeolithic. In R. Dunbar, C. Gamble, & J. Gowlett (Eds.), *Lucy to Language: The Benchmark Papers*. Oxford: Oxford University Press.

Cole, J. (2015). Handaxe symmetry in the Lower and Middle Palaeolithic: implications for the Acheulean gaze. In F. Coward, R. Hosfield, M. Pope, & F. Wenban-Smith (Eds.), *Settlement, Society, and Cognition in Human Evolution* (pp. 234-257). Cambridge: Cambridge University Press.

d'Errico, F., & Nowell, A. (2000). A new look at the Berekhat Ram figurine: Implications for the origins of symbolism. *Cambridge Archaeological Journal, 10*, 123-167.

Dauvois, M. (1981). De la simultaneite des concepts Kombewa et Levallois dans l'Acheuléen du Maghreb et du Sahara Nord-Occidental. In C. Roubet, H. Hugot, & G. Souville (Eds.), *Préhistoire africaine : Mélanges offerts au Doyen L Balout* (pp. 313-321). Paris ADPF

Foley, R., Maillo-Fernandez, J., & Lahr, M. (2013). The Middle Stone Age of the Central Sahara: Biogeographical opportunities and technological strategies in later human evolution. *Quaternary International, 300*, 153-170.

Geyh, M., & Thiedig, F. (2008). The Middle Pleistocene Al Mahruqah Formation in the Murzuq Basin, northern Sahara, Libya evidence for orbitally-forced humid episodes during the last 500,000 years. *Palaeogeography, Palaeoclimatology, Palaeoecology, 257*, 1-21. doi:1-.1016/j.palaeo.2007.07.001

Goren-Inbar, N., & Peltz, S. (1995). Additional remarks on the Berekhat Ram figurine. *Rock Art Research, 12*, 131-132.

Gowlett, J. (2011). The vital sense of proportion: Transformation, Golden Section, and 1:2 preference in Acheulean bifaces. *PaleoAnthroology, 2011*, 174-187. doi:10.4207/PA.2011.ART51

Hachi, S. (1996). L'Ibéromaurusien, découverte des fouilles d'Afalou (Bedjaia, Algérie). . *L'Anthropologie, 100*(1), 55-76.

Hachi, S. (2006). Du comportement symbolique des derniers chasseurs Mechta-Afalou d'Afrique du Nord. . *Comptes Rendus Palevol, 5*(1-2), 429-440.

Hodgson, D. (2015). The symmetry of Acheulean handaxes and cognitive evolution. *Journal of Archaeological Science: Reports, 2*, 204-208.

Hublin, J., Ben-Ncer, A., Bailey, S., Freidline, S., Neubauer, S., Skinner, M., . . . Gunz, P. (2017). New fossils from Jebel Irhoud, Morocco and the pan-African origin of *Homo sapiens*. *Nature, 546*, 289-292. doi:10.1038/nature22336

Larrasoana, J., Roberts, A., & Rohling, E. (2013). Dynamics of Green Sahara Periods and their role in hominin evolution. *PLoS One, 8*(10), e76514. doi:10.1371/journal.pone.0076514

Le Quellec, J.-L. (2008). A propos des molettes zoomorphes du Sahara central. *Sahara, 19*, 39-60.

Le Tensorer, J.-M. (2006). Les cultures acheuleennes et la question de l'emergence de la pensee symbolique chez *Homo erectus* a partir des donnees relatives a la forme symetrique et harmonique des bifaces. *Comptes Rendus Palevol, 5*, 127-135.

Le Tensorer, J.-M. (2009). L'image avant l'image: reflexions sur le colloque. *L'Anthropologie, 13*, 1005-1017.

Lefevre, D., & Raynal, J.-P. (2002). Les formations plio-plèistocènes de Casablanca et la chronostratigraphie du Quaternaire marin du Maroc revisitées. . *Quaternaire, 13*(1), 9-21.

Lubell, D. (2001). Late Pleistocene-Early Holocene Maghreb. In P. Peregrine & M. Ember (Eds.), *Encyclopedia of Prehistory*. New York: Kluwer Academic/Plenum.

Machin, A. (2009). The role of the individual agent in Acheulean biface variability. *Journal of Social Archaeology, 9*(1), 35-58.

Machin, A., Hosfield, R., & Mithen, S. (2006). Why are some handaxes symmetrical? Testing the influence of handaxe morphology on butchery effectiveness. *Journal of Archaeological Science, 34*, 883-893.

Malafouris, L. (2013). *How Things Shape the Mind: A Theory of Material Engagement*. Cambridge, Mass.: MIT Press.

Marshack, A. (1997). The Berekhat Ram figurine: a late Acheulian carving from the Middle East. *Antiquity, 71*, 327-337.

Mateu, J. (1964). Un gisement néolithique en place en bordure de l'Erg er Raoui (Sahara Nord-Occidental). . *Bulletin de la Société Préhistorique Française, 61*(6), 143-144.

Mateu, J. (1970). Un gisement néolithique des environs de Béni-Abbès (Sahara Nord-Occidental) *Libyca, 18*, 155-176.

Mateu, J., & Favergeat, G. (1965). Découverte d'une station néolithique au Tanezrouft Occidental. . *Libyca, 13*( ), 157-182.

Nougier, L.-R. (1933). Polissoir à main néolithique du Sud-Oranais. *Bulletin de la Société Préhistorique de France, 30*(7-8), 479-480.

Savary, J.-P. (1965a). A propos des sépultures néolithiques sahariennes. . *Bulletin de la Société Préhistorique Française. Etudes et travaux, 62*(1), 221-235.

Savary, J.-P. (1965b). Une nouvelle sépulture néolithique saharienne (région de Béni-Abbès). . *Bulletin de la Société Préhistorique Française, 62*(5), 178-183.

Saxon, E., Close, A., Cluzel, C., Morese, V., & Shackelton, N.-J. (1974). Results of recent investigations at Tamar Hat. . *Libyca, 22*, 49-91.

Tixier, J. (1957). *Le hachereau dans l'Acheuléen nord-africain. Notes typologiques*. . Paper presented at the Congrès Préhistorique de France.

Wynn, T. (1995). Handaxe enigmas. *World Archaeology, 27*, 10-24.

Wynn, T., & Gowlett, J. (2018). The handaxe reconsidered. *Evolutionary Anthropology: Issues, News, and Reviews, 27*(1), 21-29. doi:10.1002/evan.21552

Thomas Wynn
University of Colorado Center for Cognitive Archaeology, Colorado Springs, CO 80918, USA
twynn@uccs.edu

Mohamed Sahnouni
Centro Nacional de Investigación sobre la Evolución Humana, 09002 Burgos – Spain
and
Centre National de Recherches Prehistoriques, Anthropologiques et Historiques, 16500 Algiers, Algeria.

Tony Berlant
University of Colorado Center for Cognitive Archaeology, Colorado Springs, CO 80918, USA

Claude Douce
Musée de Préhistoires, Chateau de Sauveboeuf, Aubas 24290, France

Variable Cognition in the Evolution of *Homo*:
Biology and Behaviour in the African Middle Stone Age

Robert A. Foley and Marta Mirazón Lahr

Abstract

The emergence of the human mind is a core problem in human evolutionary studies, and many attempts have been made to describe the pattern of emergence, to characterise how it is distinctive from other forms of cognition, and to determine how and why it evolved. It is now evident that the African Middle Stone Age (MSA) is the context in which many key elements developed or came together, for this is the period in which both anatomical and behavioural modernity can first be identified. And yet, there are many unsolved problems. While the MSA is the setting for the first modern human behaviours, many of those behaviours are shared with Eurasian and African archaic hominins, suggesting either convergence or a deep shared ancestry. The MSA is also a period of at least 300 Ky, with both long periods of continuity and the recurrent appearance of novel traits, and the pattern does not match particularly closely with the evidence for the origin of anatomical modernity. In this paper we review the evidence for cognitive and behavioural evolution in the context of the African MSA, and draw the conclusion that a more sophisticated way of measuring human cognition across this period is required. We introduce the variable cognitive state model as a means of achieving this, and consider how it might be applied to the archaeological evidence.

Introduction

The modern human mind reflects the cumulative evolution of mental abilities along one hominin lineage, and key traits can be traced deep into our past through to ancestral species. But at some point in that past people thought like us – the universal mind of hunters, fishers and gatherers, of families, of externalized group identities, of different languages. The material expression of that mind is in the African Middle Stone Age (MSA). The MSA is the prehistoric technology manufactured by the first modern humans in Africa (Fleagle *et al.*, 2008; Mirazón Lahr & Foley, 2016), and the technological background for the expansion of modern humans out of Africa (Foley & Mirazon Lahr, 1997; Mirazon Lahr & Foley, 2003; Rose *et al.*, 2011; Groucutt *et al.*, 2015; Clarkson *et al.*, 2017). Yet, the MSA is not exclusively the material product of modern humans, which together with major uncertainties about the timing and structure of our lineage in Africa, makes attempts at mapping what must have been some of the most significant cognitive changes in the course of hominin evolution extremely challenging. This paper aims to discuss the conceptual and methodological challenges posed by the Middle Pleistocene African hominin and archaeological records, and explores how we can improve the visibility of cognitive evolution in recent hominin evolution.

Exploring the complexity of evolving the human mind resonates with the contributions of Gowlett to the study of early hominin cognition, particularly in relation to the makers of the Acheulean technology (Gowlett, 1984, 2014, 2015; Crompton & Gowlett, 1993; Gowlett & Crompton, 1995). One major link is that new data on the age of the genetic roots of the African *sapiens* lineage, of early MSA sites and potentially of a modern morphology have chronologically extended the 'origins of modern humans', and thus brought it closer to the Acheulean. Indeed, some might even argue that the late Acheulean is, in fact, the behavioural context where we should be seeking answers. The second is that the shift from archaic to modern humans clearly involves a cognitive change, and that the primary source of information that might signal any change – or multiple changes – would be the record of stone tools, and Gowlett has long been a pioneer in the principles and methods by which cognitive states can be inferred from technology (Gowlett, 1984, 2010; Dunbar *et al.*, 2010; Gamble *et al.*, 2011). Finally, Gowlett himself (Gowlett, 2009) identifies three revolutions in prehistory – one in the early phases of the evolution of the genus *Homo* (the basic human socio-cultural-economic package) (2.6 – 1.6 million years ago [Ma]), a second, silent one (large brains and language) between 1.5 and 0.5 Ma, and a third one (*sapiens* intensification) in the last 100 Ka. Here we examine the one on which he is himself silent – between 500,000 and 100,000 years ago (Ka).

The African record between 500 and 100 Ka

Advances in ancient genomics have extended the estimated age of the last common ancestor of modern humans and the Neanderthal-Denisovan clade to an African population at ~500-700 Ka (Meyer *et al.*, 2012; Prüfer *et al.*, 2014; Mendez *et al.*, 2016; Posth *et al.*, 2017). This age is consistent with recent simulations of the virtual ancestral morphology between the two clades (Mounier & Mirazón Lahr, 2016), but finds no clear reflection in the African Middle Pleistocene record.

The latter suffers from a limited sample of hominin fossils and chronological uncertainty. Is the matter one of poor fossil preservation and chronological resolution? Available data suggest no. Major advances in the dating of key sites and fossils have taken place in the last two decades through the refined correlation between volcanic tephras and palaeoanthropological strata (Deino & McBrearty, 2002; Brown *et al.*, 2012; Sahle *et al.*, 2014), the increasing use of luminescence dating techniques (Duller *et al.*, 2015; Dirks *et al.*, 2017a; Richter *et al.*, 2017), and the application of Uranium Series dating to fossil remains (Grün, 2006; Aubert *et al.*, 2012; Richter *et al.*, 2017). These new dates place the beginnings of the African MSA (Tryon *et al.*, 2005; Sahle *et al.*, 2014; Richter *et al.*, 2017; Potts *et al.*, 2018) and fossils with a mosaic of modern and non-modern traits (Hublin *et al.*, 2017) at *ca.* 300 Ka, while the earliest specimens formally assigned to *H. sapiens* - despite retaining some primitive traits - occur after 200 Ka (McDougall *et al.*, 2005; Fleagle *et al.*, 2008). They also establish the contemporaneous presence of non-modern hominins, ranging from singularly distinct forms, such as *Homo naledi* (Berger *et al.*, 2015; Dirks *et al.*, 2015) to fossils that most probably represent populations along an evolving *sapiens* lineage, such as Florisbad (Grün, 2006). Thus, the emerging picture is one in which two of the key features of modern human evolution - the manufacture of MSA artefacts and the evolution of a derived anatomy universal in living humans - post-date by hundreds of thousands of years the establishment of the ancestral *sapiens* African genetic lineage in the early Middle Pleistocene. This pattern is further challenged by recent genomic studies of recent and contemporaneous African populations, which suggest that the last common ancestor of all living humans also dates to *ca.* 300 Ka (Schlebusch *et al.*, 2017), thus like the MSA, also pre-dating the age of the fossils that show unique human derived morphological features. Equally fascinating, and taking us beyond the end of the Middle Pleistocene, the persistence of the MSA (Scerri *et al.*, 2017) and what are considered archaic anatomical features in the fossil of Iwo Ileru (Harvati *et al.*, 2011) in West Africa into the Pleistocene-Holocene transition, together with evidence for archaic genomic introgression into some West African populations (Hammer *et al.*, 2011; Lachance *et al.*, 2012; Skoglund *et al.*, 2017) pose questions about population history and structure within the continent, as well as the definition of 'modernity'.

This rapidly changing landscape for the evolution of modern humans in Africa raises a number of major questions that impinge on any attempt at reconstructing the processes that led to cognitive change in our ancestry. Among these, the role of the MSA as the expression of modern human behaviour, and ultimately the human mind, is key.

The African MSA

The MSA is a long phase of contrasts - it persists in a recognizable form for over 300 Ka, and yet displays enormous variation; it is technologically stable, and yet shows evidence of remarkable behavioural innovations; it occurs in every habitat in Africa and through multiple warm/wet and cold/dry phases; it is associated with both phenotypically non-modern and modern hominin species; and much of it is shared with the technology of Neanderthals. Yet, it is a fascinating phase, and one that is thought to reflect a significant technological and cognitive change in human evolution (Clark, 1989; Foley & Mirazon Lahr, 1997; Mirazon Lahr & Foley, 2001; Wynn, 2009; Barham, 2010). Here we focus on two questions – 1. Is the MSA a 'definable' entity with an identifiable beginning and end? and 2. What is the nature of the variation and diversity that occurs within these limits?

Timing of the MSA

In 1965, when Clark wrote his seminal paper on the Late Pleistocene cultures of Africa (Clark, 1965), he placed the earliest MSA at ~50 Ka. Now, ages of *ca.* 300 Ka are generally accepted - the earliest dated MSA sites are Kalambo Falls, Zambia (>300 Ka; (Duller *et al.*, 2015)), Jebel Irhoud, Morocco (~ 300 Ka; (Richter *et al.*, 2017)) Gademotta, Ethiopia (~287 Ka; (Sahle *et al.*, 2014)), Kapthurin, Kenya (~280 Ka; Tryon *et al.*, 2005), Twin Rivers, Zambia (possibly 265 Ka; (Barham & Smart, 1996) and Benzu Cave, Morocco (~250 Ka; (Richter *et al.*, 2017)). The geographical distribution includes northwestern, eastern, and southern Africa. However, to reach a consensus on the earliest occurrence of the MSA it is necessary to establish its diagnostic features.

'Middle Stone Age' was the name given by Goodwin and Van Riet Lowe in 1929 to archaeological assemblages in South Africa that lacked both the typical tools of the Early Stone Age (ESA) and Later Stone Age (LSA), such as handaxes and microliths (Goodwin & Lowe C. van Riet, 1929). This early definition emphasises the fact that the MSA lacks ubiquitous diagnostic tools as found in the ESA or LSA. While this remains the case, and is indeed one of the most interesting aspects of the MSA, researchers recognise a few technological features that, even if not universal to the MSA, occur in such a majority of cases as to be considered typical of the industry. These are core preparation (Mode 3 technology, or the pre-determination of the shape of the target flake (Clark, 1965)) and, as a proxy, flake platform faceting (Leakey, 1931; Cole, 1954) . This is not unproblematic because platform faceting can occur as an un-intentional by-product of reduction of unprepared cores (van Peer, 1992), because a form of core preparation can be found in the ESA, and because the variability in preparation techniques within the

MSA is significant and interesting (Porat *et al.*, 2002; Tryon, 2006).

Core preparation has been shown to be part of the technological repertoire of the distinctive South African ESA industry known as the Victoria West (Backwell & d'Errico, 2001; Kuman, 2001; Wilkins *et al.*, 2010; Li *et al.*, 2017). Analyses of Victoria West cores demonstrate that they were made using standardised core preparation reduction strategies that match Boëda's definition of the Levallois volumetric concept (Boëda, 1995; Sharon & Beaumont, 2006; Li *et al.*, 2017). However, they also show that they are distinct from MSA cores in form, predominance of a particular flake removal strategy, size and intended output (flakes for large cutting tools), and that they represent a relatively small percentage of the range of cores within Victoria West industries (Li *et al.*, 2017). While there is no continuity between the South African Victoria West currently dated between 1.1-0.8 Ma and the MSA found across Africa at 300 Ka, what this shows is that aspects of one of the key characterising features of MSA industries – the preparation of a core to standardise flake production – were part of the technological repertoire of some elements of the ESA.

This problem is in fact two problems. The Victoria West and other assemblages showing core preparation embedded in the Acheulean are a question of the deep roots of the MSA, and whether these are monophyletic. The mirror of this is the persistence of hand-axes and cleavers into time periods when the MSA is well established – for example, the persistence of handaxes/cleavers in Ethiopia until ~212 Ka (De La Torre *et al.*, 2014), some at Kibish ~195 Ka (Shea, 2008), some at Sai island between 223-150 Ka (Van Peer *et al.*, 2003) and some at Herto (Clark *et al.*, 2003). The question is whether these late occurrences represent an independent persistence of the Acheulean alongside the MSA, or some form of braided development of the two technologies over time.

While a better definition is needed, the roots of the MSA are clearly to be sought in the diversity of the late Acheulean. We would further argue that the 'traits' that may provide diachronic information on evolutionary patterns and cognition are not broad technical categories, such as core preparation, but variance and innovation within its particulate parts – mode and direction of flake removal, intended product, etc. – in other words, the variable state of disaggregated aspects of technology.

Notwithstanding the challenges in drawing a boundary between the late Acheulean and the early MSA, and thus in establishing its roots, few would dispute that the MSA is an entity in African prehistory that existed from at least *ca.* 300 Ka, one which includes North Africa (Scerri, 2017), and that is dominated by a particular form of prepared core techniques. The broad contemporaneity and wide spatial distribution of the earliest generally recognised MSA sites may reflect its independent origins in regional Acheulean traditions across northern, eastern and southern Africa, their relative synchrony resulting from a convergent functional response to resource pressure; alternatively, the MSA may have a single localised origin in a regional Acheulean tradition (of uncertain time-depth), and its synchronic appearance in the record after *ca.* 280 Ka be the signature of a pan-African dispersal during the warm Marine Isotope Stage (MIS) 9 interglacial. Further studies will clarify much of the present uncertainties, but whatever the resolution, the chronology of the MSA will remain significantly extended in relation to its original conceptualisation.

At the other temporal end, the disappearance of the MSA is also complex. The transition between the MSA and Later Stone Age (LSA) was generally thought to have occurred <25 Ka (Klein, 2009), largely based on stratified cave sites from South Africa. However, recent excavations and dating programmes have established the presence of LSA industries between 44-41 Ka at the South African sites of Border Cave (Villa *et al.*, 2012) and Rose Cottage Cave (Pienaar *et al.*, 2008), with suggestions that the transition may have started as early as 56 Ka (Villa *et al.*, 2012). These earlier dates are consistent with those from Enkapune Ya Muto in Kenya (Ambrose, 1998) and Goda Buticha in Ethiopia (Leplongeon, 2014; Pleurdeau *et al.*, 2014) although evidence from Mumba, Tanzania, suggests a much earlier technological shift in eastern Africa at 65 Ka (Gliganic *et al.*, 2012). Other dates in equatorial Africa have wide margins of error, but there are good grounds for placing the earliest LSA before 30 Ka (Matupi Cave (Zaire), Shum Laka (Cameroon), Kisele II, Nasera, Olduvai (Tanzania), and Prospect Farm, Lukenya (Kenya)). Yet, dates for the youngest MSA in certain areas, such as at the sites of Affad, Sudan, and Ndiayène Pendao, Senegal, are as young as 15 Ka and 11 Ka (Osypiński & Osypińska, 2016; Scerri, 2017; Scerri *et al.*, 2017) Together with evidence for multiple transitions (e.g. Mumba and Border Cave (Diez-Martin *et al.*, 2009; Villa *et al.*, 2012) and the persistence of MSA elements in local Holocene traditions (e.g., Goda Buticha (Pleurdeau *et al.*, 2014))), the diversity of dates for the MSA/LSA transition across Africa indicates that there may be much greater demographic complexity underling the relationship between the two industries, with the technology changing intermittently, locally and adaptively. Thus, the raggedness of the end of the African MSA suggests that its disappearance is likely to be evolutionarily and cognitively irrelevant.

How dynamic is the MSA?

The preceding ESA/Acheulean tradition has been (possibly unfairly) characterised as showing little geographical or temporal variation, and the MSA in contrast has been seen as more dynamic. Two elements are important here. The first of these is that the MSA displays significant and regular patterns of sub-continental spatial variation in lithic assemblages for the first time in prehistory, both typologically and technologically (Clark, 1992). This has been used to argue for the beginnings of regional and ultimately group identity, a significant element of later human social and cultural organisation. Some 50 taxonomic units have been named within the African MSA (Clark, 1965; Klein, 2009), mostly based on regional variation in technology and typology. More recent studies have suggested that there is evidence for population structure across the northern half of the continent during the MSA (Scerri *et al.*, 2014).

How much of that regional variation is also chronological is uncertain, but there is also strong evidence for considerable change during the course of the MSA. One aspect of this is that some MSA traditions are relatively short-lived, and so there is a pattern of rapid change (primarily observable in lithic technology). The best documented examples are the Stillbay and the Howieson's Poort, two very distinctive South African MSA Upper Pleistocene variants, which have been shown to be short-lived episodes of ~5,000 years duration (Jacobs *et al.*, 2008a; Jacobs & Roberts, 2017). Even if a longer chronology for these two industries is accepted (Tribolo *et al.*, 2013; Feathers, 2015), it still depicts a dynamic process of change within the MSA, which is reflected not only in the lithic industries, but also in hunting technologies (Wadley, 2010b). In North Africa, the emerging temporal boundaries of the Aterian also suggest fluctuating demographics and adaptations (Barton *et al.*, 2009; Scerri, 2012). So rather than a static picture, the evidence suggests the MSA was dynamic and adaptable, encompassing longer-term trends and patterns, as well as punctuated localized traditions. These are identifiable by distinctive traits, many of which also suggest that some previously considered important innovations were instead ephemeral (Jacobs *et al.*, 2008b; Jacobs & Roberts, 2017), making their dynamics or instability itself a focus of interest.

The dynamism of the MSA is expressed in the appearance of novel behaviours, often associated with the emergence of 'modern human behaviour', cumulative culture, and symbolism. Some of these behaviours include: *200 Ka+:* presence of ochre (Barham, 1998), microliths (Barham, 2002); *160 Ka:* a polished child's skull (White *et al.*, 2003)[39], indicating preservation and prolonged carrying; *130-80 Ka (including the Levant):* perforated shell beads (Henshilwood, 2004; Vanhaeren *et al.*, 2006; Bouzouggar *et al.*, 2007; d'Errico *et al.*, 2008); *75-60 Ka:* delicate pressure flaking (Mourre *et al.*, 2010); incised objects (ochre, ostrich eggshell) (D'Errico *et al.*, 2001; Texier *et al.*, 2010). In some cases, the function of particular innovations may be unclear – for example, whether the use of ochre was

for mastic in hafting (Lombard, 2007), or for pigmented decoration (Watts, 2002); in others, the novelty is thought to have led to increased or more demonstrable use – hafting of points being a classic example (Wilkins *et al.*, 2012) and the production of compound adhesives that gave stability to the hafted products (Wadley, 2010a); while others have very significant implications – for example, the evidence that microliths were used in composite tools that are most likely to indicate the early (70 Ka) presence of bows (D'Errico *et al.*, 2012a). It has also been argued that the MSA shows evidence of more complex hunting practices (Clark & Plug, 2008), and the first systematic use of aquatic resources (Marean *et al.*, 2007), the latter representing a shift towards defendable, dense and predictable resources that led to greater territoriality, inter-group conflict and hyper-social behaviours (Marean, 2016). These innovations often cannot be interpreted in a simple linear manner – for example, microliths may appear >250 Ka, but they do not occur again until about 75 Ka – so that while there may be directional change during the MSA, there may equally be stasis interrupted by periods of rapid change, as well as multiple trajectories of change. Thus, although individual data may be sparse, the overall signal is strong – hominins and humans were changing their behaviour and expressing those changes in the MSA record.

The 300 Ka threshold

The brief discussion of the African MSA above highlights the tension between the evidence for an extended process that establishes behaviourally and anatomically modern humans in Africa and beyond (Mirazón Lahr, 2016; Stringer, 2016), and the potential existence of one or more thresholds during that process indicative of major changes in biological and behavioural organisation. On the one hand, ~300 Ka is a major temporal threshold, from which time direct genetic ancestors of living humans are found in Africa and MSA industries become the technological context for behavioural change; on the other, the evidence suggests that substantial morphological, behavioural, and cognitive change occurs in the subsequent 300,000 years, with periods when these changes are diffused geographically through population expansion and dispersals, and others when the innovations are localised, and often short-lived (Table 1).

The process covers too great a temporal range to be treated as an archaic to modern cognitive threshold, as it is also too complex to be treated as a simple or single transition. This complexity raises questions of convergence and independent trajectories, unknown ancestral conditions, and mosaic evolution over long periods of time, all of which make attempts at understanding the evolution of the human mind even more challenging.

Table 1. Synthesis of a timeline for the evolution of modern humans in Africa.

Period	Empirical observation	Biogeographic implications
750-500 Ka	Genetic and morphological evidence for the last common ancestor (LCA) of modern humans and the Neanderthal/Denisovan clade[1,2].	• Afro-Eurasian dispersal
≥ 300 Ka	MSA[3-6]	
300-280 Ka	• Morphological evidence for hominins with mosaic derived and primitive traits (e.g., Jebel Irhoud, Florisbad)[7-8] • Genetic evidence for *sapiens* introgression into Eurasian Neanderthals[9] • Genetic evidence for the LCA of modern humans in Africa[10]	• Pan-African dispersals and establishment of regional populations of the *sapiens* lineage, one of which is ancestral to living humans • Afro-Eurasian dispersals
280-200 Ka	• Archaeological evidence for the persistence of Acheulean tools in Africa[11-14] • Archaeological evidence for novel behaviours in Africa[15-16], some shared with Eurasian hominins (e.g., ochre and engravings[17-18]) • Fossil evidence for the co-existence of multiple African hominin lineages (e.g. *Homo naledi*)[19-20]	• Localised population change
≥ 200 Ka	• Morphological evidence for derived universal *H. sapiens* traits[21-22]	
200-130 Ka	• Archaeological evidence for increased foraging breadth[23-24] and complex funerary practices[25]	• Localised population change
130-60 Ka	• Morphological evidence for *H. sapiens* throughout Africa (and the Levant)[26] • Archaeological evidence for multiple novel behaviours in different African (and Levantine) modern populations[27-34] • Archaeological evidence for localised MSA-LSA transition in East Africa[35] • Genetic evidence for last common ancestor of living African and non-African humans[36-39]	• Pan-African and Afro-Eurasian dispersals
60-11 Ka	• Archaeological evidence for expansion of LSA technology[40-43], and eventual disappearance of MSA[44-45]	• Localised population change

References: (1) (Meyer *et al.*, 2016); (2) (Mounier & Mirazón Lahr, 2016); (3) (Tryon *et al.*, 2005); (4) (Sahle *et al.*, 2014); (5) (Duller *et al.*, 2015); (6) (Richter *et al.*, 2017); (7) (Hublin *et al.*, 2017); (8) (Grün, 2006); (9) (Posth *et al.*, 2017); (10) (Schlebusch *et al.*, 2017); (11) (Clark *et al.*, 2003);(12) (Van Peer *et al.*, 2003); (13) (Shea, 2008); (14) (De La Torre *et al.*, 2014); (15) (Barham, 1998) (16) (Barham, 2002) ; (17) (Roebroeks *et al.*, 2012); (18) (Rodríguez-Vidal *et al.*, 2014); (19) (Berger *et al.*, 2015)(20) (Dirks *et al.*,2017b)(21) (McDougall *et al.*, 2005)(22) (Fleagle *et al.*, 2008)(23) (Marean *et al.*, 2007)(24) (Marean, 2016)(25) (White *et al.*, 2003)(26) (Mirazón Lahr, 2016) and references therein; (27) (Henshilwood, 2004); (28) (Vanhaeren *et al.*, 2006); (29) (Bouzouggar *et al.*, 2007); (30) (d'Errico *et al.*, 2008) (31) (D'Errico *et al.*, 2012b); (32) (Mourre *et al.*, 2010); (33) (Texier *et al.*, 2010); (34) (Wadley, 2013); (35) (Gliganic *et al.*, 2012); (36) (Malaspinas *et al.*, 2016); (37) (Mallick *et al.*, 2016); (38) (Pagani *et al.*, 2016);: (40) (Ambrose, 1998); (41) (Villa *et al.*, 2012); (42) (Leplongeon, 2014); (43) (Pleurdeau *et al.*, 2014); (44) (Osypiński & Osypińska, 2016); (45) (Scerri *et al.*, 2017).

Cognitive evolution, the MSA and modern humans

The brief discussion above highlights the fact that a number of the foundational elements of humanity came into place in Africa after 300 Ka, during the African MSA. While the spatial and temporal dimensions of this process require empirical progress, it is also necessary to have a more complex set of models of hominin cognitive evolution, ones that should make it easier to read the signals of the archaeological and fossil records.

The 'out of Africa' model had the positive impact of focusing attention on the evolution of the human mind as a critical part of the process of becoming human (Davidson & Noble, 1991; Klein, 1992; Mithen, 1997). It is generally agreed that the material record, including the

innovations associated with the MSA discussed above, reflects behaviours generated by unique cognitive capacities operating in a cultural and environmental context. New cognition generates new behaviours, and indeed new ways of generating behaviour (cumulative culture) - an individual 'does something' (e.g. makes a flake), which involves a cognitive mechanism that generates the behavioural outcome. That mechanism will comprise both innate biological cognitive traits and learning potentials derived from the demographic, social and cultural environment. The motivation for the behaviour will derive from a particular want or need, and so is shaped by environmental (including the social and cultural environment) or ecological circumstances. The flake, in this case, is therefore a product of the interaction of internal cognitive processes and external factors.

Although the application of theories of cognitive evolution to the problem of modern human behaviour opened interesting new avenues of research (Vanhaeren & d'Errico, 2006; Krause *et al.*, 2007; Wadley, 2013; d'Errico & Backwell, 2016), it also tended to throw into sharp contrast the differences between modern humans and archaic hominins (epitomized by the debate over the Middle to Upper Palaeolithic transition in Europe) (Mellars, 1991, 2005; d'Errico *et al.*, 1998; Zilhao, 2006). This paradigm of 'archaic vs modern' has shaped the nature of the cognitive, cultural and behavioural models proposed, most of which emphasise a sharp change between the old and the new. As stressed in the previous section, the complex pattern and extended process of the evolution of humans and its African MSA context must imply that such a binary view of hominin cognition is inappropriate. How, though, can one avoid the trap of dividing the cognitive world into the ancient and modern?

Models of cognitive evolution

Current models of human cognitive evolution have focused on a number of aspects, of which five are relatively well-established in the literature. These are fluid, cross-domain cognition (Mithen, 1996), cumulative culture (Tomasello, 1999), symbolic thought (Henshilwood, 2002), language (Davidson & Noble, 1991) and enhanced working memory (Coolidge *et al.*, 2013). Each of these has been identified as a key element of the modern mind. For example, 'symbolic thought' has often been seen as a trait of modern humans (with debate as to whether other hominins may have been capable of symbolic thought), and inferred from such things as the decorative use of ochre (Henshilwood *et al.*, 2009). In the past, 'culture' was also seen as an attribute of being human, but with the growth of evidence for culture in other animals, emphasis is now on 'cumulative culture', the capacity to build cumulatively on existing socially-learnt behaviours (sometimes referred to as the ratchet effect) (Tomasello, 1999). Cumulative culture is described as a distinguishing, if not unique, feature of modern humans (Boyd & Richerson, 1996). Perhaps the most widely used of all cognitive aspects is language, often seen as the fundamental basis for many other features – learning, social co-operation, planning and organisation – and certainly as a pre-requisite of modern human behaviour.

What these elements of 'modernity' share is that they are complex, made up of many components; however, they are usually bundled together conceptually. Culture, for example, is in many formulations, a set of traits (teaching, social learning, material culture, imitation, innovation, etc.), and the concept is ultimately a shorthand for those traits coming together. However, as a compound concept, it is also difficult to operationalize – seeing culture in a stone tool is virtually impossible.

Furthermore, the component attributes that constitute the compound concept may, in the past and in other hominins and populations, have been bundled together in different ways, or with only a partial set of the components. There is no reason why such compound concepts cannot be disaggregated and made more operationable (Foley, 1991), and indeed studies that disaggregate the manufacturing components of an artefact provide major insights as to the underlying behaviours and cognitive capacities (e.g. (Wadley, 2010a).

There is an advantage in doing this, as it means that we can avoid the binary, presence/absence form in which these compound concepts are often used , and so avoids the 'archaic versus modern' framework that has dominated the field since the out of Africa model became the dominant explanation. Research into cognitive traits in cognate disciplines suggests that a presence-absence dichotomy is a major over-simplification. For example, Hauser has shown that abstract thought and use of icons are quite distinct (Hauser, 2008), and so terms such as symbolic thought or symbolic behaviour are already too complex. Language is a modular entity with different components (sound, phonemes, meaning, syntax, recursion, pragmatics, etc.) that can have independent histories (Corballis, 2011). So, while the *sharp contrast models* were suitable when the African rather than multiregional evolution of *H. sapiens* was being established, they are not the right conceptual framework for exploring the spatio-temporal complexities in the African MSA which point to continuity and temporal overlap between forms of actions and types of artefacts from which archaic or modern behaviours are inferred, suggesting the changes took place over a prolonged period of time in a gradual or stepwise manner. This mismatch between existing models and current evidence for how 'modernity' came to dominate the behavioural repertoire of African hominins in the late Quaternary is a pressing problem.

Disaggregated models of cognitive evolution

Our central proposal is that human cognition is not unitary. A brief comparison with the visual system is helpful at this point. We experience the visual world as a unitary phenomenon, in real time. Cognitively, however, visual information is processed in four major systems (motion, colour, position and shape), which together create the perception we actually 'see'. This way of looking at vision can be applied to other aspects of human cognition. Instead of thinking in terms of broad concepts, such as cumulative culture or symbolic thought, we can consider more elemental capacities, which can be developed, combined and bound in different ways across the evolution of archaic hominins into modern humans. This would also be more consistent with developments in cognitive science, where viewing cognitive processing as a series of interacting systems has proved to be a useful approach (Barnard & Teasdale, 1991; Barnard *et al.*, 2007). A number of models have attempted to disaggregate the components of culture and cognition. Wynn and Coolidge's enhanced working memory model (Wynn & Coolidge, 2006) is an example; another is Hauser's four key cognitive components - recursion, promiscuous domain interfaces, mental symbols, and abstraction (Hauser, 2008); (Nowell & White, 2010) and Barham (Barham, 2010) distinguish between innovation (variants on an existing theme) and invention (novel steps), which in effect discriminate between two forms of generating novelty. Language should also be included, but in turn be divided into a number of linguistic elements - nature of phonology, syntax, morphology, and semantics and pragmatics, with different roles to play in the evolution of human cognition (Berwick *et al.*, 2013). The number of planned steps in an action, central to some of Binford's ideas (Binford, 1981), might also be included in this list, although in practice it is probably a specific outcome of enhanced executive function. Underlying much of this would be types of social learning, which forms the mechanism by which information is transmitted (imitation, teaching, etc.).

Variable cognitive state model

Breaking complex concepts down into constituent units represents a major research tool in any field; in this particular case, it permits us to consider prehistoric humans as having variable combinations of different traits, it encourages us to look at cognitive and behavioural evolution using a more continuous model (equivalent to dosage effects in genetics), and it offers the possibility to consider what might be the particular observable predictions in the archaeological record of each component.

We propose a *variable cognitive state model,* in which evolving hominin cognition is viewed as a series of

variable states across different traits. In principle, there are a very large number of potential traits that might be disaggregated and used, at all levels from that of 'outputs' through to their neurobiological basis. There must be, however, a trade-off between comprehensiveness and reductionism, and utility, particularly in terms of testability in the record.

In determining these we can start by considering a simple model, where an output behaviour is the product of an 'output' cognition, which in turn is the product of specific cognitive processes and their interaction with each other; these in turn are based on neurobiological architecture and processes (Figure 1). For the purposes of this model, it does not matter whether the priming of these is 'cultural', as argued, for example, by Heyes (Heyes, 2012) or genetic, or more likely, a combination of the two. In this basic model, for example, the use of ochre as an ethnic signal might be the behavioural outcome, occurring as a result of mental properties, such as language or symbolic thought. These, however, might rest on the inputs and interactions of a large number of cognitive processes. The variable cognitive state model (VCS Model) focuses on breaking down these cognitive processes in more detail.

A number of such processes can be proposed (Table 2). A starting point is Hauser's components which contribute to both human thought and language – *recursive operations, combinatorial operations, mental symbols, promiscuous interfaces* and *abstract thought* (Hauser, 2008). These more fundamental processes would apply both to mentalising thought (inner language), and language itself. To these can be added Wynn and Coolidge's *working memory* (Wynn & Coolidge, 2011), and the cognition associated with *causal reasoning,* the cognitive capacity to make models of causative relationships that go beyond simple correlations as an essential building block of the human mind (Buchsbaum *et al.*, 2012). An extension of causal theorizing is what can be referred to as

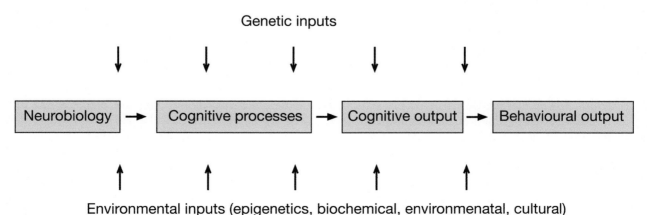

Figure 1. General model of the relationship between neurobiological processes, cognition and observed behaviour. Cognition can be measured or inferred at each of these levels, to which there may be both genetic and environmental input

Table 2. Components of hominin cognition. The variable cognitive state model attempts to break larger cognitive concepts of process down to more fine grained one, as these may evolve independently and be variable expressed in different hominin taxa and populations. Discussion and sources for the categories are in the main text.

Cognitive process	Description
Recursive and combinatorial operations	Recursion and combination is an iterative operation that allows the creation of new forms (phrases, tunes, technologies), either through bringing together separate elements, or repeating combinations.
Mental symbols	Humans convert sensory experiences and abstract thoughts into externalized symbols (words, images, objects).
Promiscuous interfaces	Combinations of representations across domains (e.g. technology and social interactions)
Abstract thought	Abstract thoughts or representations that have no direct physical association (good, evil, fairness)
Working memory	Working memory is the mind's ability to hold in attention, and process, task-relevant information in the face of interference. It consists of two subsystems, the visuospatial sketchpad and the phonological loop, operated through an episodic buffer
Causal reasoning	Humans have structured, generative, causal representations of the world (causal models), that are more than associative processes and conditioning.
Contingent decision making	An extension of causal reasoning is decision making based on modelling cognitively possible outcomes of actions, and making decisions on the basis of simulated 'what-if' scenarios.
Spatio-temporal displacement	The ability to think or communicate about phenomena that are distant in time or space; in practice this will be continuously graded, as the temporal or spatial horizon is extended. http://www.oxfordscholarship.com/view/10.1093/acprof:oso/9780195377804.001.0001/acprof-9780195377804-chapter-9
Planning	Behaviours and actions may be immediately motivated and carried out, but humans in particular have a noted capacity to plan actions either distant in time or across a number of sequential steps (planning depth).
Cultural transmission	Cultural transmission is information capable of affecting individuals' behaviour that they acquire from other members of their species through social interaction
Teaching	Teaching is the active effort to transmit information to conspecifics through social interactions
Imitation/copying	Copying or imitating behaviours and actions is an important means of social transmission, and so for the emergence of traditions or cultural properties. Byrne
Invention	*Most novelties are basically variants on existing forms or behaviours, cumulatively adding to performance; invention therefore refers to the ability to change current forms, rather than depart radically from them (Arthur)*
Innovation	In contrast to an invention, an innovation is the development of an entirely new form of behaviour, or a radical departure from existing forms. Cognitively it can be associated with some more substantial level of creativity. Arthur
Implicit metacognition	*The cognitive capacity to think about thinking, the conscious monitoring or controlling, insofar as it is possible, cognitive processes. Implicit metacognition is automatic, and humans are largely unconscious of this. Frith*
Explicit metacognition	Explicit metacognition is "concerned with generating reportable knowledge about the processes underlying our behaviour"; in other words, it is a form of cognition that is firmly part of conscious thought processes.

contingent decision making. This is, in essence, a form of mental simulation of potential actions and outcomes ('what if' scenarios), and plays a key role in planning (Buchsbaum *et al.*, 2012). *Planning* itself might be another element, although this parameter highlights an important issue, namely that planning is not an either/or, but might be considered to grade in relation to something like the number of steps involved, or the length of time over which the proposed behaviour might be executed.

Some other elements of cognitive processes fall into slightly different categories but are none the less either important or have been discussed in relation to human cognitive uniqueness. These would include: *cultural transmission* or *social learning* – that is, the acquisition of information or the capacity to carry out behaviours from conspecifics (Richerson & Boyd, 2005). For the most part this is based on an ability to *imitate* or *copy* (Byrne, 2002), this can be elevated in some cases when *teaching* is involved – in other words, the active and purposeful dissemination of information (D'Errico & Banks, 2015; Morgan *et al.*, 2015). Equally important is *theory of mind*, the ability to rationalise or intuit the thoughts of others; although this is actually another composite concept, it summarises an important element of the social and interactive nature of cognition (Whiten, 1991). Finally, the concept of *metacognition* should be included in any set of the major components of human cognition; metacognition is the monitoring and controlling by conscious thoughts of actions and their underlying non-conscious cognition – for example, error correction. Frith distinguishes between *explicit metacognition*, and *implicit metacognition* – the former a relatively slow and active process, the latter fast and largely unconscious (Frith, 2012). The concept of metacognition provides access to what can broadly be conceived as consciousness or self-consciousness, but tied into particular cognitive processes.

These add complexity to models of symbolic thought and language, or indeed mental templates, but are themselves simplifications. For example, any of these may involve either temporal or spatial aspects, which may differ; planning may be well developed spatially, less so temporally, or vice-versa. Furthermore, these can all be considered to be executive function elements of cognition, but we know that these can be integrated with and influenced by emotions; experiments have shown that, under exactly the same rational conditions (i.e. rewards and costs), individuals make different decisions in relation to varying framing or emotional conditions – becoming more or less risk-averse (Palframan *et al.*, 2006). However, these additional level considerations are not taken into account here for the sake of clarity and simplicity.

The VCS model can be visualized as a series of slider switches, ranging from less to most developed as a slide switch on a radio may go from low to high. Living humans, it can be assumed, have each of these set to full, not least because that is the only way to set a benchmark for them. The hypothesis underlying the model is that different hominin taxa would exhibit different settings. It might be, for example, that early *Homo* would have had a high setting for the level of social learning, but not for metacognition or abstract thought. Our closest living relative is also likely to vary in the extent to which these attributes approximate human cognition - theory of mind among chimpanzees, for example, is thought to be limited, but social learning and copying more well developed (Whiten & Schaik, 2007). Furthermore, theory of mind is not an either/or category, and it has been suggested that different levels of intentionality can be ascribed to successive enhancements of theory of mind, with some hominins exhibiting more than others (McNabb & Cole, 2015). In other words, the VCS model allows hominin cognition

to be measured as a series of partially independent parameters, and thus for different hominin taxa to vary cognitively in ways that are not based on the presence or absence of traits, nor simply linear developments towards the human cognition.

The advantage of the VCS model is that it moves away from binary comparisons between either humans and chimpanzees, or between modern humans and archaic hominins; it also allows avoids focusing on complex or compound traits such as language or symbolic thought that are less easily inferred and so difficult to test. It furthermore provides a means by which the cognition of different hominins is not necessarily a matter of lineal ranking, but that they may have more derived settings for some element, and less for others. Finally, it might also be considered to address some of the criticisms about the absence of suitable bridging arguments in inferring language and cognition that have been made by Botha (Botha, 2000).

Variable cognitive states and the behaviour of MSA hominins

The archaeological challenge is to determine these 'settings' for different human and hominin populations on the basis of the MSA record. In practice, this can only be done through the application of experimental approaches that determine the extent to which different outputs are conditional upon specific cognitive inputs. An example of one such experiment would be that carried out by Morgan and colleagues, which looked at the effect of teaching by using language or gestures (as opposed to just imitation) on learners of Mode 1 and Mode 2 technologies, suggesting that language would have been necessary for improving performance in the production of Mode 2 technologies (Morgan *et al.*, 2015).

However, many aspects of our prehistoric mind are not amenable to experimental testing. Yet, some preliminary predictions can be made. Mental symbols, for example (see Table 2), the product of the human ability to convert sensory experiences into externalized symbols (for example words) might be inferred if there was evidence for speech, or, more probably, images or symbols made on objects in the natural world (cave walls), or manufactured items (e.g. bone, shell, etc.). This capacity has been claimed controversially for hominins in South East Asia at 500 Ka (Joordens *et al.*, 2015), for Neanderthals at Gorham's Cave, Gibraltar (Hoffmann *et al.*, 2018.; Rodríguez-Vidal *et al.*, 2014), and more convincingly for modern humans by 120 Ka (D'Errico *et al.*, 2001; d'Errico *et al.*, 2008). However, viewing this as a continuous trait (it is unlikely that the human capacity for mental symbols emerged in one single step), the record suggests that persistent and frequent evidence for mental symbols does not occur until well into the Upper Pleistocene.

A second example might be causal reasoning - structured, generative, causal representations of the world (causal models), that are more than associative processes and conditioning (Table 2). To some extent, the development of lithic technology might be used as a measure of this capacity, as fracturing rocks in predictable ways indicates an understanding of causal mechanisms. A measure of technological complexity might thus stand evidence for various levels of causal reasoning. Another context for such an inference might be the control of fire. Gowlett has proposed a gradual process of pyrotechnology (Gowlett, 2016), seeing it as a process rather than an event, moving from opportunistic use of natural fires, through conservation and control of natural fires, to the full initiation and control of them. The development through these stages can each be seen as extensions of causal reasoning.

Figure 2 provides a speculative model for the possible character states for several of the cognitive processes shown in Table 2. Cognitive states are hypothesized for three hominin populations – makers of pre-Mode 3 technologies; makers of early Mode 3/ early MSA; and makers of MSA technologies in MIS5 and early MIS6 (in other words, the later MSA in Africa, normally associated with AMH). The primary evidence for these assessments is, in the first instance, lithic technology as this provides a thread of continuity, but also non-lithic technologies, such as portable images and adornment, and the control of fire. It is likely that mobility, settlement patterns, and subsistence-associated behaviours can be recruited to provide further evidence. It should be stressed that these are indeed hypotheses to illustrate the approach, rather than firm conclusions, which would require more detailed analysis. Figure 2 would suggest that some of the cognitive states were already well developed towards the human cognition among the makers of Acheulean assemblages (imitation, innovation, cultural transmission), while others saw a sharp increase with the early MSA (promiscuous interfaces, causal reasoning), and some only with MIS5/4 hominins (recursive and combinatorial representations, metacognition). In the model shown in Figure 2, it is also posited that in some cases, even by MIS5/early 4, states equivalent to living humans may not have been achieved, underlining the possibility for further cognitive evolution based on either ratchet effects (Tomasello, 1999) or cultural processes (Heyes, 2012), interacting with biological ones. Regardless of the accuracy of these particular hypotheses, the point to emphasise is that finding independent metrics for cognition allows for different rates of cognitive

evolution (Foley, 1996), which may be independent, or may interact with each other to lead to an accelerated rate of cognitive evolution

Conclusion

This paper has addressed the intersection between a number of major issues in current human evolutionary studies; first, the central importance of the MSA in the process by which later hominin diversity evolved and modern humans appeared as a lineage; second, the increasing recognition of the role of changing cognition – and the cultural and social capacities that are thus made possible – in this phase. The conclusion is that better models are required to allow us to link the MSA archaeological observations to cognitive inferences, and thus to evolutionary reconstructions.

The central aim of this contribution has been to argue that to tackle the complexity of the MSA and its role in the evolution of the human mind we need to develop more subtle and reductionist models of behaviour and cognition. Large compound concepts such as 'symbolic thought' or 'language' may be fine for discriminating between creatures as different as humans and chimpanzees, or even humans and australopithecines, but they are too crude to help unravel what is clearly a complex and mosaic form of transformation that occurs from the end of the Acheulean and the full establishment of global humans.

John Gowlett wrote that "that a first human revolution had been achieved by about 1.6 million years, that a second revolution, largely silent archaeologically, had occurred by about 500,000; a third revolution, of modernity, occurred thereafter" (Gowlett, 2009). Numerous models to describe the human cognitive niche have been developed, and palaeobiological and archaeological evidence employed to identify its existence in different hominin populations and taxa. Were human cognitive evolution to indeed consist of a succession of stepwise and directional revolutions, such models would be what is required. However, as Gowlett states in the same paper, these revolutions are episodes within a longer and continuous evolutionary process; to this can be added the observation that this evolutionary process is a mixture of private trajectories in lineages, convergence between them, shifting combinations of traits and their cognitive bases, and the acquisition and loss of behavioural outcomes. The challenge is not to make the case for the distinctiveness of human cognition, nor even what its character is, but to find ways of measuring its variable expression across hominin taxa and through millions of years of evolution. The variable cognitive state model is an initial attempt at this.

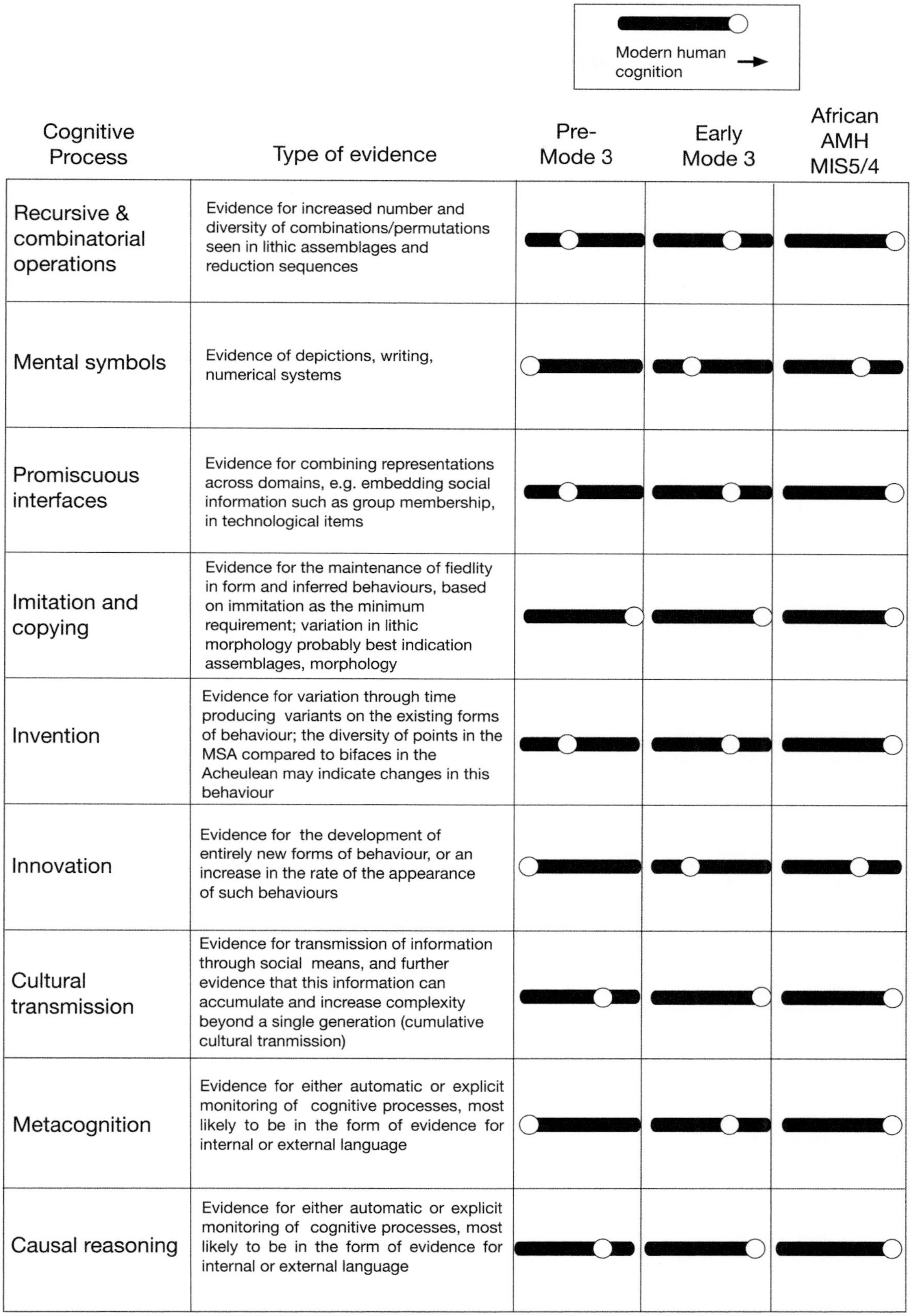

Figure 2. A selection of the component cognitive processes proposed for the variable cognitive state model (see Table 1) (column 1), and some proposed sources of evidence for these (column 2). Some hypothesised states for these for pre-Mode 3 (Acheulean) populations, early Mode 3 or MSA populations, and early Upper Pleistocene African AMH populations. Estimated states are relative to modern humans (to the right of the bars).

Acknowledgements

We would like to thank James Cole, Matt Grove and John McNabb for inviting us to contribute to this volume in honour of John Gowlett, and to John for the many years of discussion, stretching back for one of us (RAF) to Kilombe in 1973. MML's research is funded by the ERC IN-AFRICA Award No. 29590792.

References

Ambrose, S.H. 1998. Chronology of the later stone age and food production in East Africa. *J. Archaeol. Sci.* 25: 377–392.

Aubert, M., Pike, A.W.G., Stringer, C., Bartsiokas, A., Kinsley, L., Eggins, S., *et al.* 2012. Confirmation of a late middle Pleistocene age for the Omo Kibish 1 cranium by direct uranium-series dating. *J. Hum. Evol.* 63: 704–710.

Backwell, L.R. & d'Errico, F. 2001. Evidence of termite foraging by Swartkrans early hominids. *Proc. Natl. Acad. Sci. U. S. A.* 98: 1358–1363.

Barham, L. 2010. A technological fix for "Dunbar's dilemma"? *Proc. Br. Acad.* 158: 367-389+514-515.

Barham, L. 2002. Backed tools in Middle Pleistocene central Africa and their evolutionary significance. *J. Hum. Evol.* 43: 585–603.

Barham, L.S. 1998. Possible early pigment use in South-Central Africa. *Curr. Anthropol.* 39: 703–710.

Barham, L.S. & Smart, P.L. 1996. An early date for the Middle Stone Age of central Zambia. *J. Hum. Evol.* 30: 287–290.

Barnard, P.J., Duke, D.J., Byrne, R.W. & Davidson, I. 2007. Differentiation in cognitive and emotional meanings: An evolutionary analysis. *Cogn. Emot.* 21: 1155–1183.

Barnard, P.J. & Teasdale, J.D. 1991. Interacting cognitive subsystems: A systemic approach to cognitive-affective interaction and change. *Cogn. Emot.* 5: 1–39.

Barton, R.N.E., Bouzouggar, A., Collcutt, S.N., Schwenninger, J.L. & Clark-Balzan, L. 2009. OSL dating of the Aterian levels at Dar es-Soltan I (Rabat, Morocco) and implications for the dispersal of modern Homo sapiens. *Quat. Sci. Rev.* 28: 1914–1931.

Berger, L.R., Hawks, J., de Ruiter, D.J., Churchill, S.E., Schmid, P., Delezene, L.K., *et al.* 2015. Homo naledi, a new species of the genus Homo from the Dinaledi Chamber, South Africa. *Elife* 4: e09560. eLife Sciences Publications Limited.

Berwick, R.C., Friederici, A.D., Chomsky, N. & Bolhuis, J.J. 2013. Evolution, brain, and the nature of language. *Trends Cogn. Sci.* 17: 98.

Binford, L.R. 1981. *Bones: Ancient Men and Modern Myths.* Academic Press, New York & London.

Boëda, E. 1995. Levallois: a volumetric construction, methods, a technique. In: *The definition and interpretation of Levallois technology* (H. Dibble &

O. Bar-Yosef, eds), pp. 41–68. Prehistory Press, Madison, WI.

Botha, R.P. 2000. On the role of bridge theories in accounts of the evolution of human language. *Lang. Commun.* 21: 61–71.

Bouzouggar, A., Barton, N., Vanhaeren, M., d'Errico, F., Collcutt, S., HIGHAM, T.T.F.G., *et al.* 2007. 82,000-year-old shell beads from North Africa and implications for the origins of modern human behavior. *Proc. Natl. Acad. Sci.* 104: 9964–9969.

Boyd, R. & Richerson, P.J. 1996. Why culture is common, but cultural evolution is rare. *Proc. Br. Acad.* 88: 77–94.

Brown, K.S., Marean, C.W., Jacobs, Z., Schoville, B.J., Oestmo, S., Fisher, E.C., *et al.* 2012. An early and enduring advanced technology originating 71,000 years ago in South Africa. *Nature* 491: 590–593.

Buchsbaum, D., Bridgers, S., Skolnick Weisberg, D. & Gopnik, A. 2012. The power of possibility: causal learning, counterfactual reasoning, and pretend play. *Philos. Trans. R. Soc. Lond. B. Biol. Sci.* 367: 2202–12.

Byrne, R.W. 2002. Imitation of novel complex actions: what does the evidence from animals mean? *Adv. Study Behav.* 31: 77–105.

Clark, G.A. 1989. Alternative models of Pleistocene biocultural evolution: a response to Foley. *Antiquity* 63: 153–162.

Clark, J.D. 1992. African and Asian Perspectives on the Origins of Modern Humans. *Philos. Trans. R. Soc. Ser. B.* 337: 148–178.

Clark, J.D. 1965. The Later Pleistocene Cultures of Africa. *Science* 150: 833–847.

Clark, J.D., Beyene, Y., WoldeGabriel, G., Hart, W.K., Renne, P.R., Gilbert, H., *et al.* 2003. Stratigraphic, chronological and behavioural contexts of Pleistocene Homo sapiens from Middle Awash, Ethiopia. *Nature* 423: 747–752.

Clark, J.L. & Plug, I. 2008. Animal exploitation strategies during the South African Middle Stone Age: Howiesons Poort and post-Howiesons Poort fauna from Sibudu Cave. *J. Hum. Evol.* 54: 886–898.

Clarkson, C., Jacobs, Z., Marwick, B., Fullagar, R., Wallis, L., Smith, M., *et al.* 2017. Human occupation of northern Australia by 65,000 years ago. *Nature* 547: 306–310.

Cole, S. 1954. The prehistory of East Africa. *Am. Anthropol.* 56: 1026–1050.

Coolidge, F.L., Wynn, T. & Overmann, K.A. 2013. The evolution of working memory. In: *Working Memory: The Connected Intelligence* (T. P. Alloway & R. G. Alloway, eds), pp. 37–60. Psychology Press, New York.

Corballis, M.C. (ed). 2011. *The Recursive Mind: The Origins of Human Language, Thought, and Civilization.* Princeton University Press, Princeton, NJ:

Crompton, R.H. & Gowlett, J.A.J. 1993. Allometry and Multidimensional Form in Acheulean Bifaces from Kilombe, Kenya. *J. Hum. Evol.* 25: 175–199.

d'Errico, F. & Backwell, L. 2016. Earliest evidence of personal ornaments associated with burial: The Conus shells from Border Cave. *J. Hum. Evol.* 93: 91–108.

D'Errico, F., Backwell, L., Villa, P., Degano, I., Lucejko, J.J., Bamford, M.K., *et al.* 2012a. Early evidence of San material culture represented by organic artifacts from Border Cave, South Africa. *Proc. Natl. Acad. Sci.* 109: 13214–9.

D'Errico, F. & Banks, W.E. 2015. The Archaeology of Teaching: A Conceptual Framework. *Cambridge Archaeol. J.* 25: 859–866.

D'Errico, F., García Moreno, R. & Rifkin, R.F. 2012b. Technological, elemental and colorimetric analysis of an engraved ochre fragment from the Middle Stone Age levels of Klasies River Cave 1, South Africa. *J. Archaeol. Sci.* 39: 942–952.

D'Errico, F., Henshilwood, C. & Nilssen, P.J. 2001. An engraved bone fragment from c. 70,000-year-old Middle Stone Age levels at Blombos Cave, South Africa : implications for the origin of symbolism and language. *Antiquity* 75: 309–318.

d'Errico, F., Vanhaeren, M. & Wadley, L. 2008. Possible shell beads from the Middle Stone Age layers of Sibudu Cave, South Africa. *J. Archaeol. Sci.* 35: 2675–2685.

d'Errico, F., Zilhao, J., Julien, M., Baffier, D. & Pelegrin, J. 1998. Neanderthal acculturation in western Europe? A critical review of the evidence and its interpretation. *Curr. Anthropol.* 39: S1–S44.

Davidson, I. & Noble, W. 1991. The evolutionary emergence of modern human behaviour: language and its archaeology. *Man* 26: 223–254.

De La Torre, I., Mora, R., Arroyo, A. & Benito-Calvo, A. 2014. Acheulean technological behaviour in the Middle Pleistocene landscape of Mieso (East-Central Ethiopia). *J. Hum. Evol.* 76: 1–25.

Deino, A.L. & McBrearty, S. 2002. 40Ar/39Ar dating of the Kapthurin Formation, Baringo, Kenya. *J. Hum. Evol.* 42: 185–210.

Diez-Martin, F., Domínguez-Rodrigo, M., Sánchez, P., Mabulla, A.Z.P., Tarriño, A., Barba, R., *et al.* 2009. The Middle to Later Stone Age technological transition in East Africa. New data from Mumba Rockshelter Bed V (Tanzania) and their implications for the origin of modern human behavior. *J. African Archaeol.* 7: 147–173.

Dirks, P.H., Roberts, E.M., Hilbert-Wolf, H., Kramers, J.D., Hawks, J., Dosseto, A., *et al.* 2017a. The age of *Homo naledi* and associated sediments in the Rising Star Cave, South Africa. *Elife* 6: 1–59.

Dirks, P.H.G.M., Berger, L.R., Roberts, E.M., Kramers, J.D., Hawks, J., Randolph-Quinney, P.S., *et al.* 2015. Geological and taphonomic context for the new hominin species Homo naledi from the Dinaledi Chamber, South Africa. *Elife* 4.

Dirks, P.H.G.M., Roberts, E.M., Hilbert-Wolf, H., Kramers, J.D., Hawks, J., Dosseto, A., *et al.* 2017b. The age of homo naledi and associated sediments in the rising star cave, South Africa. *Elife* 6.

Duller, G.A.T., Tooth, S., Barham, L. & Tsukamoto, S. 2015. New investigations at Kalambo Falls, Zambia: Luminescence chronology, site formation, and archaeological significance. *J. Hum. Evol.* 85: 111–125.

Dunbar, R., Gamble, C. & Gowlett, J. 2010. The social brain and the distributed mind. *Proc. Br. Acad.* 158: 3–15.

Feathers, J. 2015. Luminescence dating at Diepkloof Rock Shelter - new dates from single-grain quartz. *J. Archaeol. Sci.* 63: 164–174.

Fleagle, J.G.J.G., Assefa, Z., Brown, F.H.F.H. & Shea, J.J.J.J. 2008. Paleoanthropology of the Kibish Formation, southern Ethiopia: Introduction. *J. Hum. Evol.* 55: 360–365.

Foley, R.A. 1991. How useful is the culture concept in early human studies. In: *The origins of human behaviour* (R. A. Foley, ed), pp. 25–38. Unwin Heinmann, London.

Foley, R.A. 1996. Measuring the cognition of extinct hominids. In: *Modelling the Early Human Mind* (P. Mellars & K. Gibson, eds), pp. 57–66. MacDonald Institute for Archaeological Research Monograph, Cambridge.

Foley, R.A. & Mirazon Lahr, M. 1997. Mode 3 technologies and the evoltion of modern humans. *Cambridge Archaeol. J.* 7: 3–36.

Frith, C.D. 2012. The role of metacognition in human social interactions. *Philos. Trans. R. Soc. B Biol. Sci.* 367: 2213–2223.

Gamble, C., Gowlett, J. & Dunbar, R. 2011. The social brain and the shape of the palaeolithic. *Cambridge Archaeol. J.* 21: 115–135.

Gliganic, L.A., Jacobs, Z., Roberts, R.G., Domínguez-Rodrigo, M. & Mabulla, A.Z.P.P. 2012. New ages for Middle and Later Stone Age deposits at Mumba rockshelter, Tanzania: Optically stimulated luminescence dating of quartz and feldspar grains. *J. Hum. Evol.* 62: 533–547.

Goodwin, A. & Lowe C. van Riet. 1929. The stone age cultures of South Africa. *Ann. South African Museum* 27: 1–289.

Gowlett, J.A.J. 2010. Firing Up the Social Brain. *Proc. Br. Acad.* 158: 341–366.

Gowlett, J.A.J. 1984. Mental abilities of early man: a look at some hard eveidence. In: *Hominid evolution and comunity ecology* (R. A. Foley, ed), pp. 167–192. Academic Press, New York & London.

Gowlett, J.A.J. 2016. The discovery of fire by humans: a long and convoluted process. *Philos. Trans. R. Soc. B Biol. Sci.* 371: 20150164.

Gowlett, J.A.J. 2014. The Elements of Design Form in Acheulean Bifaces. In: *Lucy to Language: The Benchmark Papers* (R. I. M. Dunbar, C. Gamble, & J. A. J. Gowlett, eds), pp. 409–426. Oxford University Press, Oxford.

Gowlett, J.A.J. 2009. The longest transition or multiple revolutions? Curves and steps in the record of human origins. In: *A Sourcebook of Palaeolithic Transitions: Methods, Theories and Interpretations* (M. Camps & P. Chauhan, eds), pp. 65–78. Springer Verlag, Berlin.

Gowlett, J.A.J. 2015. Variability in an early hominin percussive tradition: the Acheulean versus cultural variation in modern chimpanzee artefacts. *Philos. Trans. R. Soc. B Biol. Sci.* 370: 20140358.

Gowlett, J.A.J. & Crompton, R.H. 1995. Kariandusi: Acheulean morphology and the question of allometry. *African Archaeol. Rev.* 12: 3–42.

Groucutt, H.S., White, T.S., Clark-Balzan, L., Parton, A., Crassard, R., Shipton, C., *et al.* 2015. Human occupation of the Arabian Empty Quarter during MIS 5: evidence from Mundafan Al-Buhayrah, Saudi Arabia. *Quat. Sci. Rev.* 119: 116–135.

Grün, R. 2006. Direct dating of human fossils. *Yearb. Phys. Anthropol.* 49: 2–48.

Hammer, M.F., Woerner, A.E., Mendez, F.L., Watkins, J.C. & Wall, J.D. 2011. Genetic evidence for archaic admixture in Africa. *Proc. Natl. Acad. Sci.* 108: 15123–15128.

Harvati, K., Stringer, C., Grün, R., Aubert, M., Allsworth-Jones, P. & Folorunso, C.A. 2011. The Later Stone Age Calvaria from Iwo Eleru, Nigeria: Morphology and Chronology. *PLoS One* 6: e24024.

Hauser, M. 2008. *The seeds of humanity.* Princeton University Press, Princeton, N.J.

Henshilwood, C. 2004. Middle Stone Age Shell Beads from South Africa. *Science* 304: 404–404.

Henshilwood, C.S. 2002. Emergence of Modern Human Behavior: Middle Stone Age Engravings from South Africa. *Science* 295: 1278–1280.

Henshilwood, C.S., d'Errico, F. & Watts, I. 2009. Engraved ochres from the Middle Stone Age levels at Blombos Cave, South Africa. *J. Hum. Evol.* 57: 27–47.

Heyes, C. 2012. Grist and mills: on the cultural origins of cultural learning. *Philos. Trans. R. Soc. B Biol. Sci.* 367: 2181–2191.

Hoffmann, D.L., Standish, C.D., García-Diez, M., Pettitt, P.B., Milton, J.A., Zilhão, J., *et al.* n.d. U-Th dating of carbonate crusts reveals Neandertal origin of Iberian cave art. , doi: 10.1126/science.aap7778.

Hublin, J.J., Ben-Ncer, A., Bailey, S.E., Freidline, S.E., Neubauer, S., Skinner, M.M., *et al.* 2017. New fossils from Jebel Irhoud, Morocco and the pan-African origin of Homo sapiens. *Nature* 546: 289–292.

Jacobs, Z., Roberts, R., Galbraith, R. & Deacon, H. 2008a. Ages for the Middle Stone Age of southern Africa: Implications for human behavior and dispersal. *Science* 322: 733–735.

Jacobs, Z. & Roberts, R.G. 2017. Single-grain OSL chronologies for the Still Bay and Howieson's Poort industries and the transition between them: Further analyses and statistical modelling. *J. Hum. Evol.* 107: 1–13.

Jacobs, Z., Wintle, A.G., Duller, G.A.T., Roberts, R.G. & Wadley, L. 2008b. New ages for the post-Howiesons Poort, late and final Middle Stone Age at Sibudu, South Africa. *J. Archaeol. Sci.* 35: 1790–1807.

Joordens, J.C.A.A., D'Errico, F., Wesselingh, F.P., Munro, S., de Vos, J., Wallinga, J., *et al.* 2015. Homo erectus at Trinil on Java used shells for tool production and engraving. *Nature* 518: 228–231.

Klein, R.G. 1992. The archaeology of modern human origins. *Evol. Anthropol.* 1: 5–14.

Klein, R.G. 2009. *The Human Career.* Chicago University Press, Chicago.

Krause, J., Lalueza-Fox, C., Orlando, L., Enard, W., Green, R.E., Burbano, H.A., *et al.* 2007. The Derived FOXP2 Variant of Modern Humans Was Shared with Neandertals. *Curr. Biol.* 17: 1908–1912.

Kuman, K. 2001. An Acheulean factory site with prepared core technology near Taung, South Africa. *South African Archaeol. Bull.* 56: 8–22.

Lachance, J., Vernot, B., Elbers, C.C.C., Ferwerda, B., Froment, A., Bodo, J.-M., *et al.* 2012. Evolutionary history and adaptation from high-coverage whole-genome sequences of diverse African hunter-gatherers. *Cell* 150: 457–469.

Leakey, L. 1931. *The stone age cultures of Kenya.* Cambridge University Press, Cambridge.

Leplongeon, A. 2014. Microliths in the Middle and Later Stone Age of eastern Africa: New data from Porc-Epic and Goda Buticha cave sites, Ethiopia. *Quat. Int.* 343: 100–116.

Li, H., Kuman, K., Lotter, M.G., Leader, G.M. & Gibbon, R.J. 2017. The Victoria West: earliest prepared core technology in the Acheulean at Canteen Kopje and implications for the cognitive evolution of early hominids. *R. Soc. open Sci.* 4: 170288. The Royal Society.

Lombard, M. 2007. The gripping nature of ochre: The association of ochre with Howiesons Poort adhesives and Later Stone Age mastics from South Africa. *J. Hum. Evol.* 53: 406–419.

Malaspinas, A.-S., Westaway, M.C., Muller, C., Sousa, V.C., Lao, O., Alves, I., *et al.* 2016. A genomic history of Aboriginal Australia. *Nature* 538: 207–214. Nature Research.

Mallick, S., Li, H., Lipson, M., Mathieson, I., Gymrek, M., Racimo, F., *et al.* 2016. The Simons Genome Diversity Project: 300 genomes from 142 diverse populations. *Nature* 538: 201–206. Nature Publishing Group.

Marean, C.W. 2016. The transition to foraging for dense and predictable resources and its impact on the evolution of modern humans. *Philos. Trans. R. Soc. Lond. B. Biol. Sci.* 371: 160–169.

Marean, C.W., Bar-Matthews, M., Bernatchez, J., Fisher, E., Goldberg, P., Herries, A.I.R., *et al.* 2007. Early human use of marine resources and pigment in South Africa during the Middle Pleistocene. *Nature* 449: 905–908.

McDougall, I., Brown, F.H. & Fleagle, J.G. 2005. Stratigraphic placement and age of modern humans from Kibish, Ethiopia. *Nature* 433: 733–736.

McNabb, J. & Cole, J. 2015. The mirror cracked: Symmetry and refinement in the Acheulean handaxe. *J. Archaeol. Sci. Reports* 3: 100–111.

Mellars, P. 1991. Cognitive changes and the emergence of modern humans in Europe. *Cambridge Archaeol. J.* 1: 63–76.

Mellars, P. 2005. The impossible coincidence. A single-species model for the origins of modern human behavior in Europe. *Evol. Anthropol.* 14: 12–27.

Mendez, F.L., Poznik, G.D., Castellano, S. & Bustamante, C.D. 2016. The divergence of Neandertal and Modern Human Y chromosomes. *Am. J. Hum. Genet.* 98: 728–734.

Meyer, M., Arsuaga, J.-L.L., de Filippo, C., Nagel, S., Aximu-Petri, A., Nickel, B., *et al.* 2016. Nuclear DNA sequences from the Middle Pleistocene Sima de los Huesos hominins. *Nature* 531: 504–507.

Meyer, M., Kircher, M., Gansauge, M.-T.M.-T.M.-T.T., Li, H., Racimo, F., Mallick, S., *et al.* 2012. A High-Coverage Genome Sequence from an Archaic Denisovan Individual. *Science* 338: 222–226.

Mirazón Lahr, M. 2016. The shaping of human diversity: filters , boundaries and transitions. *Philos. Trans. R. Soc. London* 371: 1–12.

Mirazon Lahr, M. & Foley, R.A. 2003. Demography, dispersals and human evolution in the Last Glacial Period. In: (T. H. van Andel & W. Davies, eds), pp. 241–256. McDonald Institute for Archaeological Research, Cambridge.

Mirazon Lahr, M. & Foley, R.A. 2001. Mode 3, Homo helmei, and the pattern of human evolution in the Middle Pleistocene. In: *Human Roots: Africa and Asia in the Middle Pleistocene* (L. Barham & K. Robson Brown, eds), pp. 23–39. Western Academic & Specialist Press, Bristol.

Mirazón Lahr, M. & Foley, R.A. 2016. Human evolution in Late Quaternary Eastern Africa. In: *Africa from MIS 6-2: Population Dynamics and Paleoenvironments* (S. C. Jones & B. A. Stewart, eds), pp. 215–231. Springer, Dordrecht.

Mithen, S. 1996. *The prehistory of the mind: a search for the origins of art, religion and science.* Thames and Hudson, London.

Mithen, S. 1997. The prehistory of the mind. *Cambridge Archaeol. J.*

Morgan, T.J.H., Uomini, N.T., Rendell, L.E., Chouinard-Thuly, L., Street, S.E., Lewis, H.M., *et al.* 2015. Experimental evidence for the co-evolution of hominin tool-making teaching and language. *Nat. Commun.* 6: 6029.

Mounier, A. & Mirazón Lahr, M. 2016. Virtual ancestor reconstruction: Revealing the ancestor of modern humans and Neandertals. *J. Hum. Evol.* 91: 57–72.

Mourre, V., Villa, P. & Henshilwood, C.S. 2010. Early Use of Pressure Flaking on Lithic Artifacts at Blombos Cave, South Africa. *Science* 330: 659–662.

Nowell, A. & White, M. 2010. Growing Up in the Middle Pleistocene: Life History Strategies and Their Relationship to Acheulian Industries. In: *Stone Tools and the Evolution of Human Cognition* (I. Davidson & A. Nowell, eds), pp. 67–82. Cambridge, Cambridge.

Osypiński, P. & Osypińska, M. 2016. Optimal adjustment or cultural backwardness? New data on the latest Levallois industries in the Nile Valley. *Quat. Int.* 408: 90–105.

Pagani, L., John Lawson, D., Jagoda, E., Mörseburg, A., Eriksson, A., Mitt, M., *et al.* 2016. Genomic analyses inform on migration events during the peopling of Eurasia. *Nat. Publ. Gr.* 538: 238–242.

Palframan, W.J., Meehl, J.B., Jaspersen, S.L., Winey, M. & Murray, A.W. 2006. Anaphase inactivation of the spindle checkpoint. *Science (80-.).* 313: 680–684.

Pienaar, M., Woodborne, S. & Wadley, L. 2008. Optically stimulated luminescence dating at Rose Cottage Cave. *S. Afr. J. Sci.* 104: 65–70.

Pleurdeau, D., Hovers, E., Assefa, Z., Asrat, A., Pearson, O., Bahain, J.-J., *et al.* 2014. Cultural change or continuity in the late MSA/Early LSA of southeastern Ethiopia? The site of Goda Buticha, Dire Dawa area. *Quat. Int.* 343: 117–135.

Porat, N., Chazan, M., Schwarcz, H. & Horwitz, L.K. 2002. Timing of the Lower to Middle Paleolithic boundary: new dates from the Levant. *J. Hum. Evol.* 43: 107–122.

Posth, C., Wißing, C., Kitagawa, K., Pagani, L., van Holstein, L., Racimo, F., *et al.* 2017. Deeply divergent archaic mitochondrial genome provides lower time boundary for African gene flow into Neanderthals. *Nat. Commun.* 8: 16046.

Potts, R., Potts, R., Behrensmeyer, A.K., Faith, J.T., Tryon, C.A., Brooks, A.S., *et al.* 2018. Environmental dynamics during the onset of the Middle Stone Age in eastern Africa. 2200: 1–8.

Prüfer, K., Racimo, F., Patterson, N., Jay, F., Sankararaman, S., Sawyer, S., *et al.* 2014. The complete genome sequence of a Neanderthal from the Altai Mountains. *Nature* 505: 43–49.

Richerson, P.J. & Boyd, R. 2005. *Not by Genes Alone: How Culture Transformed Human Evolution.* University of Chicago Press, Chicago.

Richter, D., Grün, R., Joannes-Boyau, R., Steele, T.E., Amani, F., Rué, M., *et al.* 2017. The age of the hominin fossils from Jebel Irhoud, Morocco, and the origins of the Middle Stone Age. *Nature* 546: 293–296.

Rodríguez-Vidal, J., d'Errico, F., Pacheco, F.G., Blasco, R., Rosell, J., Jennings, R.P., *et al.* 2014. A rock engraving made by Neanderthals in Gibraltar.

Roebroeks, W., Sier, M.J., Nielsen, T.K., De Loecker, D., Pares, J.M., Arps, C.E.S., *et al.* 2012. Use of red ochre by early Neanderthals. *Proc. Natl. Acad. Sci.* 109: 1889–1894.

Rose, J.I., Usik, V.I., Marks, A.E., Hilbert, Y.H., Galletti, C.S., Parton, A., *et al.* 2011. The Nubian complex of Dhofar, Oman: An African Middle Stone Age industry in Southern Arabia. *PLoS One* 6: e28239.

Sahle, Y., Morgan, L.E., Braun, D.R., Atnafu, B. & Hutchings, W.K. 2014. Chronological and behavioral contexts of the earliest middle stone age in the gademotta formation, main ethiopian rift. *Quat. Int.* 331: 6–19.

Scerri, E. 2017. The Stone Age Archaeology of West Africa. *Oxford Res. Encycl. African Hist.* 1–37.

Scerri, E.M.L. 2012. The Aterian and its place in the North African Middle Stone Age. *Quat. Int.* 1–20.

Scerri, E.M.L., Blinkhorn, J., Niang, K., Bateman, M.D. & Groucutt, H.S. 2017. Persistence of Middle Stone Age technology to the Pleistocene/Holocene transition supports a complex hominin evolutionary scenario in West Africa. *J. Archaeol. Sci. Reports* 11: 639–646.

Scerri, E.M.L., Drake, N.A., Jennings, R. & Groucutt, H.S. 2014. Earliest evidence for the structure of Homo sapiens populations in Africa. *Quat. Sci. Rev.* 101: 207–216.

Schlebusch, C.M., Malmström, H., Günther, T., Sjödin, P., Coutinho, A., Edlund, H., *et al.* 2017. Southern African ancient genomes estimate modern human divergence to 350,000 to 260,000 years ago. *Science* 358: 652–655.

Sharon, G. & Beaumont, P.B. 2006. 2006. Victoria West: a highly standardized prepared core technology. In: *Axe age: Acheulian tool making from quarry to discard* (N. Goren-Inbar & G. Sharon, eds), pp. 181–199. Equinox, London, UK.

Shea, J.J. 2008. The Middle Stone Age archaeology of the Lower Omo Valley Kibish Formation: Excavations, lithic assemblages, and inferred patterns of early Homo sapiens behavior. *J. Hum. Evol.* 55: 448–485.

Skoglund, P., Thompson, J.C., Prendergast, M.E., Mittnik, A., Sirak, K., Hajdinjak, M., *et al.* 2017. Reconstructing Prehistoric African Population Structure. *Cell* 171: 59–71.e21.

Texier, P.-J.J., Porraz, G., Parkington, J., Rigaud, J.-P.P., Poggenpoel, C., Miller, C., *et al.* 2010. A Howiesons Poort tradition of engraving ostrich eggshell containers dated to 60,000 years ago at Diepkloof Rock Shelter, South Africa. *Proc. Natl. Acad. Sci. U. S. A.* 107: 6180–6185.

Tomasello, M. 1999. The human adaptation for culture. *Annu. Rev. Anthropol.* 28: 509–529.

Tribolo, C., Mercier, N., Douville, E., Joron, J.L., Reyss, J.L., Rufer, D., *et al.* 2013. OSL and TL dating of the Middle Stone Age sequence at Diepkloof Rock Shelter (South Africa): A clarification. *J. Archaeol. Sci.* 40: 3401–3411.

Tryon, C.A. 2006. "Early" Middle Stone Age Lithic Technology of the Kapthurin Formation (Kenya). *Curr. Anthropol.* 47: 367–375.

Tryon, C.A., McBrearty, S. & Texier, P.J. 2005. Levallois lithic technology from the Kapthurin Formation, Kenya: Acheulian origin and Middle Stone Age diversity. *African Archaeol. Rev.* 22: 199–229.

van Peer, P. 1992. *The Levallois Reduction Strategy*. Prehistory Press, Madison.

Van Peer, P., Fullagar, R., Stokes, S., Bailey, R., Moeyersons, J., Steenhoudt, F., *et al.* 2003. The Early to Middle Stone Age Transition and the Emergence of Modern Human Behaviour at site 8-B-11, Sai Island, Sudan. *J. Hum. Evol.* 45: 187–193.

Vanhaeren, M. & d'Errico, F. 2006. Aurignacian ethno-linguistic geography of Europe revealed by personal ornaments. *J. Archaeol. Sci.* 33: 1105–1128.

Vanhaeren, M., D'Errico, F., Stringar, C., Jamas, S.L., Todd, J.A., Mienis, H.K., *et al.* 2006. Middle Paleolithic Shell Beads in Israel and Algeria. *Science (80-.).* 312: 1785–1788. American Association for the Advancement of Science.

Villa, P., Soriano, S., Tsanova, T., Degano, I., HIGHAM, T.F.G., D'Errico, F., *et al.* 2012. Border Cave and the beginning of the Later Stone Age in South Africa. *Proc Nat Acad Sci USA* 109: 13208–13213.

Wadley, L. 2010a. Compound-Adhesive Manufacture as a Behavioral Proxy for Complex Cognition in the Middle Stone Age. *Curr. Anthropol.* 51: S111–S119.

Wadley, L. 2013. Recognizing Complex Cognition through Innovative Technology in Stone Age and Palaeolithic Sites. *Cambridge Archaeol. J. Cambridge Archaeol. J. 163183 Cambridge Archaeol. J.* 23: 163–83.

Wadley, L. 2010b. Were snares and traps used in the Middle Stone Age and does it matter? A review and a case study from Sibudu, South Africa. *J. Hum. Evol.* 58: 179–192.

Watts, I. 2002. Ochre in the Middle Stone Age of Southern Africa: ritualised display or hide preservative. *Source South African Archaeol. Bull.* 57: 1–14.

White, T.D.T., Asfaw, B., DeGusta, D., Gilbert, H., Richards, G.D.G., Suwa, G., *et al.* 2003. Pleistocene Homo sapiens from Middle Awash, Ethiopia. *Nature* 423: 742–747.

Whiten, A. 1991. *Natural theories of mind: evolution, development, and simulation of everyday mindreading.* Wiley Blackwell, London.

Whiten, A. & Schaik, C.P. van. 2007. The evolution of animal 'cultures' and social intelligence. *Philos. Trans. R. Soc. London B Biol. Sci.* 362: 603–620.

Wilkins, J., Pollarolo, L. & Kuman, K. 2010. Prepared core reduction at the site of Kudu Koppie in northern South Africa: Temporal patterns across the Earlier and Middle Stone Age boundary. *J. Archaeol. Sci.* 37: 1279–1292.

Wilkins, J., Schoville, B.J., Brown, K.S. & Chazan, M. 2012. Evidence for Early Hafted Hunting Technology. *Science* 338: 942–946.

Wynn, T. 2009. Hafted spears and the archaeology of mind. *Proc. Natl. Acad. Sci. U. S. A.* 106: 9544–9545.

Wynn, T. & Coolidge, F.L. 2006. The effect of enhanced working memory on language. *J. Hum. Evol.* 50: 230–231.

Wynn, T. & Coolidge, F.L. 2011. The Implications of the Working Memory Model for the Evolution of Modern Cognition. *Int. J. Evol. Biol.* 2011: 1–12. Hindawi.

Zilhao, J. 2006. Neandertals and moderns mixed, and it matters. *Evol. Anthropol.* 15: 183–195.

Robert A. Foley
Leverhulme Centre for Human Evolutionary Studies
Department of Archaeology
University of Cambridge

Marta Mirazón Lahr
Leverhulme Centre for Human Evolutionary Studies
Department of Archaeology
University of Cambridge

Initial Source Evaluation of Archaeological Obsidian from Middle Stone Age Site Kilombe GqJh3 West 200, Kenya, East Africa

Sally Hoare, Stephen Rucina and John A.J. Gowlett

Abstract

The new Middle Stone Age locality of Kilombe GqJh3 West 200 lies approximately 2 km west of the main Acheulean site of Kilombe. The site, which was first excavated in 2012, comprises a large surface collection of long obsidian points and several in situ MSA obsidian flake tools. The latter lie at levels within and overlying a primary tuff currently dated by $^{40}Ar/^{39}Ar$ to ca. 120,000 Ka. This paper presents the preliminary results of a small pilot study (n 19 artefacts) using portable XRF (Niton Gold 3) to determine the sources and hence transport distances of the raw materials present in this assemblage. PXRF has proven a useful tool in determining sources of obsidian on archaeological sites based on the use of the trace elements Rb, Zn, Sr, Y, Zr, and Nb. Preliminary results here suggest a source north of Lake Naivasha for part of the surface collection and in situ flake tools. Other sources are apparent within this data set but are more difficult to determine using the PXRF method. These data suggest the long-distance transportation of obsidian over 80 km away from north Naivasha to Kilombe at ca. 120,000 Ka.

Introduction

This paper reports initial sourcing studies on obsidian artefacts from a new Middle Stone Age site on the southern flanks of Kilombe volcano in central Kenya. The results are preliminary, but the study has yielded important new information and will justify further efforts to refine the dating and characterisation of the site, as well as to trace other sources.

The late Middle and Late Pleistocene of East Africa are important periods for understanding the evolution of modern humans. The period charts a succession from Acheulean to Middle Stone Age, but recent finds and new dating have stressed the complexity of associations between hominin species and archaeological industries. The approximate contemporaneity between *Homo naledi* in South Africa and early *Homo sapiens* at Jebel Irhoud in North Africa, at around 300,000 ka, emphasises this point (Berger *et al.* 2017; Hublin *et al.* 2017; Richter *et al.* 2017). The earliest fossils found in an East African context are dated to ca. 195 ka from Omo-Kibish and from ca. 160 ka from Herto (Clark *et al.* 2003; White *et al.* 2003; McDougall *et al.* 2005). The Middle Stone Age (MSA)

in Africa is well known for documenting increased technological diversity such as the appearance of hafted tools (Barham 2013) and other markers of behavioural complexity (McBrearty and Brooks, 2000). Larger hand-held tools such as Acheulean handaxes were generally replaced by a much smaller and more diverse tool kit including smaller cores and flakes, points and hafted tools (McBrearty 1999; Johnson and McBrearty 2010; Barham, 2013), but the process appears to occur in mosaic fashion, with both early appearances of MSA and late appearances of Acheulean (McBrearty, 2003; Wilkins and Chazan 2012; Wilkins *et al.* 2012). Along with appearance of more diverse technologies the MSA also has been widely attested to document changes in patterns of mobility, subsistence strategies and symbolic behaviours (Ambrose 2001, 2002; McBrearty and Brooks 2000; Tryon & Faith, 2013).

Raw material provenance studies can be used to investigate long-distance contact and/or transport of lithic raw materials, which arguably give documentation to some of the main aspects and trends concerning the emergence of modern human behaviour (Féblot-Augustins 1999; McBrearty and Brooks 2000; Moutsiou 2014). Characterization through geochemical analysis of lithics and their potential geological sources is one long-known approach that can be taken to investigate long distance contact and/or transport lithic raw materials (Merrick and Brown 1984; Feblot-Augustins 1999). In recent years raw material provenance studies of archaeological obsidians have received an increasing amount of attention across Africa (Negash and Shackley, 2006; Nash *et al.*, 2013; Brown *et al.*, 2013; Zipkin *et al.* 2015; Blegen 2017; Shackley and Sahle, 2017). The geochemical composition of obsidian from a single eruptive event is normally highly distinctive, enabling the attribution of lithic artefacts to a specific volcanic source based on similarity in their geochemical compositions (Glascock *et al.* 1998; Negash *et al.* 2006).

In Kenya numbers of studies demonstrate the utility of elemental analysis on archaeological obsidians in this regard (Merrick and Brown 1984; Brown *et al.* 2013). Geological sources of obsidian in East Africa tend to be restricted geographically and can therefore be geochemically finger-printed in relation to the sourcing of artefacts (Merrick & Brown, 1984; Merrick

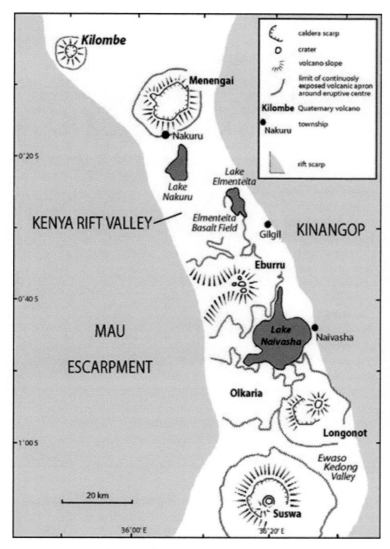

Figure 1. Map of the Central Rift Valley showing the site of Kilombe, and Eburru, source of MSA group 3 artefacts.

Materials and methods

MSA Site 200

The new Middle Stone Age locality of Kilombe GqJh3 West 200 lies approximately 2 km west of the main Achuelean site of Kilombe, in the Central Rift Valley, Kenya (Figures 1, 2a and b). The site was discovered during field walking in July 2011 (in the British Academy-supported Mobility and Links project between University of Liverpool and National Museums of Kenya); and small test trenches where excavated during the 2012 and 2013 field seasons. The site which is described in Hoare *et al.* (in prep) lies above an unconformity with earlier Pleistocene sediments. Artefacts which had eroded out and mainly fallen down the exposure scarp were found on the surface include long MSA obsidian points and flakes. The step trenches established the origin of these finds, with several artefacts found in situ. The in-situ flakes were found directly above but also within a primary tuff, named the Kibberenge tuff, for which there is an $^{40}Ar/^{39}Ar$ date of 0.120 +/- 0.03 Ma (Hoare *et al.* in prep).

MSA 200 local stratigraphy

Figures 2 a and b show the local stratigraphy of MSA 200 which comprises five lithologically distinct units (MSA 200 Units 1-5, from the stratigraphically lowest to highest). Unit 1 is a brown clay, which is approximately 3 metres thick at this locality, and is stratigraphically the same unit as a brown clay (Unit 2) on Kilombe main site (Gqh1), where it records a palaeomagnetic excursion determined by Herries (in Gowlett *et al.* 2015) to be the Jaramillo reversal (0.990 Ma). Unit 1 is capped by Unit 2, the three-banded tuff (3BT), primary airfall tuff, also dated by $^{40}Ar/^{39}Ar$ to 1.03 +/- 0.05 Ma (Hoare *et al.* in prep). Above the 3BT is an unconformity which is overlain by 3 metres of later sediments, labelled as MSA 200 Units 3-5. Unit 3 is a light brown silty clay. Above Unit 4 is primary tuff MSA Unit 4, the Kibberenge tuff, which is dated by $^{40}Ar/^{39}Ar$ to 0.120 +/- 0.03 Ma. Unit 5 which contains in-situ artefacts is also a reddish brown silty-clay (Figure 2 a and b). Most artefacts lie directly above Unit 4 but one was found within the tuff itself.

Excavation of MSA site 200

Pilot excavations of MSA 200 were confined to two step trenches, following discovery of surface artefacts. Plans to conduct a larger excavation were interrupted by the

et al. 1994; Brown *et al.* 2013). Obsidian artefacts are rarely found in Acheulean contexts in the Early and Middle Pleistocene (Kariandusi is a rare exception: Merrick and Brown 1984; Gowlett and Crompton 1994) and studies on transport distances indicate averages of little more than 30 km (Negash *et al.* 2006; Merrick *et al.* 1994; Merrick & Brown, 1984). The MSA in East Africa documents an increase in the long-distance transportation of obsidian as a lithic raw material during both the Middle and Late Pleistocene (Ambrose 2002). Distances of 250 km are documented from several east African sites including Karungu and Porc Epic (Negash & Shackley, 2006; Faith *et al.* 2015). The most comprehensive and published reference resource characterising the elemental compositions of obsidian sources in Kenya comes from Brown *et al.* 2013 who provide data from 90 localities for the Northern, Central and Southern Kenyan Rift for 84 compositionally distinct groups using electron microprobe, XRF and LA-ICP-MS.

Figure 2A. The local stratigraphy for MSA site 200.
Unit 1 - brown claystone, Unit 2 - three-banded tuff (3BT), Unit 3 - light brown silty clay, Unit 4 - the Kibberenge tuff and Unit 5, reddish brown silty clay. The in-situ MSA artefacts lie directly above and within the Kibberenge Tuff.

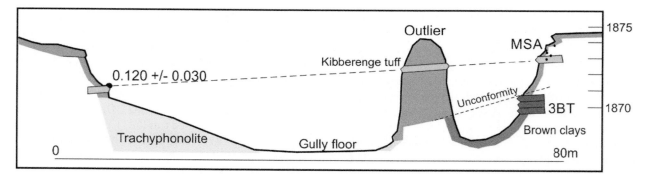

Figure 2B: Measured section of the GqJh3 West Gully from North to South (approximately), indicating position of the dating sample.

unforeseen construction of a road on and immediately west of the find site: subsequent erosion of bull-dozed spoil indicates that some of the site is still available for future work. The trenches were excavated by removal of 10 -15 cm spits right down through to Unit 1, in case any Acheulean material was present in the older sediments but no finds were made at the lower levels (a hand-axe was found on the surface less than 50 m away).

MSA 200 artefacts

A substantial surface collection of obsidian debitage and tools was made from the gully floor below the exposures, see Figure 3. It includes numbers of long MSA points as well as flakes and small debitage. The material is exclusively obsidian, in contrast to finds of MSA in older levels at nearby Moricho, where lava was also used for points.

analysis. The main aims were to identify whether more than once source was evident in the in situ and surface collections. Due to limitations, both of field time and of taking readings in the field we were not able to use the PXRF to analyse geological source materials, and have made comparisons based on the published data of Brown *et al.* (2013).

Results

Geochemical analysis MSA 200

The results of the geochemical analysis for characterisation of the artefacts and geological sources for MSA 200 are interesting. The geochemical composition of the obsidian artefacts is presented in Table 1 with one element, Fe (Fe_2O_3) expressed as wt%, and four as ppm (Nb, Zr, Sr and Rb). Table 1, and Figures 4 and 5 show the PXRF results of the elemental composition of the obsidian artefacts for MSA 200. Three groups of compositionally distinct obsidians were identified at MSA site 200 by PXRF (n19) from the surface and in situ collection and based on the bivariate plots of Zr, Nb and Rb, and Fe_2O_3/Rb. These are assigned to three groups labelled G1, G2 and G3 for ease of reference. Middle Stone Age group 3 is particularly distinctive based on exceptionally high levels of Zr, Nb and Rb.

Figure 3. Four obsidian points from MSA site 200. As the lightly weathered obsidian does not show detail in photographs this illustration has been prepared from casts in a lighter-coloured material. Group G3 shows closet affinity with MER70 from Mount Eburru.

Obsidian analysis

All artefacts were prepared for analysis by cleaning with de-ionised water. Only pieces > 5cm diameter were selected for analysis by X-Ray Fluorescence (XRF). A Niton XL3T 950 He GOLDD+ XRF Analyser was used, which generates X-rays via a miniaturised 50 kV, 200 µA tube with a silver anode. This instrument has a silicon drift detector with a resolution of better than 155 eV. The obsidian artefacts were analysed for trace elements (Nb, Zr, Sr and Rb), and the major element Fe via the oxide Fe_2O_3, using mining mode. Measurements were taken for 120s each, and each reading was repeated 3 times. Standard references were run after every 3 readings. Only those elements returning measurements above the limits of detection have been employed in the

Table 1. Individual MSA obsidian artefacts analysed for geological sourcing. Major oxide by wt% and trace elements ppm.

Sample n19	Fe wt%	Rb ppm	Sr ppm	Zr ppm	Nb ppm
MSA 200 62	3.6	115	3	647	176
MSA 200 63	3.6	120	4	621	147
MSA 200 108	3.8	117	4	619	177
MSA 200 112	3.74	113	3	605	170
MSA 200 114	3.69	115	4	620	173
MSA 200 22	3.81	117	4	628	147
MSA 200 119	3.73	111	5	625	173
MSA 200 110	3.28	119	3	602	166
MSA 200 109	3.66	117	4	623	175
MSA 200 61	8.1	165	10	1598	324
MSA 200 20	7.98	170	7	1607	322
MSA 200 118	7.74	148	7	1610	307
MSA 200 111	7.81	140	2	1596	319
MSA 200 116	7.64	142	1	1614	314
MSA 200 117	7.83	149	6	1609	309
MSA 200 19	7.75	356	5.6	2775	429
MSA 200 13	7.91	354	6.1	2769	429
MSA 200 21	7.76	349	5.4	2771	423
MSA 200 15	7.92	347	4.4	2773	431

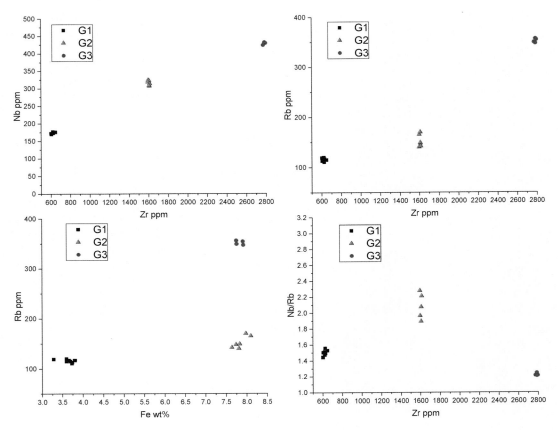

Figure 4. Bivariate plots of major and trace elements for the MSA 200 obsidian artefacts.

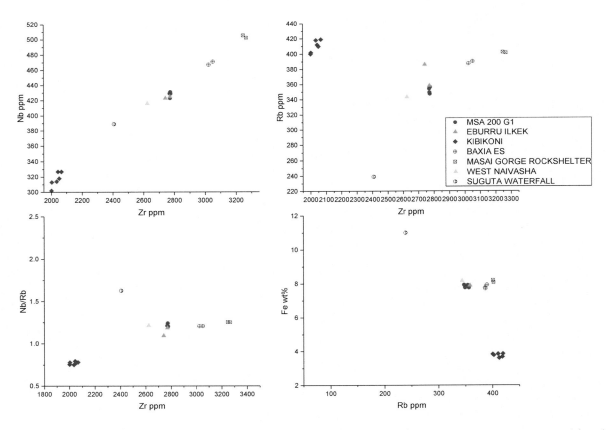

Figure 5. Bivariate plots of major and trace elements for the MSA 200 group 3 and geological sources from Brown et al (2013).

Initial source evaluation

The MSA 200 obsidian artefacts can potentially be assigned to their geological sources by using the geochemical composition of their major and trace elements and comparison with source rock data available in (Brown *et al.* 2013). MSA 200 group 3 is particularly distinctive based on extremely high levels of Zr. In Kenya only a few known sources have elevated levels of Zr < 2000 ppm, including Suguta in northern Kenya, and Eburru mountain and Masai Gorge Rockshelter in the Central Rift. Figure 5 shows a comparison of major and trace elements of all Kenya type obsidian sources with high levels of Zr. The MSA 200 obsidian type group 3 shows the closest correlation with sample MER 70 from Eburru, Ilkek, (Brown *et al.* 2013). This source which is located over 80 km south of MSA 200, in an area just north of lake Naivasha (Figure 1). A single in situ flake and 3 MSA points from the surface collection gives matches with the geochemistry of this source (see Figure 3). The other two MSA 200 groups cannot be assigned to sources at present, as they show no overall affinity to published sources from Brown *et al* (2013) - suggesting that they belong to a yet unidentified or unpublished Kenyan sources. These are unlikely to be in the Naivasha area, as the profiles are outside the range of the many sources identified there based on Zr values of 600 ppm and low values of Nb for MSA G1, but equally they do not match with known nearer sources in the rift between Nakuru and Baringo. Middle Stone Age group 2 shows some similarity in elemental composition to Mount Eburru sources (MER 114-119) based on the trace elements of Zr, Nb and Rb, however, Fe_2O_3 is ca. 5% lower in these samples.

Discussion

Distances of raw material transport from site to source can provide one of the strongest proxies concerning the size of both physical and social landscapes of hominin populations. When compared to hominin groups in the Early Stone Age, Middle Stone Age groups appear to show an increased use of finer grained rocks e.g. obsidian which was transported over much greater distances than those documented for ESA groups (Ambrose 2002; Moutsiou 2014). The MSA site 200 obsidian data demonstrate the long-distance transportation of raw materials for lithic artefacts from over 80 km at ca. 120,000 ka (Figure 5). Greater precision is desirable in the dating, but the current date gives approximately an 85% chance that the age is greater than 90,000 years. There is thus a likelihood that extensive channelling in the area at these levels belongs to MIS5 sensu lato. The data show that MSA populations at this site were utilising at least three different sources of obsidian for lithic manufacture. Although only one source can be

identified in this pilot study, it is sufficient to show that long distance material transport networks were already in use at this period. Mount Eburru sources are considered to represent a high-quality source of obsidian for artefact manufacture due to their flaking properties, and are therefore a commonly used source in both Middle and Late Pleistocene sites (Merrick *et al.* 1994). Blegen (2017) has recently demonstrated the use of this source for artefact manufacture from the MSA Sibolo School Road Site, Baringo at ca. 200 ka, with transportation distance of 140 km.

Long-distance transportation of obsidian from multiple sources is now well documented during the African Middle and Later Stone Ages (McBrearty, 1981; Merrick and Brown, 1984; Mehlman, 1989; Merrick *et al* 1994; Faith *et al* 2015). In southern Africa transport distances for silcrete of 220 km have been documented by MSA populations from Tsodilo Hills in Botswana (Nash et al. 2013). In East Africa, northern Tanzanian sites such as the Nasera and Mumba rockshelters demonstrate the transportation of obsidian over distances of 240 and 325 km respectively (Merrick & Brown, 1984; Mehlman, 1989; Merrick *et al.* 1994). The site of Porc Epic, Ethiopia also demonstrates the use of obsidian from multiple sources in conjunction with the long-distance transportation of raw materials from 150-250 km (Negash and Shackley, 2006). In Kenya, several sites from the Central Rift Valley also yield evidence of both obsidian derived from multiple sources and the long-distance transportation of these materials from source to site (Michels *et al.* 1983; Blegen 2017). The obsidian assemblage from the site of Prospect Farm, in the Nakuru-Elmenteita basin, dated to between 120 – 50 ka, shows the use of multiple sources both locally from Eburru ca. 40 km and from Kisanana at ca. 75 km (Michels *et al.* 1983). The earliest evidence for the long-distance transportation of obsidian as a raw material now comes from the site of Sibilo School Road, Baringo at 200 ka, with three different sources including Eburru, Ilkek and distances of 25 km, 140 km and 166 km (Blegen 2017). Blegen also noted that 43% of the Sibilo lithic assemblage is manufactured from obsidian which is comparable in frequency observed at much later MSA sites such as Lukenya Hill.

Conclusion

Recent excavations and analysis of obsidian artefacts from Middle Stone Age site Kilombe GqJh3 West 200, dating to ca. 120,000 ka, demonstrate new evidence of the long-distance transport of archaeological obsidian from a secure and dated early late Pleistocene Middle Stone Age context. The geochemical composition of the MSA 200 obsidians using the PXRF method attributes the lithic artefacts to three distinct geological sources. When compared to WD-XRF data from Brown *et al.* (2013), it is only possible to identify only one of the sources utilised for artefact manufacture by the MSA 200 population. The source identified is Eburru, Ilkek, which is a high-quality source of obsidian, and is located north of lake Naivasha over 80 km south of Kilombe. Regardless of only one geographical source being attributed so far, the MSA 200 data clearly demonstrate the use of multiple sources of obsidian in conjunction with the long-distance transport of high-quality raw materials for lithic artefact manufacture, and provide justification for further research to gain more detailed knowledge of the site and its raw materials. The distance of 80 km is suggestive of a highly mobile hunter-gatherer population, and tallies with similar evidence from a number of other sites.

Acknowledgments

The research was carried out within a Mobility and Links project between University of Liverpool and National Museums of Kenya, funded by the British Academy. We thank NACOSTI and KNM and Nakuru County for research permissions and assistance, and the British Institute in East Africa for material assistance. We thank also our colleagues James Brink, Laura Basell and Jason Hall.

References

Ambrose, S.H. 2001. Paleolithic technology and human evolution. Science 291, 1748-1753.

Ambrose, S.H. 2002. Small Things Remembered: Origins of Early Microlithic Industries in Sub-Saharan Africa. In Elston, R.G. and Kuhn, S.L. (Eds) *Thinking Small: Global Perspectives on Microlithization. Archeological Papers of the American Anthropological Association* 12, 9-29.

Barham, L. 2013 *From hand to handle: the first industrial revolution*. Oxford, UK: Oxford University Press.

Berger, L.R., Hawks, J.,Dirks, P.H.G.M., Elliott, M., & Roberts, E.M. 2017. *Homo naledi* and Pleistocene hominin evolution in subequatorial Africa. eLife 2017;6:e24234 doi: 10.7554/eLife.24234

Blegen, N. 2017. The earliest long-distance obsidian transport: Evidence from the 200 ka Middle Stone Age Sibilo School Road Site, Baringo, Kenya. *Journal of Human Evolution* 103: 1-19

Brown, F.H., Nash, B.P., Fernandez, D.P., Merrick, H.V., Thomas, R.J. 2013. Geochemical composition of source obsidians from Kenya. *Journal of Archaeological Science*. 40: 3233-3251.

Clark, J.D., ed. 2001.. *Kalambo Falls, Vol 3*. Cambridge: Cambridge University Press.

Clark, J.D., Beyene, Y., WoldeGabriel, G., Hart, W.K., Renne, P.R., Gilbert, H., Defleur, A., Suwa, G., Katoh, S., Ludwig, K.R., Boisserie, J.-R., Asfaw, B.,White, T.D., 2003. Stratigraphic, chronological and behavioural contexts of Pleistocene Homo sapiens from Middle Awash, Ethiopia. Nature 423: 747–752.

Deino, A., and S. McBrearty. 2002. [40]Ar/[39] Ar dating of the Kapthurin Formation, Baringo, Kenya. *Journal of Human Evolution* 42: 185-210.

Faith, J.T., Tryon, C.A., Peppe, D.J., Beverly, E.J., Blegen, N., Blumenthal, S., Chritz, K.L., Driese, S.G. and Patterson, D. (2015). Paleoenvironmental context of the Middle Stone Age record from Karungu, Lake Victoria Basin, Kenya, and its implications for human and faunal dispersals in East Africa. Journal of Human Evolution, 83, 28-45.

Féblot-Augustins J. 1999 Raw material transport patterns and settlement systems in the European Lower and Middle Palaeolithic: continuity, change and variability. In *The Middle Palaeolithic occupation of Europe* (eds W Roebroeks, C Gamble), pp. 193-214. Leiden: European Science Foundation and University of Leiden.

Glascock, M., Braswell, G.E., and Cobean, R.H. 1998. A Systematic Approach to Obsidian Source Characterization. In *Archaeological Obsidian Studies: Method and Theory*. Shackley, M.S. Ed. Plenum Press: New York pp 15-65.

Gowlett, J.A.J. 2009. The longest transition or multiple revolutions? Curves and steps in the record of human origins. In *Sourcebook of Paleolithic transitions: methods, theories and interpretations*, edited by M. Camps, and P. Chauhan, 65-78. Dordrecht: Springer Verlag.

Gowlett, J.A.J., J.S. Brink, A.I.R. Herries, S. Hoare, I. Onjala, and S.M.Rucina. 2015. At the heart of the African Acheulean: the physical, social and cognitive landscapes of Kilombe. In *Settlement, Society and Cognition in human evolution: Landscapes in Mind*, edited by F.Coward, R.Hosfield, and F. Wenban-Smith, 75-93. Cambridge: Cambridge University Press.

Gowlett, J.A.J. & Crompton, R.H. 1994. Kariandusi: Acheulean morphology and the question of allometry. *African Archaeological Review*, 12, 3-42.

Hoare, S, et al. (in prep). New [40]Ar/[39]Ar (Argon Argon) dates from Kilombe Volcano and archaeological sites, Kenya: A whole Pleistocene series.

Hublin, J.-J., Abdelouahed, B.-N., Bailey, S.E., Freidline, S.E., Neubauer, S., Skinner, M.M., Bergmann, I., Le Cabec, A., Nenazzi, S., harvarti, K. and Gunz, P. 2017. New fossils from Jebel Irhoud, Morocco and the pan-African origin of *Homo sapiens*. Nature 546: 289–292. doi:10.1038/nature22336

Jennings, D.J. 1971. *Geology of the Molo area*. Nairobi: Ministry of Natural Resources, Geological Survey of Kenya, Report No. 86.

Johnson, S.R. and S. McBrearty. 2010. 500, 000-year-old blades from the Kapthurin Formation, Kenya. *Journal of Human Evolution* 58: 193–200.

Jones, W.B. and S.J. Lippard. 1979. New age determinations and geology of the Kenya Rift-Kavirondo Rift junction, W Kenya. *Journal of the Geological Society, London* 136: 693-704.

Lane, C.S., B.T. Chorn, and C.T Johnson. 2013. Ash from the Toba supereruption in Lake Malawi shows no volcanic winter in East Africa at 75 ka. *Proceedings of the National Academy of Sciences of the United States of America* 110: 8025–8029.

Leakey, Margaret, P.V. Tobias, J.E. Martyn, and R.E.F. Leakey.1969. An Acheulian industry with prepared core technique and the discovery of a contemporary hominid at Lake Baringo, Kenya. *Proceedings of the Prehistoric Society* 35 : 48-76.

McBrearty, S. 1981. Songhor : a Middle Stone Age site in Western Kenya. *Quaternaria.* 23 :171-190.

McBrearty, S. 1999. The Archaeology of the Kapthurin formation. In *Late Cenozoic environments and hominid Evolution: a tribute to Bill Bishop*, edited by P. Andrews, and P Banham, 143-156. London: Geological Society.

McBrearty, S and Brooks, A.S. 2000. The revolution that wasn't: a new interpretation of the origin of modern human behaviour. *Journal of Human Evolution.* 39: 453-563.

McBrearty S. 2003. Patterns of technological change at the origin of Homo sapiens. *Before Farming* 3:1-6

McDougall I., Brown F. H. & Fleagle J. G. 2005. Stratigraphic placement and age of modern humans from Kibish, Ehtiopia. *Nature* 433:733 - 736.

McCall, G.J.H. 1964. Kilombe caldera, Kenya. *Proceedings of the Geologists' Association* 75: 563-572.

Mehlman, M.J. 1989. Later Quaternary archaeological sequences in northern Tanzania. PhD. Dissertation. University of Illinois at Urbana- Champaign.

Merrick, H.V., Brown, F.H., 1984. Obsidian sources and patterns of source utilization in Kenya and Tanzania: some initial findings. *African Archaeological Review.* 2: 129-152.

Merrick, H.V., Brown, F.H., Nash, W.P., 1994. Use and movement of obsidian in the Early and Middle Stone ages of Kenya and northern Tanzania. In: Childs, S. (Ed.), Society, Culture, and Technology in Africa. MASCA Res. Pap. in Sci. and Archaeol., Philadelphia. Supplement to vol. 11: 29-43.

Michels, J.W., Tsong, I.S., Nelson, C.M., 1983. Obsidian dating and East African archaeology. *Science* 219: 361-266

Moutsiou, T. 2014. The obsidian evidence for the scale of social life during the Palaeolithic. Archaeopress, BAR-IS 2613.

Nash, D.J., Coulson, S., Staurset, S., Smith, M.P., & Ullyott, J.S. Babutsi, M., Hopkinson, L., Smith, M. P. 2013. Provenancing of silcrete raw materials indicates long-distance transport to Tsodilo Hills, Botswana during the Middle Stone Age *Journal of Human Evolution* 64:280-288

Negash, A., & Shackley, M.S. 2006. Geochemical provenance of obsidian artifacts from the MSA site of Porc Epic, Ethiopia. *Archaeometry* 48: 1–12.

Negash, A., Shackley, M.S., and Alene, M. 2006. Source provenance of obsidian artifacts from the Early Stone Age (ESA) site of Melka Konture, Ethiopia. *Journal of Archaeological Science.* 33: 1647-1650

Potts, R. 1998. Variability selection in hominid evolution. *Evolutionary Anthropology* 7: 81-96.

Richter, D., Grün, R., Joannes-Boyau, R., Steele, T.E., Amani, F., Rué, M., Fernandes, P., Raynal, J.-P., Geraads, D., Abdelouahed, B.-N., Hublin, J.-J., and McPherron, S.P. 2017. The age of the hominin fossils from Jebel Irhoud, Morocco, and the origins of the Middle Stone Age. *Nature* 546: 293–296.

Rightmire, G.P. 2012. The evolution of cranial form in mid-Pleistocene *Homo. South African Journal of Science.* 108(3/4)

Shackley, M.S., & Sahle, Y. 2017. Geochemical Characterization of Four Quaternary Obsidian Sources and Provenance of Obsidian Artifacts from the Middle Stone Age Site of Gademotta, Main Ethiopian Rift. *Geoarchaeology* 32: 302-310.

Tryon, C.A., and Faith, J.T. 2013. Variability in the Middle Stone Age of Eastern Africa. *Current Anthropology.* 54: s234-s254.

White, T.D., Asfaw, B., DeGusta, D., Giblbert, H., Richards, G.D., Suwa, G., Howell, F.C., 2003. Pleistocene Homo sapiens from Middle Awash, Ethiopia. Nature 423:742–747.

Wilkins, J., and M. Chazan. 2012. Blade production ~500 thousand years ago at Kathu Pan 1, South Africa: support for a multiple origins hypothesis for early Middle Pleistocene blade technologies. *Journal of Archaeological Science* 39: 1883-1900.

Wilkins, J., B.J. Schoville, K.S. Brown, and M. Chazan. 2012. Evidence for early hafted hunting technology. *Science* 338: 942-946.

Zipkin, A.M., Hanchar, J.M., Brooks, A.S., Grabowski, M.W., Thompson, J.C., & Gomani-Chindebvu, E. 2015. Ochre fingerprints: Distinguishing among Malawian mineral

Sally Hoare
ACE, HLC, University of Liverpool, L69 3BX, UK

Stephen Rucina
Palynology and Paleobotany section, National Museums of Kenya, P.O. Box 40658-00100, Nairobi, Kenya

John A.J. Gowlett
ACE, HLC, University of Liverpool, L69 3BX, UK

The Eternal Triangle of Human Evolution

Clive Gamble

John Gowlett, palaeoanthropologist extraordinary

When John and I started as undergraduates at Cambridge in 1969 the map of the Palaeolithic was unrecognisable from today's satellite image. We were taught a chronology of four ice ages and their pluvial equivalents in low latitudes, stone tool typology rather than lithic technology and a simple ancestral line that led from *Homo habilis* to Charles McBurney's office where we were inducted into the details of Dorothy Garrod's excavations at Mt Carmel. Our textbooks were Kenneth Oakley's (1969) *Frameworks for Dating Fossil Man*, the joint venture by John Coles and Eric Higgs (1969) *The Archaeology of Early Man*, the excellent *Guide to Fossil Man* by Michael Day (1967) and the largely incomprehensible *The Old Stone Age* by François Bordes (1968).

The gendered language of these titles, not forgetting Clark Howell's (1965) *Early Man*, instantly marks John's archaeological career as starting a long time ago. But elsewhere a new map was already being drawn up. At the Godwin Institute Nick Shackleton was revolutionising quaternary chronology and climatology with the first deep sea wiggle curves. Absolute dating in the form of radiocarbon (¹⁴C) was pushing back into the Upper Palaeolithic while potassium argon (K-Ar) had defined a baseline for human evolution through its application to the tuffs at Olduvai Gorge. Although that still left Glynn Isaac's muddle-in-the-middle untethered to a reliable chronology (Butzer and Isaac 1975) there was an optimistic mood that this would change. Under the urging of Eric Higgs, faunal remains were kept and studied for information on human diet, and the catchments from which they came, rather than thrown away or used simply as chronological markers. Excavation in caves and open locales had moved on from Garrod's rapid digging days. Hallam Movius Jnr was slowly excavating the Abri Pataud, Gerhard Bosinski was plotting in great detail the Magdalenian house at Gönnersdorf – an exercise which matched the ethnographic approach of André Leroi-Gourhan at Pincevent – and Isaac was pioneering his scatters and patches approach in African conditions. And in one massive volume on her Olduvai excavations Mary Leakey (1971) set the gold-standard for Lower Palaeolithic archaeology. Metrication and computer analysis had arrived, principally in the form of Derek Roe's (1968b; 1968a) method for analysing handaxes and his seminal Gazetteer of the British Lower Palaeolithic, while at UCLA, Lewis and Sally Binford (1966) published their factor analysis claiming functional variability among Bordes' Mousterian assemblages. McBurney promised us a session on this 'provocative' case, but it never happened.

Much, however, was missing. There was no archaeogenetics to worry about. Computing was restricted to mainframes. Even using a photocopier was a student's distant dream. Furthermore, the concept of the 'anatomically modern *Homo sapiens*' (Brose and Wolpoff 1971:1183) did not appear until our final year. Radiocarbon accelerators which transformed Palaeolithic chronologies were some years away (Gowlett and Hedges 1986), while Bayes was something that covered billiard tables.

The biggest changes that underpin John's career as the leading archaeologist of early humans in Europe and Africa relate neither to hardware nor software but the shaping of a new interdisciplinary approach. Archaeologists, and particularly those working in the Palaeolithic, have always collaborated with colleagues in the sciences. But in the last fifty years a distinctive kind of interdisciplinarity has been forged which is partially covered by the concomitant rise of the term Palaeoanthropology. This interdisciplinary approach to human evolution grew from the American four-field tradition and achieved early prominence in such syntheses as *The Social life of Early Man* (Washburn 1961) and the *Man the Hunter* symposium (Lee and DeVore 1968). The range of participating disciplines was on show in another Cambridge text book *Science in Archaeology* (Brothwell and Higgs 1969) then in a second, fatter, edition. In the subsequent decades the number of disciplines with which Palaeolithic archaeologists interdigitate has kept on growing. This trend continued when John linked-up with psychologist Robin Dunbar when they were both at the University of Liverpool and out of which grew the British Academy Centenary Project *From Lucy to language: the archaeology of the social brain* (Dunbar et al. 2010; Gowlett et al. 2012; Dunbar et al. 2013; Gamble et al. 2014).

I regard John as the embodiment of the success of the palaeoanthropological approach and the interdisciplinarity it thrives upon. It can only succeed with a generosity of spirit, an interest in discovery and a commitment to share results; values which John has followed in his work across two continents facilitated by another innovation of the 1970s, the jumbo jet, which made international fieldwork a reality for under-funded British archaeologists.

The eternal triangle

The need for an interdisciplinary approach is encapsulated in Gowlett's (2010: 357) eternal triangle of human evolution and where its three points are diet change, social collaboration and detailed environmental knowledge.

The development of fire use and management by hominins illustrates his perspective. Fire acts as that external stomach allowing the trade-off between a smaller gut and a larger brain to proceed (Aiello 1998). Those larger hominin brains, according to predictions from the social brain model of apes and monkeys, required social collaboration to be re-purposed. This involved novel forms of communication that included language (Aiello and Dunbar 1993) and signs incorporated into material culture (Gowlett 1984).

Social collaboration is often only considered as a Palaeolithic issue in the context of the act of communal hunting. However, selection for higher energy diets to support larger more expensive brains changed the way that landscapes were exploited through the flexible hominin tactics of fission and fusion. Hominins are not the only animals to aggregate and separate as resources vary. During their evolutionary history hominins fashioned from the mammalian heritage of a distributed cognition, an extended mind that allowed them to live apart yet stay in touch.

The three points of Gowlett's triangle are central to the evolutionary process where a distinctive human community emerged from a primate *bauplan*.

The eternal triangle also reminds us that environmental knowledge is not only acquired but remembered and passed on. Those landscapes are not just the ecological ranges where food is intercepted and collected but also the landscapes of the imagination; the stories shared around the well-kindled fire as the hominin social day was extended deep into the socially unproductive night of the primates (Gowlett 2016).

Hearths such as those John has excavated at Chesowanja, Kenya, (Gowlett et al. 1981) and Beech's Pit, England, (Gowlett et al. 2005) provided a new structure for hominin interaction and re-shaped hominin biology. These features are a classic example of a constructed ecological niche reminding us that hominins did not simply interact *with* their environments but instead were inseparable from them. A hominin from any Palaeolithic period can only be understood if, as Gowlett has shown, they are considered as one with the environment they created, lived in and transformed.

A focus on fire and its application to the triangle that structured hominin evolution serves as an example of the palaeoanthropological approach in action. In the rest of this paper I will take another nexus where material, social and environmental connections effected a similar re-purposing to that of fire. This nexus is the collection and accumulation of stuff. Sometimes stuff is arranged to form a pattern of associations that suggests a dwelling or a food store. At others it is a set of objects, fashioned or not, either similar or dissimilar in composition, shape and size, that has been brought together as a demonstration of association and knowledge (Scott and Shaw 2018). Such assemblages form a nexus of connections ranging from the individual to the community, the locale to the landscape and the single track to the maze of overlapping itineraries (Brody 1981).

These acts of accumulation produce Isaac's (1980) patches among the continuous scatters of artefacts across the landscape. In other contexts caves form capture points that arrest the flows of people and stuff as they circulate through the landscape. Such archaeological 'sites', however, have to stand the test of taphonomic scrutiny to become evidence of hominin agency rather than any number of other forces which bring materials together. Hominins are not the only animals that accumulate sets of material; wolves, hyenas, beavers and birds are all well known for their abilities to accumulate and construct dens, lodges and nests, while bears made their own hibernation cemeteries in the caves of Pleistocene northern Europe (Musil 1980-1). But none of these examples are anthropological *stuff* (Miller 2012). Primates, and particularly great apes, are less conspicuous when it comes to such behaviour. Chimpanzees makes nests at nightfall but their tropical existence and avoidance of caves does not result in significant accumulations of materials. This was Isaac's (1978) insight in his home base model where food sharing among hominins formed a platform for further cultural developments.

Stuff defines humans because it is the way we mediate, organise and imaginatively transcend what is otherwise an impersonal Darwinian existence. Our immediate ancestors, those anatomically modern humans (AMH) who first appeared in 1971, were once thought to be the originators of such stuff. It was their symbolic use of materials and the inception of a new suite of material culture – jewellery, figurative art, composite artefacts, status burials, musical instruments – that marked a human revolution in creative taste and value. Previously stuff dealt with necessity and hominin imagination was limited to function.

La Cotte case study

In what follows I will take Gowlett's triangle and apply it to a Middle Pleistocene case study where he learned his excavation skills. The granite headland cave of La

Cotte de St Brelade, Jersey, was excavated by McBurney between 1961 and 1978 (Callow and Cornford 1986; McBurney and Callow 1971; Scott 1980). Following McBurney's death in 1979, the report edited by Paul Callow and Jean Cornford outlined the richness of the site and the depth and complexity of the stratigraphy that contains ten archaeological layers. Three aspects from this work have ensured that La Cotte is known as an important Middle Pleistocene locale. First there was the recovery and interpretation by Kate Scott (1986a) of two piles of megafauna bones in the North Ravine. These were plausibly accounted for as two separate hunting episodes with the cliff edge serving as a game drive. Second the work by Callow (1986) on the changing raw materials meant that La Cotte would not be seen in isolation but instead as a nexus in a landscape of stone opportunities dependent upon fluctuating sea level and hence global climate. The third element was Cornford's (1986) study of the re-sharpening flakes among the lithic assemblage that pointed to a strategy that managed lithic resources in a variety of ways.

It is tempting to see these three studies as an inspiration for Gowlett's triangle of diet, collaboration and environmental knowledge. I will now examine them in more detail

Diet change, climate and brains

Diet change is represented by the climatic changes La Cotte witnessed during the time that its deposits were forming. If we include also the Upper Pleistocene sequence, which McBurney did not investigate in any detail, then there are at least two cold stages (MIS6 and MIS4) present, two interglacials, MIS7 and MIS5e, each of which is sub-divided into temperate and cooler conditions, and the variable stadial-interstadial conditions of MIS5a-d and MIS3 (Van Andel and Davies 2003). Hominin occupation is restricted to phases of low sea-level during the colder parts of the locale's climatic history. The sediments from which new OSL samples could be obtained are behind a protective concrete wall that McBurney built when his excavations ended. Exactly what condition the sediments it encases are in cannot be ascertained. This means that the only date for the Middle Pleistocene deposits is the TL age estimate for a sample of six burnt flints from layers C and D of 238 ±35ka BP (Huxtable 1986). In those parts of the sequence which are available for OSL dating a stratigraphically ordered sequence of Upper Pleistocene ages has recently been obtained (Bates et al. 2013). What we do know is that the La Cotte sequence in the Middle Pleistocene covers the appearance of reindeer (*Rangifer tarandus*) in Western Europe (Cordy 1992: Figure 1). This herd ungulate was of critical importance to the ecology and diet of later hominins and humans. Reindeer is absent in the two bone heaps in levels 6 and 3 (Scott 1986b; Julien pers. comm.) and poorly represented in

the rest of the fauna from the locale, even though the cooler climatic conditions would have suited it. The evidence from the fauna at La Cotte does not suggest that during this long sequence hominin prey selection changed significantly.

By the time la Cotte was occupied in the Late Middle Pleistocene the costs of energetically expensive large brains had been mitigated by diet change. Mary Stiner (2002; 2009) has pointed to the archaeozoological evidence that by 400,000 years ago a change in predation had occurred that now targeted prime-aged herd animals. This change coincided with a change in the steepness of the brain growth curve for hominins (Gowlett 2009: Fig.1; Rightmire 2004). And it is in this broader context of diet change/encephalisation/social collaboration that La Cotte must be placed.

There are no hominin crania from La Cotte. However, in the time range of 300 – 250ka hominin fossil crania from Europe were at least 1300cms^3 in size, just below Neanderthal and AMH values (Gowlett 2009: Fig.1; Rightmire 2004). Using the social brain graph this suggests a personal network size of 132 (slightly below Dunbar's number of 150 for contemporary humans) for such large brained hominins (Gamble 2013: Table 4.2 for values). If social interaction was non-verbal, reliant instead on the primate pattern of finger-tip grooming, then the La Cotte people would have spent up to 37 per cent of their day attending to others (Dunbar 1992). This time budget is clearly unsustainable and supports a prediction that by this stage rapid communication in the form of speech had evolved to solve a time-budgeting issue that arose from an individual's larger social network (Gamble et al. 2014: 138-148). La Cotte lacks the hearths around which Neanderthal story-telling took place during the long nights of MIS7. But there is abundant evidence for burning and well-tended hearths are known elsewhere at this time (Alperson-Afil and Goren-Inbar 2010; Gowlett 2010; 2016).

Social collaboration

Kate Scott's game drive hypothesis for Jersey (Scott 1986a) speaks of social collaboration directed towards the food quest. The two bone heaps as described by Scott are also remarkable because they are preserved in a granite ravine system that is normally inimical to the survival of bone. This preservation suggested to the excavators that the skulls and long bones of mammoth and rhino had been dragged by hominins under the overhang where they were then mantled by loess. This wind-blown material is high in calcium carbonate and is the factor that aided preservation. It might well be, however, that the heaps of bone originally extended beyond the overhang and the subsequent erosion of the loess unintentionally 'sculpted' the faunal assemblage into its final form as two distinct heaps of bones.

The taphonomy of these two assemblages is therefore critical. In her original analysis Scott comments on the rarity of carnivore damage visible on the bone surfaces. Subsequent cleaning of these surfaces has not only confirmed her original assessment (Scott 1980) but through the recent work of Marie-Anne Julien (mss) revealed stone tool marks on a significant number of long bones from mammoth, rhino and bison. This is surprising given that cut marks are generally extremely rare on megafaunal bones as Schreve (2012) found in her study of the 11 mammoths from the MIS3 Lynford hunting and processing locale. Furthermore the bone breakage and cut mark evidence at La Cotte needs to be added to other instances of hominin involvement with megafaunal bones; for example the mammoth rib driven intentionally through a mammoth skull in Level 3, the lower bone 'heap' (Scott 1986a: 175 and Fig18.15).

Recent claims (Smith 2015) for substantial carnivore rather than hominin alteration to the la Cotte fauna are not supported by the evidence. Carnivore damage to bones is rare. A detailed assessment of bone surfaces (Julien et al. mss: Figure 2) shows that levels A and B, stratigraphically below the bone 'heaps', are the only ones that have any reliable carnivore modification and these comprise 0.5 and 0.6 per cent of the number of identified specimens (NISP). By comparison, in those same levels human modification is found on 7.2 and 9.3 per cent of NISP. The La Cotte fauna has no hyena bones, coprolites or traces of their distinctive gnawing. Morever, claims of carnivore damage to bone surfaces can be parsimoniously ascribed to trowel marks (Julien et al. mss).The game-drive hypothesis for the two bone heaps at la Cotte has recently been re-assessd (Scott et al. 2014). This analysis confirms hominins as responsible for the faunal collections but questioned on topographic grounds the suitability of the headland for a game drive. An alternative is proposed that moves the hunting from the top of the headland to the currently submerged granite canyons that lead towards La Cotte. These have been revealed by underwater bathymetric survey. They would have provided the opportunity for close quarter hunting where topography was used to disadvantage prey (Churchill 1993; 2014).

La Cotte is a hominin locale as indicated by the >100,000 lithic artefacts that McBurney excavated. Carnivores are present but in very low numbers for a Middle/Upper Pleistocene cave in northern Europe where their remains frequently dominate collections (Gamble 1995) and their gnawing is omni-present. The faunal assemblage from La Cotte contradicts this general pattern. There is hominin modification of megafaunal bones as well as the assembly of stuff in layers and on two occasions, *pace* taphonomy, into 'heaps' of bone. Social collaboration in the pursuit of food is still an inference that can be made from the data; an activity that required co-operation if only to lift mammoth heads weighing up to 400kg, and even more depending on the size of the tusks that were attached.

Detailed environmental knowledge and the accumulation of stuff

The last point of Gowlett's triangle can be explored at La Cotte through the acquisition and management of lithic raw materials. Indeed when viewed through this lens the locale emerges as a stable place within a changing landscape (Scott and Shaw 2018). The shallow continental shelf in the Normano-Breton Gulf was repeatedly exposed and submerged revealing raw materials of different flaking quality and abundance at distances of at least 20kms from the cave. When combined with the large number of lithic artefacts, more than 100,000 in ten archaeological layers, then it is possible to characterise La Cotte as a persistent place (Shaw et al. 2016). As described by Scott and Shaw (2018), 'A persistent place is in part a long-lived capture point – but it is not only that: precisely because particular places can preserve the residues of human behaviour over at least one glacial-interglacial cycle, so they can also reflect the different ways in which changing human action continually made, remade and encultured the landscape.' They show how detailed landscape knowledge around La Cotte can be traced through the exploitation of raw materials as they are revealed and lost by changing sea levels. They show how life is far from local at a locale so persistently occupied. Instead the cave captures the heterogeneous qualities of gatherings that brought together people, tools, sounds, animals and other places. The landscape is not simply a set of ecological affordances but rather a source of sensations that are made meaningful at both the emotional and rational level to the people who experienced their human lifespans against the backdrop of La Cotte's 300,000 year-long persistence.

Detailed environmental knowledge is evident also in Cornford's analysis of a large sample of sharpening flakes. The majority were long sharpening flakes with a smaller number of transversal examples. In her analysis she pays particular attention to the changing proportions of flint in the lithic assemblage (Cornford 1986: 343 and Fig.29.3). As flint declines as a proportion of all lithics through the sequence so the ratio of sharpening flakes to flint tools increases. This trend is reversed in layer 5, the latest she examined in her study. Cornford regards the overall trend as a response to flint shortage and the situation in layer 5 as a fundamental change in behaviour (Cornford 1986:344).

This behaviour is further explored by Scott and Shaw (2018). Their study of stone tool refits in layer 5 shows two distinct patterns. Local materials, from less than 5km distant (quartz and schist), were brought into La Cotte and worked there as shown by refitting. Nodules of feldspathic sandstones were acquired from sources 10-15km away and the initial phases of working were carried out elsewhere. They were then carried to La Cotte for further use.

These two patterns of transport were however surpassed by the use of bedrock flint whose nearest source is 20km from the cave. Artefacts in this raw material were brought into La Cotte as finished artefacts and then further modified on-site. Many of the artefacts are broken. As Scott and Shaw (2018) conclude the persistent, but not continuous use of La Cotte, points to versatility and change in hominin patterns of movement across the landscape of the exposed Normano-Breton Gulf. The attritional cost of distance is, in this instance, not met by a strategy that appears to economise the use of flint as a higher value raw material. The accumulation of stuff, for example the worked out discoidal cores that are common in layer 5, continues at la Cotte but the paths and tracks by which the materials were obtained, transported and accumulated now varies compared to earlier layers E and A (See Scott and Shaw (2018) for a full review).

Discussion

John Gowlett's eternal triangle draws attention to the co-evolutionary character of hominin evolution. This relationship is shown clearly with the role fire played in transforming hominin agents and in turn being transformed by them (Gowlett 2016). Change is always a two way process involving human and non-human agents, animate and inanimate. I have argued in this contribution that a similar co-evolutionary path is discernible in the assemblages of stuff. Traditionally the appearance of sets and nets of material have been identified as an Upper Palaeolithic phenomenon and more broadly used to infer a Human revolution (Mellars and Stringer 1989). While the evidence points to a late appearance of assemblages that take the classic form of a well-structured 'Childean' culture, I have argued elsewhere that this is simply the extension of a Neolithic model of deep human history back into the Pleistocene (Gamble 2007). However, the eternal triangle with its co-evolutionary perspective demands that we look further back in deep history to see when and how change occurred. Gowlett (2009:65) has referred to this as a Middle Pleistocene silent revolution; one where archaeologists have to work harder when not dazzled by the flashiness of the last 50,000 years (Gowlett et al. 2012). And it is here that the assembly of stuff at a persistent place such as La Cotte both challenges the simple pathways archaeologists often reconstruct for hominin evolution and calls into question the idea of a Human revolution. Persistent places are found across the Late Middle Pleistocene world (Pope et al. 2018; Pope 2018). They represent a threshold in behaviour that is crossed about 400,000 years ago but which is

done silently (Pope 2018; Gamble 2018). This was not the world of busy modern *Homo sapiens*, so it is a mistake to expect it to be garlanded by art, symbols and explained by metaphors of containment, order and the imposition of structure. In this regard the Late Middle Pleistocene belongs to the Bricoleur rather than the Engineer (Gamble 2007:138-9). Both agents create assemblages of stuff. The latter makes sense to us because the cultural bricks they build with result in the familiar shapes we live in and express our understanding of the world; houses, villages, towns and cities, containers both of people and the concepts we apply to the world. By contrast the Bricoleur creates social life by bringing together what seem to us disparate things. This is the activity preserved at La Cotte in the two 'heaps' and the set of re-sharpening flakes. Comparable creative acts of bricolage are found both earlier than La Cotte at the Sima de los Huesos (Carbonell et al. 2003) where a body 'heap' was created, and later inside Bruniquel Cave where Neanderthals constructed speleofact 'heaps' in annular patterns (Jaubert et al. 2016). The meaning of these assemblages is obscure to the Archaeologist as Engineer and neither is the task of the Archaeologist as Bricoleur an easy one. There is always the thought that our job is to translate, like an anthropologist, the socially unfamiliar into the familiar, the old into the new and the exotic into the everyday. Sometimes we need to accept that archaeologists encounter in their evidence a different understanding of humanity. Gowlett's silent revolution and his cardinal points suggest thresholds being passed long before we are confident in calling them revolutions. Silent they may be but we are fortunate to have an extraordinary Palaeoanthropologist as our expert guide through the long corridors of deep human history.

Acknowledgements

I am grateful to Andy Shaw, Beccy Scott, Matt Pope, Marie-Anne Julien and John McNabb for comments on this paper and invaluable advice during the *Crossing the Threshold* project that was funded by the Arts and Humanities Research Council (AHRC-AH/K00338X/1). I would like also to acknowledge the support the project received from Jon Carter and Olga Finch of Jersey Heritage, John Renouf and the Société Jersiaise who are the custodians of La Cotte.

References

Aiello, L. and R. Dunbar, 1993. Neocortex size, group size and the evolution of language. *Current Anthropology*, 34, 184-93.

Aiello, L.C., 1998. The 'expensive tissue hypothesis' and the evolution of the human adaptive niche: a study in comparative anatomy, in *Science in archaeology: an agenda for the future*, ed. J. Bayley.London: English Heritage, 25-36.

Alperson-Afil, N. and N. Goren-Inbar, 2010. *The Acheulian site of Gesher Benot Ya'aqov volume II: ancient flames and controlled use of fire,* Dordrecht: Springer.

Bates, M., M. Pope, A. Shaw, B. Scott and J.-L. Schwenninger, 2013. Late Neanderthal occupation in North-West Europe: rediscovery, investigation and dating of a last glacial sediment sequence at the site of La Cotte de Saint Brelade, Jersey. *Journal of Quaternary Science, 28,* 647-52.

Binford, L.R. and S.R. Binford, 1966. A preliminary analysis of functional variability in the Mousterian of Levallois facies. *American Anthropologist, 68(2),* 238-95.

Bordes, F., 1968. *The old stone age,* London: Weidenfeld and Nicholson.

Brody, H., 1981. *Maps and Dreams,* Vancouver: Douglas and McIntyre.

Brose, D. and M. Wolpoff, 1971. Early Upper Paleolithic man and Late Middle Palaeolithic tools. *American Anthropologist, 73,* 1156-94.

Brothwell, D. and E.S. Higgs (eds.), 1969. *Science in archaeology,* London: Thames and Hudson.

Butzer, K.W. and G.L. Isaac (eds.), 1975. *After the Austalopithecines,* The Hague: Mouton.

Callow, P., 1986. Raw materials and sources, in *La Cotte de St. Brelade 1961-1978: excavations by C.B.M. McBurney,* eds. P. Callow and J.M. Cornford.Norwich: Geo books, 203-13.

Callow, P. and J.M. Cornford (eds.), 1986. *La Cotte de St. Brelade 1961-1978. Excavations by C.B.M.McBurney,* Norwich: Geo Books.

Carbonell, E., M. Mosquera, A. Ollé, X.P. Rodríguez, R. Sala, J.M. Vergès, J.L. Arsuaga and J.M. Bermúdez de Castro, 2003. Les premiers comportments funéraires auraient-ils pris place à Atapuerca, il y a 350,000 ans? *L'Anthropologie, 107,* 1-14.

Churchill, S.E., 1993. Weapon technology, prey size selection, and hunting methods in modern hunter-gatherers: implications for hunting in the Palaeolithic and Mesolithic, in *Hunting and animal exploitation in the later Palaeolithic and Mesolithic of Eurasia,* eds. G.L. Peterkin, H. Bricker and P.A. Mellars.Archaeological Papers of the American Antropological Association 4, 11-24.

Churchill, S.E., 2014. *Thin on the Ground: Neandertal Biology, Archeology and Ecology,* London: Wiley-Blackwell.

Coles, J.M. and E.S. Higgs, 1969. *The archaeology of early man,* London: Faber and Faber.

Cordy, J.-M., 1992. Apport de la paléomammologie à la paléoanthropologie en Europe, in *Cinq millions d'années, l'aventure humaine,,* ed. M.Toussaint.Liège: ERAUL, 77-94.

Cornford, J.M., 1986. Specialised resharpening techniques and evidence of handedness, in *La Cotte de St. Brelade 1961-1978: excavations by C.B.M. McBurney,* eds. P. Callow and J.M. Cornford.Norwich: Geo books, 337-53.

Day, M.H., 1967. *Guide to fossil man,* London: Cassell.

Dunbar, R., C. Gamble and J.A.J. Gowlett (eds.), 2010. *Social brain and distributed mind,* Oxford: Oxford University Press, *Proceedings of the British Academy* 158.

Dunbar, R., C. Gamble and J.A.J. Gowlett (eds.), 2013. *Lucy to language the benchmark papers,* Oxford: Oxford University Press.

Dunbar, R.I.M., 1992. Time: a hidden constraint on the behavioural ecology of baboons. *Behavioural ecology and sociobiology, 31,* 35-49.

Gamble, C., 2018. Thresholds in hominin complexity during the Middle Pleistocene: a persistent places approach, in *Crossing the human threshold: dynamic transformation and persistent places during the Middle Pleistocene,* eds. M. Pope, J. McNabb and C. Gamble. London: Routledge, 3-23.

Gamble, C.S., 1995. Large mammals, climate and resource richness in Upper Pleistocene Europe. *Acta Zoologica Cracovensis, 38(1),* 155-75.

Gamble, C.S., 2007. *Origins and revolutions: human identity in earliest prehistory,* New York: Cambridge University Press.

Gamble, C.S., 2013. *Settling the Earth: the archaeology of deep human history,* Cambridge: Cambridge University Press.

Gamble, C.S., J.A.J. Gowlett and R. Dunbar, 2014. *Thinking big: the archaeology of the social brain,* London: Thames and Hudson.

Gowlett, J.A.J., 1984. Mental abilities of early man: a look at some hard evidence, in *Hominid evolution and community ecology,* ed. R.A.Foley.London: Academic Press, 167-92.

Gowlett, J.A.J., 2009. The longest transition or multiple revolutions? Curves and steps in the record of human origins, in *A sourcebook of Palaeolithic transitions: methods, theories and interpretations,* eds. M. Camps and P. Chauhan.Berlin: Springer Verlag, 65-78.

Gowlett, J.A.J., 2010. Firing up the social social brain, in *Social brain and distributed mind,* eds. R. Dunbar, C. Gamble and J.A.J. Gowlett.Oxford: Oxford University Press, 341-66.

Gowlett, J.A.J., 2016. The discovery of fire by humans: a long and convoluted process. *Philosophical Transactions of the Royal Society B, 371,* 20150164.

Gowlett, J.A.J., C. Gamble and R. Dunbar, 2012. Human evolution and the archaeology of the social brain. *Current Anthropology, 53,* 693-722.

Gowlett, J.A.J., J. Hallos, J.S. Hounsell, V. Brant and N.C. Debenham, 2005. Beeches Pit - archaeology, assemblage dynamics and early fire history of a Middle Pleistocene site in East Anglia, UK. *Eurasian Prehistory, 3,* 3-38.

Gowlett, J.A.J. and R.E.M. Hedges (eds.), 1986. *Archaeological results from Accelerator Dating,* Oxford: Alden Press.

Gowlett, J.A.J., J.W.K.Harris, D.Walton and B.A.Wood, 1981. Early archaeological sites, hominid remains and traces of fire from Chesowanja, Kenya. *Nature, 294,* 125-9.

Howell, F.C., 1965. *Early man,* London: Time Life Books.

Huxtable, J., 1986. The thermoluminescence dates, in *La Cotte de St. Brelade 1961-1978: excavations by C.B.M. McBurney,* eds. P. Callow and J.M. Cornford.Norwich: Geo books, 145-51.

Isaac, G., 1978. The food sharing behaviour of proto-human hominids. *Scientific American,* 238, 90-108.

Isaac, G., 1980. Stone age visiting cards: approaches to the study of early land use patterns, in *Pattern of the Past: studies in honour of David Clarke,* eds. N.Hammond, G.Isaac and I.Hodder.Cambridge: Cambridge University Press., 131-55.

Jaubert, J., S. Verheyden, D. Genty, M. Soulier, H. Cheng, D. Blamart, C. Burlet, H. Camus, S. Delaby, D. Deldicque, R.L. Edwards, C. Ferrier, F. Lacrampe-Cuyaubère, F. Lévêque, F. Maksud, P. Mora, X. Muth, É. Régnier, J.-N. Rouzaud and F. Santos, 2016. Early Neanderthal constructions deep in Bruniquel Cave in southwestern France. *Nature,* 534(7605), 111-4.

Julien, M.A., A. Shaw, M. Pope, B. Scott, C. Gamble and J. McNabb, mss. Taphonomy and Neanderthal behaviour at La Cotte de St Brelade. *Journal of Human Evolution.*

Leakey, M.D., 1971. *Olduvai Gorge: excavations in Beds I and II 1960 - 1963,* Cambridge: Cambridge University Press.

Lee, R.B. and I. DeVore (eds.), 1968. *Man the hunter,* Chicago: Aldine.

McBurney, C.B.M. and P. Callow, 1971. The Cambridge excavations at la Cotte de St Brelade, Jersey - a preliminary report. *Proceedings of the Prehistoric Society,* 37, 167-207.

Mellars, P.A. and C. Stringer (eds.), 1989. *The Human Revolution: behavioural and biological perspectives on the origins of modern humans,* Edinburgh: Edinburgh University Press.

Miller, D., 2012. *Stuff,* Cambridge: Polity Press.

Musil, R., 1980-1. *Ursus spelaeus. Der Hohlenbar,* Weimar: Museum fur Ur und Fruhgeschichte Thuringens.

Oakley, K.P., 1969. *Frameworks for dating fossil man,* London: Weidenfeld and Nicolson.

Pope, M., 2018. Thresholds in behaviour, thresholds of visibility: landscape process, assymmeteries in landscape records and niche construction in the formation of the Palaeolithic record, in *Crossing the human threshold: dynamic transformation and persistent places during the Middle Pleistocene,* eds. M. Pope, J. McNabb and C. Gamble.London: Routledge, 24-39.

Pope, M., J. McNabb and C. Gamble (eds.), 2018. *Crossing the human threshold: dynamic transformation and persistent places during the Middle Pleistocene,* London: Routledge.

Rightmire, G.P., 2004. Brain size and encephalization in early to Mid-Pleistocene Homo. *American Journal of Physical Anthropology,* 124(2), 109-23.

Roe, D.A., 1968a. British Lower and Middle Palaeolithic handaxe groups. *Proceedings of the Prehistoric Society,* 34, 1-82.

Roe, D.A., 1968b. *A gazetteer of British Lower and Middle Palaeolithic sites,* London: Council for British Archaeology Research Report 8.

Schreve, D., 2012. The vertebrate assemblage from Lynford: taphonomy, biostratigraphy and implications for Middle Palaeolithic subsistence strategies, in *Neanderthals among mammoths: excavations at Lynford Quarry, Norfolk,* eds. W.A. Boismier, C.S. Gamble and F. Coward.London: English Heritage Monographs, 157-205.

Scott, B., M. Bates, R. Bates, C. Conneller, M. Pope, A. Shaw and G. Smith, 2014. A new view from La Cotte de St Brelade, Jersey. *Antiquity,* 88, 13-29.

Scott, B. and A. Shaw, 2018. La Cotte de St Brelade: place making, assemblage and persistence in the Normano-Breton Gulf, in *Crossing the human threshold: dynamic transformation and persistent places during the Middle Pleistocene,* eds. M. Pope, J. McNabb and C. Gamble.London: Routledge, 123-41.

Scott, K., 1980. Two hunting episodes of middle Palaeolithic age at La Cotte de Saint-Brelade Jersey (Channel Islands). *World Archaeology,* 12, 137-52.

Scott, K., 1986a. The bone assemblage from layers 3 and 6, in *La Cotte de St. Brelade 1961-1978. Excavations by C.B.M.McBurney,* eds. P. Callow and J.M. Cornford. Norwich: Geo Books, 159-84.

Scott, K., 1986b. The large mammal fauna, in *La Cotte de St. Brelade 1961-1978: excavations by C.B.M. McBurney,* eds. P. Callow and J.M. Cornford.Norwich: Geo books, 109-39.

Shaw, A., M. Bates, C. Conneller, C. Gamble, M.-A. Julien, J. McNabb, M. Pope and B. Scott, 2016. The archaeology of persistent places: the palaeolithic case of La Cotte de St Brelade, Jersey UK. *Antiquity,* 90, 1437-53.

Smith, G.M., 2015. Neanderthal megafaunal exploitation in Western Europe and its dietary implications: a contextual reassessment of La Cotte de St Brelade (Jersey). *Journal of Human Evolution,* 78, 181-201.

Stiner, M.C., 2002. Carnivory, coevolution, and the geographic spread of the genus *Homo. Journal of Archaeological Research,* 10, 1-64.

Stiner, M.C., R. Barkai and A. Gopher, 2009. Cooperative hunting and meat sharing 400-200kya at Qesem Cave, Israel. *Proceedings of the National Academy of Sciences,* 106, 13207-12.

Van Andel, T. and W. Davies (eds.), 2003. *Neanderthals and modern humans in the European landscape during the last glaciation,* Cambridge: McDonald Institute Monographs.

Washburn, S.L. (ed.) 1961. *Social life of early man,* New York: Wenner-Gren.

Clive Gamble
Department of Archaeology, University of Southampton, Avenue Campus, Southampton, SO17 1BF, UK.
clive.gamble@soton.ac.uk

Climate, Fire and the Biogeography of Palaeohominins

Robin I.M. Dunbar

While the exact timing of the first occupation of Eurasia may be subject to change in the light of new fossil discoveries, there seems little doubt that this radical shift in biogeography did not take place before the appearance of the genus *Homo.* The australopithecines and their allies remained resolutely confined to continental Africa, and mostly its eastern and southern extent at that. Not only is there no answer as to why australopithecines were confined to Africa, I am not even sure that the question has ever actually been asked. Australopithecine biogeographic distribution has simply been taken as a fact, and left at that.

By the same token, even after early *Homo* managed to break out of Africa and occupy Eurasia, the occupation of high latitudes remained elusive in the face of the Ice Ages. Archaic humans, notably the Neanderthals, had, of course, occupied southern Europe, but their northward extension remained constrained by an inability to cope with a more challenging thermal environment. As the ebb and flow of the Ice Ages moved isotherms up and down the Eurasian landmass, so the Neanderthals seem to have moved up and down in concert. It is only with the appearance of anatomically modern humans in Europe that the higher latitudes were permanently occupied.

In this contribution, I want to suggest that these particular cases can be explained by a general phenomenon that has largely remained below the radar in palaeoanthropology, namely how well animals cope with thermally challenging environments. I shall argue that this has been much more important than people realise.

Why were australopithecines confined to Africa?

Palaeoanthropologists have been much concerned with the details of the particular environments where australopithecines were found. However, they never seem to have asked the wider biogeographic question of where australopithecines lived in Africa, and why they remained confined to the continent. At best, there seems to be an implicit assumption that they were more or less confined to eastern and southern Africa, but nothing more beyond that. Of course, there has always been the implicit assumption that what fossil sites we have is just the bad luck of preservation, especially during the pre-*Homo* period: hominins were everywhere, but many of these sites are in places characterised by poor fossilisation conditions. This is always a risk, a more

careful examination of sites where hominins do and don't occur might tell us something we didn't know. In fact, all the sites at which australopithecines have been found are actually at moderate to high altitudes, and there is no evidence of their presence at low altitude sites even when those have good preservation of other mammals and primates. And may provide us with a clue to the problem they faced and the reason why they remained trapped in Africa, and, indeed, trapped in certain limited parts of Africa.

In a seminal series of papers published during the 1980s, Wheeler (1984, 1991a, 1992) introduced the suggestion that bipedal locomotion was an adaptation for thermoregulation (his 'stand tall, stay cool' hypothesis). His claim was based on extrapolating from well established thermal physiology equations to the body size and environment of (nominally) east African australopithecines. The essence of his claim was that by standing upright, especially when travelling, the animals would expose less of their body surface to direct sunlight than an animal who was walking quadrupedally, especially at midday when the tropical sun is directly overhead and at its hottest. He included both direct incident radiation as a function of time of day and the indirect effects of conduction of heat from the air, together with the cooling effects of wind speed on furred and naked bodies with different standing heights in relation to the surrounding vegetation. He concluded that while an animal of the body size of a baboon would gain little thermal advantage by standing bipedally, one the size of an australopithecine would, and hence that bipedalism would have been selected for only in this lineage of primates. Chimpanzees and other apes did not enter into the picture because, by and large, they were (and still are) confined to heavily forested habitats and were thereby protected from the thermal effects of direct sunlight.

The central issue hinges around the sensitivity of brain tissue to overheating: brain tissue has a very narrow thermal tolerance, and if it is heated up outside this range, it dies. Many savannah animals have anatomical mechanisms such as complex nasal retes (antelope: Maloiy et al. 1988) or long muzzles (baboons: Hiley 1976) that allow cerebral blood to be cooled by panting. Hominins lack these adaptations, and, Wheeler argued, would have found the open savannahs a significant challenge. Later, he showed that a naked animal would gain a significant further advantage from sweating that a furred animal would not, such that a bipedal hairless individual would be able to travel approximately twice as far on a litre of water as a quadrupedal furred one (Wheeler 1991b).

Subsequently, Ruxton & Wilkinson (2011a,b) pointed out that Wheeler had ignored the endogenous thermal costs of muscular activity during active travel. They claimed that introducing this additional cost removed any advantage of bipedalism and hairloss. However, Ruxton and Wilkinson made several rather unfortunate assumptions that directly affect their calculations.

First, they assumed that animals travel continuously throughout the day, and so calculated endogenous heat production from travel based on a constant rate of continuous travel across the whole day. In fact, no animal travels continuously; most tropical animals actually go to rest during the middle of the day when temperatures are at the hottest, and this is true of antelope (e.g. reedbuck: Roberts & Dunbar 1991), primates such as baboons (Hill 2006) that make regular use of open savannahs and even forest primates (Korstjens et al. 2010). It is even true of hunter-gatherers and post-industrial human societies in and near the tropics (Yetish et al. 2015; Monsivais et al. 2017). On average, primates spend only ~20% of their day travelling (Dunbar et al. 2009). Chimpanzees actually spend slightly less time travelling (17.1%: Lehmann et al. 2008a) than most primates, and this is because travel is particularly costly for great apes. As a result, the biogeographical distributions in Africa of both *Pan* and *Gorilla* are determined mainly by the costs of travel (Lehmann et al. 2007a, 2008a,b). Indeed, so intrusive is travel time that chimpanzees are only able to exist in most of the habitats where we now find them because they can exploit a fission-fusion form of sociality (Lehmann et al. 2007a), and this in turn severely limits the range of habitats they can occupy because of the risk of predation that they then incur by travelling in small parties (Lehmann & Dunbar 2009). Travel is much less problematic for baboons, whose distribution is limited instead mainly by the time required for feeding (Bettridge et al. 2010). Even so, it is significant that baboon troops that occupy forest edge and riverine habitats similar to those occupied by the australopithecines (Copeland et al. 2011) typically travel to and from foraging areas in the morning and late afternoon (Hill 2006). These are precisely the times when the Ruxton-Wilkinson model suggests that australopithecines might have had sufficient spare thermal capacity to allow travel. Estimates of australopithecine time budgets using baboon and chimpanzee models suggest that they probably spent about 16% of their day travelling (Bettridge 2010).

The second problem with the Ruxton-Wilkinson model is that they assumed that australopithecines lived at sea level. In fact, not only are *none* of the fossil sites

where australopithecines have been found at sea level, very few were even below ~1000m asl (allowing for the tectonic movement that has occurred over the past three million years, especially in the East African Rift Valley floor which is estimated to be ~1000m below its original position: Partridge 1997; Bonnefille et al. 1987) (Fig. 1a). Chimpanzees today live at somewhat lower altitudes (mean = 820m asl) than australopithecines did (means of 1488m and 1313m, respectively, in southern and eastern Africa), but this is only because mean global temperatures are now around 2°C lower than they were at 3-4 Ma due to the climatic downturn that occurred around 2.5 Ma. This is obvious from the actual mean temperatures that characterise these sites: in reality, the australopithecines lived under very similar thermal regimes to those occupied today by chimpanzees (Fig. 1b). These environments were considerably cooler than the Ruxton-Wilkinson model assumes (on average, by approximately 10°C!). It is worth noting, in addition, that none of the habitats occupied by either australopithecines or chimpanzees are as hot (or as cool) as those occupied by baboons. Baboons are able to cope with a much wider range of habitats, from sea level to dry woodland and even montane habitats, but chimpanzees cannot (and australopithecines could not).

Some palaeo-evidence for this is provided by the fact that, although there are low altitude sites dated between 2-4 Ma that have yielded fossil monkeys (implying that they are suitable for primates), none of them have yielded hominins. The Chiwondo beds of Lake Malawi (~500 m asl), for example, contain many papionins, but the high altitude specialist *Theropithecus* is rare (Frost & Kullmer 2008) despite being common at most of the contemporary classic higher altitude australopith sites in both eastern and southern Africa. Hominins (*Paranthropus bosei* and *Homo rudolfensis*) only appear at these low latitude sites after the 2.5 Ma climate downturn (Sandrock et al. 2007; Frost & Kullmer 2008; Bocherens et al. 2011). Other low altitude sites that contain monkeys but no australopithecines include Ahl al Oughlam in Morocco (100 m asl, dated to ~2.5 Ma; Alemseged & Geraads 1998), the late Miocene Libyan site of Qasr as Sahabi (0 asl; Boaz et al. 2008) and the Miocene/Pliocene boundary site of Langebaanweg in the Western Cape (30 m asl; Hendey 1981). Additionally, of course, there is the fact that ungulates (but not australopithecines) migrated via coastal routes back and forth between Eurasia and sub-Saharan Africa during this period (e.g. Bibi 2007), but hominins did not. In stark contrast, the appearance of *Homo* around 2.8 Ma marks a sea change in hominin biogeography. There is extensive evidence for the presence of *Homo* fossils as well as Olduwan-type industries at various coastal sites around the Mediterranean (Ternifine: Geraads et al. 2008; Kocabas, Turkey: Kappelman et al. 2008; various littoral sites in the Magreb and coastal Morocco

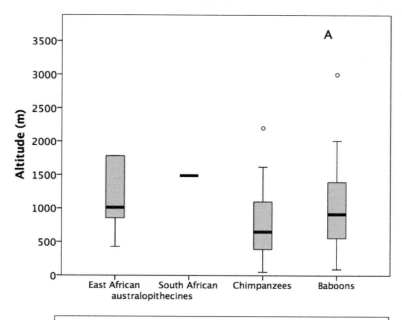

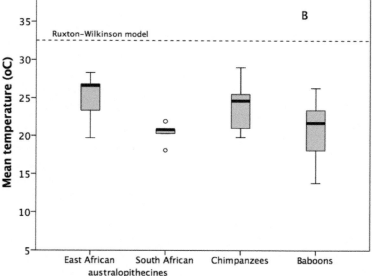

Figure 1. Median (±50% and 95% ranges) (a) altitude (m) and (b) mean ambient temperature (°C) for australpithecine fossil sites in eastern and southern Africa, together with data for contemporary chimpanzee and baboon populations. Fossil altitudes are calculated from current altitude adjusted for the extent to which the location has dropped in altitude over the last 3 million years. Temperatures for fossil sites are determined by equating them with contemporary sites on the basis of faunal profiles; the equivalent sites are given by Bettridge & Dunbar (2012b). Chimpanzee data are from Lehmann et al. (2008); baboon data are from Bettridge et al. (2010).

~1000m asl in East Africa, and ground frost is common at altitudes above 2500m asl. Cool nights are beneficial in that they provide an opportunity when heat accumulated by the body during the day can be dumped back into the environment by conduction, thus allowing the organism to cope with moderately high daytime temperatures. However, if night time temperatures drop too low, this can result in the animal losing heat too fast, with the risk of having to use energy to keep warm so as to avoid freezing to death.

Dàvid-Barrett & Dunbar (2016) introduced these three corrections into the Ruxton-Wilkinson model. The key results are shown in Fig. 2. Three important conclusions can be drawn from these results. First, Ruxton & Wilkinson (2011a,b) were right in claiming that there was no advantage to bipedalism if australopithecines were completely furred and living at sea level. However, had they checked it out, they would have found that their model also predicts that chimpanzees couldn't exist either, as they would also go into fatal thermal overload under these conditions. Had they noticed this, Ruxton & Wilkinson (2011a,b) might have been prompted to look more closely at why it was that chimpanzees seemingly bucked the trend of their model. That might have prompted them to notice where chimpanzees actually live, and, had they appreciated that present day temperatures occur at lower altitudes than they did at ~4 Ma, they might have been prompted to draw to the right conclusion. The important lesson here, perhaps, is that models are only as good as the modellers' knowledge of the animals' ecology and environmental conditions. When animals live on continents we are not familiar with, mistakes are almost inevitable. In a word, there is no substitute for firsthand knowledge.

Nonetheless, Ruxton & Wilkinson (2011a,b) unwittingly provided us with the clue as to why australopithecines remained confined to Africa. Being confined to moderate to high altitude plateau and montane habitats meant that australopithecines could not descend into the warm, humid coastal regions in order to be able to escape from continental Africa into Eurasia. In short, they were stuck where they were, unable to cope with the high heat loads in

and Algeria, including Ahl al Oughlam: Biberson 1961; Sahnouni 2006). And, of course, there is the indirect evidence for early *Homo* migrations out of Africa into Eurasia implied by the presence of fossils in Asia.

The third problem is that Ruxton and Wilkinson only considered the daytime. At moderate to high altitudes in the tropics, the nights are cool, even cold: minimum temperatures can fall as low as 5°C at altitudes of

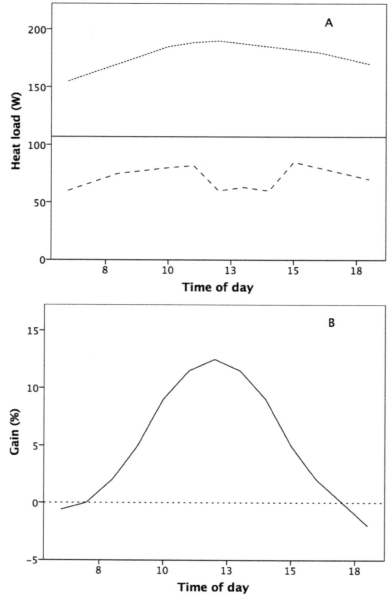

Figure 2. (a) Heat load (in watts, W) for a fully furred female bipedal australopithecine over the course of the day as predicted by the Ruxton-Wilkinson model (dotted line) and the Dàvid-Barrett/Dunbar model (dashed line). The solid line marks the maximum heat dissipation limit (the maximum rate at which the animal can lose heat = 107W) as calculated by Ruxton & Wilkinson (2011a). The Dàvid-Barrett/Dunbar model adds three corrections to the Ruxton-Wilkinson model: mean temperature corrected to 1000m asl, travel limited to 16% of time averaged across the day, and the hottest part of the day (1200-1459 h) spent resting in shade (reflected in the drop in heat load between 1200-1400 hrs in the dashed line). All three lines would be slightly higher for the males (male heat dissipation rate = 160W), but the pattern is identical. Individuals with a maximum of 15% hair cover would be below the maximum dissipation level for *both* models (i.e. even the Ruxton-Wilkinson model predicts that bipedalism would be advantageous for a hairless australopithecine). Redrawn from Dàvid-Barrett & Dunbar (2016). (b) The relative advantage in switching from quadrupedal to bipedal locomotion for a female australopithecine across the hours of the day, according to the Dàvid-Barrett/Dunbar model. The overall advantage is ~5%, but the benefit would be greatest (12%) during the middle of the day when ambient temperatures are highest and direct radiation greatest. The difference due to hairlessness is marginal. Redrawn from Dàvid-Barrett & Dunbar (2016).

the coastal lowlands that lapped, like the sea beyond, around their montane fastnesses.

The second conclusion that Dàvid-Barrett & Dunbar (2016) drew was that, at the altitudes where australopithecines did live (i.e. between ~800 and 1800m asl), bipedal australopithecines would have gained a 12% advantage over quadrupedal australopithecines in terms of heat load during the crucial hottest parts of the day. Averaged across the whole day, the selection advantage was ~5%. In terms of genetic fitness, this is a considerable advantage: most genetic selection ratios in natural populations are only in the order of 5-10% (Kingsolver et al. 2001). In short, the Wheeler hypothesis is right – at the altitudes where australopithecines actually lived. This is not, by the way, to suggest that australopithecines were adapted to open savannah plains. They almost certainly were not equipped to deal with those kinds of habitats, but then nor are baboons. Both taxa are best described as woodland specialists. Nonetheless, even these habitats are subject to significant radiant effects from sunlight, as anyone who has actually worked in them knows well.

Of course, the initial trigger for bipedalism may well have been arboreal feeding, as Hunt (1994), Thorpe et al. (2007), Crompton et al. (2008) and others have suggested. Nonetheless, it is clear that the real impetus for the evolution of the more efficient form of bipedalism that characterised australopithecines must have required something more substantial by way of a selection effect, and that selection effect had to have been explicitly terrestrial. Clambering around in trees will not promote the kind of fully plantar locomotion that characterises the hominin lineage from its earliest phase. It may have provided a transition point, but no more. Even *Papio* baboons have retained an opposable big toe in a way that their more exclusively terrestrial allies, the gelada, have not. Australopithecines were clearly adapting to a more explicitly terrestrial style of locomotion.

The third conclusion from the revised model was the fact that the australopithecines must have still been fully furred. Had they already been hairless, they would have been under a level of thermal challenge at night that would have been too intense to cope with. Dàvid-Barrett & Dunbar (2016) estimated that, even at 1000m altitude, the overnight heat loss for animals with just 15% hair cover would have left them in significant energy deficit by the morning, and the deficit would have got worse with increasing altitude. Females would have needed an extra 3500 kcals, and adult males 5600 kcals, a day to offset the deficit – set against a normal daily energy intake of just 1250 kcals for females and 1740 kcals for males (Aiello & Wells 2002). With a time budget that had absolutely no spare capacity (Bettridge 2010), this is not within the realm of ecological possibility. The bottom line is that the australopithecines *must* have been fully furred. Had they not been, they simply would have frozen to death overnight. The implication of this is that human hairlessness is unlikely to have evolved before the appearance of the genus *Homo*.

The australopithecines seem to have been adapted to a very specific kind of environment – that associated with a C4 niche (with or without termites) based on lacustrine and riverine floodplain-plus-gallery-forest habitats, where the forest strip provides safe night time refuges and the flood plain immediately beyond provides access to rich food sources. All known australopithecine fossil sites are associated with these kinds of habitats (Copeland et al. 2011). Here, the key food source seems to have been underground storage organs such as roots and tubers (Sponheimer & Lee-Thorp 2003; Sponheimer et al. 2005; Laden & Wrangham 2005; Ungar et al. 2006; Cerling et al. 2011; Ungar & Sponheimer 2011). These habitats are widespread but sparsely distributed in Africa, and are almost exclusively associated with the cooler, high altitude veldt-type regions of eastern and southern Africa (hence, no doubt, the fact that australopithecines were not more widely distributed across the continent). Nonetheless, there is a strong suggestion that australopithecines may have been dependent on caves for night time shelters: most of the early fossils, in southern Africa especially, are associated with cave deposits. Aside from providing a refuge from nocturnal predators, caves provide a significant thermal advantage: measurements from both South African and northwest European cave sites suggest that minimum temperatures are around 5°C higher inside the cave than immediately outside the cave (Barrett et al. 2004; Dunbar & Shi 2013).

In this context, it is important to appreciate how significant predation risk is for primates: most of their biogeographic distributions are in fact determined by the density of predators, even for large bodied species like chimpanzees and baboons (Lehmann & Dunbar 2009; Bettridge & Dunbar 2012a). Australopithecines could not have survived in the kinds of treeless habitats represented by floodplains had they (a) not been able to forage in large groups (terrestrial mammals' principal defence against predators) and (b) not had ready access to trees of substantial size (the only sure protection from cursorial predators, especially overnight). For baboons, the three most important predators that most strongly influence both their grouping patterns and biogeography are leopard, lion and hyaena (Cowlishaw 1994), all of whom are at their most dangerous at night when primates are most vulnerable because of poor night vision. The importance of the threat offered by hyaena is not widely appreciated: they are formidable predators and are the more threatening because they are more likely to chance an attack (and when they do, to attack in large groups) than leopard or lion.

The appearance of *Homo* after ~2.5 Ma is marked by two important factors relevant to this. One is the fall in global temperatures, which would have shifted the habitable zone downwards in altitude by 300m or so, thereby bringing the coastal zones within physiological striking distance. However, notice that even contemporary chimpanzees cannot access the coast itself, and so remain locked within Africa. So for early *Homo* to make the final step to sea level (and Eurasia) something else was required. This seems to have required five important changes: an increase in body size (and hence stature), hairlessness, the capacity to sweat copiously, a more efficient striding form of bipedal locomotion and a significant selection pressure against the original australopithecine niche.

The first three of these would have allowed them to occupy warmer lower altitude habitats. A larger body can store a greater heat load, and hairlessness combined with sweating would have helped reduce heat load. Sweating and hairlessness have to go together because, as Wheeler (1991b) pointed out, sweating with fur simply keeps the tips of the fur cool and has no effect on core body temperature. To benefit from sweating, the sweat must evaporate off the skin itself. Wheeler (1991b) showed that the capacity to sweat copiously would have allowed bipedal hominins to travel twice as far on a litre of water as animals that don't sweat (as is the case for other primates). An increase in body size has two separate advantages: larger body size reduces exposure to predators, but it also allows cursorial mammals to travel faster (or cover more ground in the same time period). All mammals move their limbs at the same constant speed when moving: what determines their speed of travel is the length of the stride they can take (Heglund et al. 1974). So an increase in limb length (essentially a third power of body size) would have allowed *Homo* to travel proportionately further in the same amount of time. Taken together, these would have allowed early *Homo* to search much more widely for food sources and yet still be within comfortable travel distance of safe night time refuges.

What exactly selected against the australopithecine niche is not clear, though shifts in altitudinal bands at which preferred habitats lie may well have been the problem. What is clear, however, is that although the robust australopithecines and the habilines survived for at least another half million years, the entire taxonomic group had died out by about 1.5 Ma.

Fire and the occupation of high latitudes

While early *Homo* were successful in colonising Eurasia, the fossil evidence suggests that they were able to colonise high latitudes, in Europe in particular, only during warm spells. Once the current Ice Ages got under way from ~800 ka, *Homo erectus* was forced southwards towards the Tropics. It was not until the appearance of archaic humans, and in particular *Homo neanderthalensis* and their sister taxon the Denisovans, that serious in-roads were made once more into high latitude Europe. Even then, there is evidence from areas like the British Isles that occupancy of higher latitudes was spasmodic and mainly coincided with warmer intervals (White 2006; White & Pettit 2011). Since Europe was still in the grip of the Ice Ages at this point and even the warm period temperatures at the latitudes of Britain were decidedly cool (White & Pettitt 2011), something must have changed to allow them to do this. I want to suggest that what made the difference was control over fire.

The problem that humans face at high latitudes in Ice Age Europe and Asia was the energetic cost of coping with low temperature. The energetic cost of maintaining thermoneutrality (constant body temperature) increases steeply as environmental temperature declines, even in large-bodied mammals (Mount 1979). During the process of development, the ability to invest in foetal brain growth depends on the mother's capacity to spare resources from somatic maintenance (including thermoregulation). If large quantities of energy have to be diverted to maintaining thermoneutrality, then less will be available to be invested in foetal brain growth, both by the mother during pregnancy and lactation and by the infant itself once it becomes nutritionally independent of the mother. Since in humans the bulk of brain growth takes place before the infant is completely nutritionally independent (at about 5 years of age in modern humans), brain development is

particularly sensitive to the mother's environmental circumstances.

This seems to have acted as an important constraint on archaic humans' capacity to grow large brains. For early *Homo*, the difference in cranial volume between high and low latitude populations is marginal, and, if anything, slightly in favour of high latitude populations (perhaps reflecting a demand for better visual processing under low ambient light conditions: see Pearce et al. 2013) (Fig. 3). By comparison, early archaic humans (pre-400 ka) show a very striking difference with high latitude populations having much smaller cranial volumes – in a context where Pearce et al. (2013) might lead us to expect the reverse. This difference is much reduced but still present in later archaics (post-400 ka), who also show a striking increase in cranial volume, especially at high latitudes. In stark contrast, anatomically modern humans exhibit the reverse pattern, with high latitude populations having larger cranial volumes than low (tropical) latitude populations. Much of this difference can be attributed directly to the demand for more acute visual processing in the low light level conditions at high latitudes (something that is also present in both historical and contemporary human populations: Pearce & Dunbar 2012; Pearce & Bridge 2013). Lower ambient light levels require a large retina (and hence a large orbit to house this) and a proportionately larger visual cortex (occipital lobe) to process the incoming signal in order to maintain equivalence of visual acuity (Pearce & Dunbar 2012).

I have suggested elsewhere (Dunbar 2014) that the dramatic shift in cranial volume at around 400 ka may be due to archaic humans gaining control over

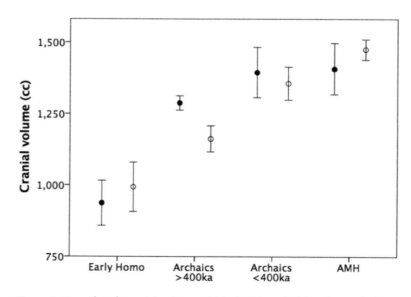

Figure 3. Mean (±2se) cranial volume within (solid symbols) and outside (open symbols) 43° latitude for early *Homo*, archaic humans (dated at before and after 400 ka) and fossil anatomically modern humans (AMH). Source of data: de Miguel & Henneberg (2001).

fire: it was this that gave later archaics and AMH the advantage they needed to colonise high latitudes outside the Tropics. The evidence for control of fire at around this time is uncontroversial. An extensive review of fire use as indexed by the presence of hearths in the archaeological record (Dunbar & Gowlett 2014; see also Roebucks & Villa 2011) indicates that, while there is some evidence for fire from ~1 Ma, it is *very* sporadic, and there are very long gaps. However, as Fig. 4 suggests, from 400 ka, the evidence for hearths increases dramatically, with a consistent pattern of ~50% of all archaeological sites having hearths or other evidence for fire (see also Shahack-Gross et al. 2014; Shimelmitz et al. 2014). Archaeologists tend to emphasise first occurrences, but the only criterion that is ecologically meaningful is when a trait becomes the norm (see also Shimelmitz et al. 2014). Casual use of fire cannot be a solution to a persistent ecological problem. There may well be evidence of fire as early as 1.6 Ma (notably at Chesowanja, Koobi Fora and Swartkrans: Gowlett et al. 1981; Clark & Harris 1985; Brain 1993) and more convincing evidence from sites dated around 1.0 Ma (Gesher Benot Ya'aqov in Israel and Wonderwerk in southern Africa: Goren-Inbar et al. 2004; Berna et al. 2012), but the reality is that regular use of fire does not emerge until around 400 ka (Roebroeks & Villa 2011; Gowlett & Dunbar 2014). The contrast before and after 400 ka is simply too dramatic to be explained away as being due to taphonomy or excavation effort.

More importantly, there is a significant correlation between cranial volume and the percentage of sites with hearths among these archaic human populations (Fig. 4: r=0.734, p=0.024). These data strongly suggest a phase shift rather than a continuum: cranial volume increases dramatically and suddenly at just the time when hearth use switches from negligible to being common. Interestingly, the variance in cranial volume among both archaic and modern humans is high after 400 ka, suggesting, perhaps, that not everyone had control over fire (as indeed seems to be the case, at least among Neanderthals: Sandgathe et al. 2011). This is precisely what one might expect of a culturally transmitted trait.

Fire has three important implications for hominins: warmth, cooking and as a source of light. Cooking is the function that has received most attention, thanks to Wrangham's (2010; Wrangham et al. 1999; Carmody et al. 2011) proposal that cooking played a seminal role in facilitating hominin brain evolution. Wrangham's argument is perfectly plausible: cooking increases the digestibility of certain key foods (specifically tubers and red meat) by about 50%. Since meat obviously became increasingly important to humans over time, this is *prima facie* evidence to support the hypothesis. The substantive issue, however, is timing: did cooking occur early (at ~2.0 Ma), as Wrangham has argued, or late (specifically, as late as 400 ka)? I suggest that the archaeological evidence for fire implies a late date.

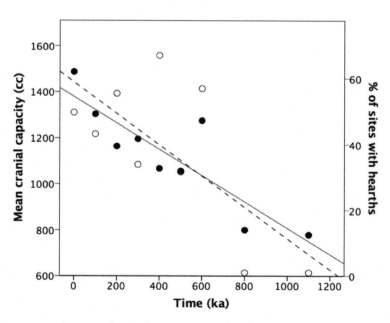

Figure 4. The correlation between cranial volume in archaic humans (Heidelbergs and Neanderthals only) and the percentage of sites with evidence for hearths, as a function of time. Sites are grouped by 100,000-year intervals. Hearth data are given as three-point rolling mean of the percentage of archaeological sites that have evidence for hearths. Sources: cranial volume data from de Miguel & Henneberg (2001); hearth data from Roebroeks & Villa (2011).

A second reason for thinking that a late date is likely for cooking is that estimates of foraging time demand suggest that archaic human time budgets were under *much* more pressure than was the case for australopithecines or early *Homo* (Dunbar 2014). Fig. 5 plots the percentage of the day that different hominin species would have had to devote both to foraging and for their total time budget adding in social time and time required for travel and rest (based on the calculations given by Dunbar 2014). Foraging time is calculated by using the ratio of body mass and brain mass (in each case, using metabolic mass, calculated as mass$^{0.75}$) to scale up the estimate for australopithecine foraging time calculated by Bettridge (2010) on the basis of great ape time budgets (see Dunbar 2014). Social time requirements are based directly on cranial volumes, using these to calculate neocortex ratio and from this social group size and then social time requirement from the relevant equation for primates (for details, see Dunbar 2014). Travel time and enforced rest time (time that has to be

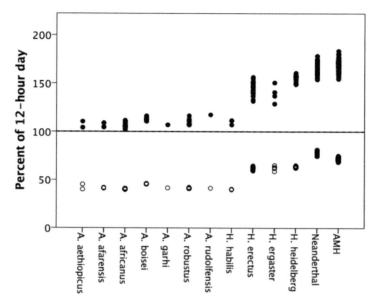

Figure 5. Time budgets of fossil hominin species. Open circles: foraging time; solid circles: total time budget, including fixed travel (16%) and rest (32%), plus social time, calculated for individual fossil specimens. Travel time and rest time are those calculated for australopithecines by Bettridge (2010) using ape time budget models, and for convenience are treated as fixed. Social time is calculated by estimating group size from cranial volume, and then using this to calculate required social time using the primate grooming time equation given by Lehmann et al. (2007b). Neanderthal social time is adjusted for the fact that their frontal lobe was smaller than would be expected for cranial volume (due to differential investment in the visual system), and hence their group sizes would have been lower (for details, see Pearce et al. 2013).

spent resting, mainly because of high environmental heat loads: see Korstjens et al. 2010) are taken as fixed throughout, using the values of 16% and 32%, respectively, calculated for australopithecines by Bettridge (2010). (For a brief summary of these models, see Appendix.)

The impact of increased body size and brain size really only becomes significant with the appearance of *Homo*, but it becomes especially intense with Neanderthals and modern humans for whom ~75% of the day would have had to be devoted to foraging had they had the same diet as australopithecines. Our problem is that if cooking solved early *Homo*'s time budgeting crisis, what was left to solve archaic humans' even worse crisis? More importantly, cooking is not a complete panacea: cooking has little effect on the digestibility of foods like white meat, fish, fruits, leaves, honey, etc. The real benefit is only in respect of tubers and red meat, and these typically account, on average, for only about 45% of modern hunter-gatherer diets (Cordain et al. 2000), with the possibility of a trade-off between these two dietary categories between low and high latitude populations (Grove 2010). In effect, cooking will reduce overall foraging time demand by only 45% x 50% = 22.5%. This would reduce early *Homo* and archaic human foraging time to ~48% of the day and that

of Neanderthals and modern humans to ~58% of the day. These are helpful contributions in both cases, but don't solve the whole problem: they leave early *Homo* and archaic human time budgets adrift by 30 percentage points and Neanderthal and AMH time budgets adrift by a substantial 50 percentage points. The former is easily taken care of by modest contributions from the expensive tissue hypothesis (Aiello & Wheeler 1990) and a shift to more nutrient-rich meat-based diet (without cooking: cooking doesn't make meat digestible, it simply increases its digestibility), combined with the use of a more efficient form of social bonding based on chorusing in the form of laughter (Dunbar 2012, 2014). The problem faced by Neanderthals and AMH, by comparison, is *much* more serious.

Fire has two further advantages that don't really become a major issue before the appearance of archaic humans. The heat generated by fire has a dramatic impact on the air temperature immediately around the hearth, especially if the hearth is within a cave. In low temperature habitats, such as those outside the Tropics, the combination of caves and fire would have introduced very considerable savings in terms of the costs of thermoregulation. For every 1°C that ambient temperature is below the animals' thermoneutral zone (37°C in the case of humans), it has to expend (and therefore acquire by foraging) a proportional amount of energy (Mount 1979). Raising night time temperatures, in particular, by 10-15°C, as would be the case from having fires inside caves, would offer massive savings in terms of energy acquisition.

In addition, by lighting up the evening, fire effectively extends the working day and allows at least some activities to be shifted into the night, thereby freeing up time during the day for other essential activities. The drawback is that the circle of light cast by a hearth is quite modest, and the quality of light is poor. However, even if the light quality is too poor to make tools by (especially if that requires fine work), it is good enough to cook and eat food by, and is perfect for social engagement. Extending the working day by 3-4 hours (from dusk) adds a significant amount of time onto the 12-hour tropical day that other primates are limited to, and especially so at high latitudes in winter when day length is appreciably shorter. If the day's activity schedule is restructured so that most of the business of eating and socialising is done in the evening, three hours can be freed off during the daytime for foraging. In fact, the average amount of time contemporary humans spend in social interaction is almost exactly 3 hours (Dunbar 1998). That would be sufficient to allow for 58% of the day to be devoted to foraging for food, which, with an additional 16% for travel, would still allow a full 30% to be devoted to resting without breaking the time budget bank. The crucial difference that allowed humans to balance their activity budget at this point is the 17 percentage points that cooking would have saved on the foraging time budget.

Some evidence to support this suggestion is the fact that, among the !Kung-San hunter-gatherers, most social conversations (those involving story telling and socially-relevant comments, as opposed to the exchange of factual or economic information) take place in the evening rather than during the day (Weissner 2014). This has important implications for the evolution of language. If most conversational activity takes place in the evening, then it puts a premium of the vocal as opposed to a gestural channel: gestures are not easy to see in the half-light of a fire, never mind elsewhere in the camp, whereas vocal communication carries well beyond the circle of firelight.

Conclusions

My aim in this chapter has been to emphasise the singular importance of environmental temperature and thermoregulation to the story of human evolution. While palaeoanthropologists invariably pay lip service to temperature in the context of the European Ice Ages, they radically underestimate how important thermoregulation is even in the Tropics. It turns out to be the single most important factor determining biogeographic distribution in Old World monkey and apes (Dunbar et al. 2009). Peter Wheeler is clearly the exception in this respect, but then he was a physiologist rather than a palaeontologist by background. At the other end of the time scale, John Gowlett, with his interest in fire, has been an equally lone voice. My interest in the topic has come directly from studying primates and antelope in their natural habitats in Africa, and learning to experience the environment as they experience it. It became obvious to me that living under the conditions in which most of these species, including hominins, lived was no mean feat – not simply because they had to find enough to eat each day to satisfy their physical requirements, but because time is extremely limited and they have many more important things to do than just eat. There are parts of the day when they simply cannot do anything except rest (even social grooming is energetically too expensive), they have to travel to and from feeding sites, and they have to devote sufficient time to social interaction to maintain the coherence of their social communities (Dunbar et al. 2009). And all this has to be done within the constraints imposed by significant daily variation in temperature and incident radiation load.

I have offered two very different examples of how this might have constrained hominin evolution. One involved the way high daytime and low night time temperatures seem to have constrained where australopithecines could live – in ways that palaeoanthropologists seem not to have appreciated. It confined them to Africa. The exodus from Africa depended partly on global climate change lowering the altitudes at which early hominins could survive. It also partly depended, however, on finding additional solutions to the problems of overheating. Hairlessness and sweating was clearly part of that, as may have been the ability to travel further in less time (i.e. having longer legs). The second example I discussed concerned later humans' ability to colonise high latitudes in Eurasia once the Ice Ages set in. I have suggested that low ambient temperatures imposed a significant constraint on later hominin brain evolution – the very many claims for a simple continuous increase in brain size over the course of human evolution notwithstanding. Developing control over fire seems to have made the crucial difference, and I suggest that the archaeological evidence implies that this happened quite late – around 400 ka. I highlighted the fact that control over fire gave us much more than just cooking, however important that was in managing nutrient throughput. Allowing the working day to be extended into the night must have played a seminal role by making it possible to shift most social activity into the evening – a period that is effectively dead time for all primates.

One further point is worth emphasising. Much of what I have argued here has depended on quantitative analyses using models. Palaeoanthropologists tend to shy away from mathematical modelling of this kind (unless it is concerned with anatomical shape). I want to stress its heuristic value. In an important sense, models provide proof of concept by showing whether or not it is possible for the explanation we have in mind to work. This is not to underestimate the difficulty of working with models of this kind. Mathematical models require us to spell out the explicit assumptions that we make in constructing our explanations. That means we really have to understand the biology and ecology of the animals concerned. If we don't, we risk what is known as the GIGO [Garbage In, Garbage Out] form of modelling (a fine example of which is provided by the recent González-Forero & Gardner (2018) analysis of hominin brain evolution – although, inadvertently, they do at least prove that the social brain hypothesis has nothing to do with social cooperation, as we have known for a long time). Avoiding such pitfalls requires that we be more engaged with the lives that our organisms actually lived. That's a big ask, but largely it is just a matter of investing time and effort in familiarising ourselves with the kinds of environments they lived in and how the animals that live in them now cope with the exigencies they face. The archaeological record tells us what those environments were like: the rest is up to us.

Appendix

As part of the *Lucy to Language* Project, we developed a series of time budget models for individual primate genera (Dunbar et al. 2009). To date, genus-specific models have been developed (by us and others) for 11 primate genera (including all three great apes and, most recently, the hylobatids) and one ungulate (feral goats). The models use data on the known time budgets of individual study groups of a given genus to derive equations for the impact of climate and demographic variables on feeding, travel and resting time. These equations are then used to determine the taxon's time budget for these three core components at a given location, and hence how much spare capacity it will have for social interaction; this spare capacity is used to determine the largest possible group size the taxon could manage as a coherent social entity at that location using the general equation relating grooming time to group size for primates as a whole. The value of group size predicted by the model is the maximum possible group size that the species can maintain in a given location. More details can be found in Dunbar et al. (2009), and the papers referenced therein.

The models are tested by determining how well they predict the biogeographic distribution of the genus on a continental scale (e.g. the whole of sub-Saharan Africa in the case of an African genus). Presence/absence is typically predicted with an accuracy of 75-85%. These models are at least as accurate in predicting a species' biogeographic distribution as conventional climate envelope models (these simply relate presence/absence to local climate variables), which, with the same datasets, typically have an accuracy of ~70% (Willems & Hill 2009).

These models tell us whether a species could live at a given site (i.e. whether they could maintain a group of sufficient minimum size). Not only do the models tell us where a taxon could and couldn't live (and what aspect of their time budget prevents them living outside that range), but they have the additional merit of telling us what group sizes the taxon can manage at these locations and how much ecological stress they are under at that site. This is because they include an extra step (time budgets and group size) in the equation relating climate to biogeography.

These models have been extended to fossil baboons (Dunbar 1993; Bettridge & Dunbar 2012b). Bettridge (2010) also developed a model for australopithecines using the *Papio* and *Pan* models as a template to determine what equations were required to predict the presence and absence of australopiths at fossil sites in the time span 2-4 Ma. Neither of the individual taxon models was at all successful in predicting where australopiths actually occurred: the baboon model had them living everywhere, and the chimpanzee model confined them to a narrow range around the equator. However, by manipulating combinations of the two models' equations, it was possible to arrive at a model that gave a conventionally reasonable fit to the observed distribution of australopithecine fossil sites. This model yields a typical time budget and a typical maximum group size for the taxon. An important finding from this analysis was that the australopithecine time budget had absolutely no spare capacity: the sum of the four predicted time budget activities was 100% of available daytime.

I used the mean values for feeding, travel and resting time predicted by this model as a baseline against which to evaluate the changing demand across individual *Homo* populations as a function of changes in body and brain size. For computational simplicity, I assumed that travel and resting time remain constant. Feeding time was adjusted by multiplying the australopithecine value by the ratio of metabolic body weight of any given *Homo* species relative to the australopithecine average, including an adjustment for the higher metabolic costs of brain matter. Social time was determined by using the ape social brain equation to predict social group size for a given species, and then using the primate grooming time equation to determine how much time would need to be devoted to social grooming (assuming that all social bonding was done by grooming). The details and results can be found in Dunbar (2014).

The aim is to calculate the bottom line time budget requirement relative to that for the australopithecine baseline. Since this will always be greater than that for the australopithecines, and hence above 100% of available time (if only because most *Homo* species had bigger bodies), this gives us some sense of the challenge that *Homo* populations faced in order to survive. Given that they clearly did survive, they must have made some kind of adjustments to behaviour or diet in order to bring their time budgets down into line (i.e. below 100% of daytime). Our task is to figure out what they could have changed and what impact such a shift would have had on their time budgets. The full preliminary analysis can be found in Dunbar (2014), but I stress that those analyses are very preliminary. More detailed analyses are in progress.

Acknowledgments

Fire was not something I had thought much about, but I was stimulated to think about its implications for hominin evolution by a long series of discussions with John Gowlett during the course of the British Academy's Lucy Project (2003-2010).

References

Aiello, L.C. & Wells, J.C.K. (2002). Energetics and the evolution of the genus Homo. *Ann. Rev. Anthropol.* 31, 323-338.

Alemseged, Z. & Geraads, D. (1998). *Theropithecus atlanticus* (Thomas, 1884) (Primates: Cercopithecidae) from the late Pliocene of Ahl al Oughlam, Casablanca, Morocco. *J. Human Evol.* 34, 609-621.

Barrett, L., Gaynor, D., Rendall, D., Mitchell, D. & Henzi, S.P. (2004). Habitual cave use and thermoregulation in chacma baboons (*Papio hamadryas ursinus*). *J. Human Evol.* 46, 215-222.

Berna, F., Goldberg, P., Horwitz, L.K., Brink, J., Holt, S., Bamford, M. & Chazan, M. (2012). Microstratigraphic evidence of in situ fire in the Acheulean strata of Wonderwerk Cave, Northern Cape Province, South Africa. *Proceedings of the National Academy of Sciences, USA*, 111: 14027-14035.

Bettridge, C.M. (2010). *Reconstructing Australopithecine Socioecology Using Strategic Modelling Based on Modern Primates.* DPhil thesis, University of Oxford.

Bettridge, C. & Dunbar, R. (2012a). Predation as a determinant of minimum group size in baboons. *Folia Primat.* 83: 332-352.

Bettridge, C. & Dunbar, R.I.M. (2012b). Modelling the biogeography of fossil baboons. *Int. J. Primatol.* 33: 1278-1308.

Bettridge, C., Lehmann, J. & Dunbar, R.I.M. (2010). Trade-offs between time, predation risk and life history, and their implications for biogeography: a systems modelling approach with a primate case study. *Ecol. Modelling* 221: 777-790.

Biberson, P. (1961). *Le Paléolithique inférieure du Maroc Altantique.* Services des Antiquités du Maroc, Rabat.

Bibi, F. (2007). Origin, paleoecology, and paleobiogeography of early Bovini. *Palaeogeogr. Palaeocl.* 248, 60-72.

Boaz, N.T., El-Arnauti, A., Augusti, J., Bernor, R.L., Pavlakis, P.P. & Rook, L. (2008). Temporal, lithographic, and biochronologic setting of the Sahabi formation, north-central Libya. *Geol. E. Libya* 3, 959-872.

Bocherens, H., Sandrock, O., Kullmer, O. & Schrenk, F. (2011). Hominin palaeoecology in Late Pliocene Malawi: First insights from isotopes (13C, 18O) in mammal teeth. *S. Afr. J. Sci.* 107. #331.

Bonnefille, R., Vincens, A. & Buchet, G. (1987). Palynology, stratigraphy and palaeoenvironment of a Pliocene hominid site (2.9-3.3 M.Y.) at Hadar, Ethiopia. *Palaeogeogr. Palaeocl.* 60, 249-281.

Brain, C.K. (1993). The occurrence of burnt bones at Swartkrans and their implications for the control of fire by early hominids. In: C.K. Brain (ed.) *Swartkrans: A Cave's Chronicle of Early Man*, pp.229-242. Pretoria: Transvaal Museum.

Brain, C.K. & Sillen, A. (1988). Evidence from the Swartkrans cave for the earliest use of fire. *Nature* 336: 464-466.

Carmody R.N., Weintraub G.S. & Wrangham R.W. (2011). Energetic consequences of thermal and nonthermal food processing. *Proceedings of the National Academy of Science, USA,* 108: 19199-19203.

Cerling, T., Mbua, E., Kirera, F., Manthi, F., Grine, F., Leakey, M., Sponheimer, M. & Uno, K. (2011). Diet of *Paranthropus boisei* in the early Pleistocene of East Africa. *Proceedings of the National Academy of Sciences, USA* 108: 9337-41.

Clark, J.D. & Harris, J.W.K. (1985). Fire and its roles in early hominid lifeways. *African Archaeological Review* 3: 3-27.

Copeland, S., Sponheimer, M., de Ruiter, J., Lee-Thorp, J., Codron, D., le Roux, P., Grimes, V. & Richards, M P. (2011). Strontium isotope evidence for landscape use by early hominins. *Nature* 474: 76–9.

Cordain, L., Miller, J.B., Eaton, S.B., Mann, N., Holt, S.H.A. & Speth, J.D. (2000). Plant-animal subsistence ratios and macronutrient energy estimations in worldwide hunter-gatherer diets. *American Journal of Clinical Nutrition* 71: 682-692.

Cowlishaw, G. (1994). Vulnerability to predation in baboon populations. *Behaviour* 131, 293-304.

Crompton, R.H., Vereecke, E.E., Thorpe, S.K.S., 2008. Locomotion and posture from the common hominoid ancestor to fully modern hominins, with special reference to the last common panin/hominin ancestor. *J. Anat.* 212, 501-543.

Dàvid-Barrett, T. & Dunbar, R. (2016). Bipedality and hair loss in human evolution revisited: the impact of altitude and activity scheduling. *J. Human Evol.* 94: 72-82.

De Miguel, C. & Henneberg, M. (2001). Variation in hominin brain size: how much is due to method? *Homo* 52: 3-58.

Dunbar, R.I.M. (1993). Behavioural ecology of the extinct theropiths. In: N. Jablonski & R.A. Foley (eds) *Theropithecus: The Rise and Fall of a Primate Genus*, pp. 465-486. Cambridge: Cambridge University Press.

Dunbar, R.I.M. (1998). Theory of mind and the evolution of language. In: J. Hurford, M. Studdart-Kennedy & C. Knight (eds) *Approaches to the Evolution of Language*, pp. 92-110. Cambridge: Cambridge University Press.

Dunbar, R.I.M. (2012). Bridging the bonding gap: the transition from primates to humans. *Phil. Trans. R. Soc. Lond.* 367B: 1837-1846.

Dunbar, R.I.M. (2014). *Human Evolution.* Harmondsworth: Pelican Press and New York: Oxford University Press.

Dunbar, R.I.M. & Gowlett, J.A.J. (2014). Fireside chat: the impact of fire on hominin socioecology. In: R.I.M. Dunbar, C. Gamble & J.A.J. Gowlett (eds) *Lucy to Language: the Benchmark Papers*, pp. 277-296. Oxford: Oxford University Press.

Dunbar, R.I.M. & Shi, J.B. (2013). Time as a constraint on the distribution of feral goats at high latitudes. *Oikos* 122, 403-410.

Dunbar, R.I.M., Korstjens, A.H. & Lehmann, J. (2009). Time as an ecological constraint. *Biol. Rev.* 84, 413-429

Frost, F.R., Kullmer, O. (2008). Cercopithecidae from the Pliocene Chiwondo beds, Malawi-rift. *Geobios* 41, 743-749.

Geraads, D., Hublin, J.-J., Jaeger, J.-J., Tong, H., Sen, S. & Toubeau, P. (2008). The Pleistocene hominid site of Ternifine, Algeria: New results on the environment, age, and human industries. *Quaternary Research* 25, 380-386.

González-Forero, M. & Gardner, A. (2018). Inference of ecological and social drivers of human brain-size evolution. *Nature* 557, 554-557.

Goren-Inbar N., Alperson N., Kislev M.E., Simchoni O., Melamed Y., Ben-Nun A. & Werker E. (2004). Evidence of Hominin Control of Fire at Gesher Benot Ya`aqov, Israel. *Science* 304: 725-727.

Gowlett, J.A.J., Harris, J.W.K., Walton, D. & Wood, B.A. (1981). Early archaeological sites, hominid remains and traces of fire from Chesowanja, Kenya. *Nature* 294: 125-129.

Grove, M. (2010). Logistical mobility reduces subsistence risk in hunting economies. *Journal of Archaeological Science* 37: 1913-1921.

Heglund, N.C., Taylor, C. R. & McMahon, T.A. (1974). Scaling stride frequency and gait to animal size: mice to horses. *Science* 186: 1112-1113.

Hendey, Q.B. (1981). Palaeoecology of the Late tertiary fossil occurrences in 'E' quarry, Langebaanweg, South Africa, and a reinterpretation of their geological context. *Ann. S. Afr. Mus.* 84, 1-104.

Hiley, P.G. (1976). The thermoregulatory responses of the galago (*Galago crassicaudatus*), the baboon (*Papio cynocephalus*) and the chimpanzee (*Pan satyrus*) to heat stress. *J. Physiol.* 254, 657-671.

Hill, R.A. (2006). Thermal constraints on activity scheduling and habitat choice in baboons. *Am. J. Phys. Anthropol.* 129, 242-249.

Hunt, K.D., 1994. The evolution of human bipedality: ecology and functional morphology. *J. Human Evol.* 26, 183-202.

Kappelman, J., Alçiçek, M.C., Kazanci, N., Schultz, M., Ôzkul, M. & Sen, S. (2008). First Homo erectus from Turkey and implications for migrations into temperate Eurasia. *Amer. J. Phys. Anthrop.* 135, 110e116.

Kingsolver, J.G., Hoekstra, H.E., Hoekstra, J.M., Berrigan, D., Vignieri, S.N., Hill, C.E., Hoang, A., Gibert, P., Beerli, P., 2001. The strength of phenotypic selection in natural populations. *Am. Nat.* 157, 245-261.

Korstjens, A., Lehmann, J. & Dunbar, R.I.M. (2010). Resting time as an ecological constraint on primate biogeography. *Anim. Behav.* 79: 361-374.

Laden, G. & Wrangham, R.W. (2005). The rise of the hominids as an adaptive shift in fallback foods: Plant underground storage organs (USOs) and australopith origins. *Journal of Human Evolution* 49, 482-498.

Lehmann, J. & Dunbar, R.I.M. (2009). Implications of body mass and predation for ape social system and biogeographical distribution. *Oikos* 118: 379-390.

Lehmann, J., Korstjens, A.H. & Dunbar, R.I.M. (2007a). Fission-fusion social systems as a strategy for coping with ecological constraints: a primate case. *Evol. Ecol.* 21, 613-634.

Lehmann, J., Korstjens, A. & Dunbar, R.I.M. (2007b). Group size, grooming and social cohesion in primates. *Anim. Behav.* 74: 1617-1629.

Lehmann, J., Korstjens, A. & Dunbar, R.I.M. (2008a). Time and distribution: a model of ape biogeography. *Ethol. Ecol. Evol.* 20: 337-359.

Lehmann, J., Korstjens, A. & Dunbar, R.I.M. (2008b). Time management in great apes: implications for gorilla biogeography. *Evol. Ecol. Research* 10: 515-536.

Maloiy, G.M.O., Rugangazi, B.M. & Clemens, E.T. (1988). Physiology of the dik-dik antelope. *Comp. Biochem. Physiol. A Physiol.* 91, 1-8.

Monsivais, D., Bhattacharya, K., Ghosh, A., Dunbar, R.I.M. & Kaski, K. (2017). Seasonal and geographical impact on human resting periods. *Scientific Reports* 7: 10717.

Mount, L.E. (1979). *Adaptation to Thermal Environment: Man and his Productive Animals.* London: Edward Arnold.

Partridge, T.C. (1997). Late Neogene uplift in Eastern and Southern Africa and its paleoclimatic implications. In: Ruddiman, W.F. (Ed.) *Tectonic Uplift and Climate Change*, pp. 63-86. Hamburg: Springer.

Pearce, E. & Bridge, H. (2103). Is orbital volume associated with eyeball and visual cortex volume in humans? *Ann. Human Biol.* 40, 531-540.

Pearce, E. & Dunbar, R.I.M. (2012). Latitudinal variation in light levels drives human visual system size. *Biol. Lett.* 8: 90-93

Pearce, E., Stringer, C. & Dunbar, R.I.M. (2013). New insights into differences in brain organisation between Neanderthals and anatomically modern humans. *Proc. R. Soc. Lond.* 280B: 1471-1481.

Roberts, S.C. & Dunbar, R.I.M. (1991). Climatic influences on the behavioral ecology of Chanler mountain reedbuck in Kenya. *Afr. J. Ecol.* 29, 316-329.

Roebroeks, W. & Villa, P. (2011). On the earliest evidence for the use of fire in Europe. *Proceedings of the National Academy of Sciences, USA*, 108: 5209–5214.

Ruxton, G.D. & Wilkinson, D.M. (2011a). Avoidance of overheating and selection for both hair loss and bipedality in hominins. *Proceedings of the National Academy of Sciences, USA*, 108, 20965-20969.

Ruxton, G.D. & Wilkinson, D.M. (2011b). Thermoregulation and endurance running in extinct hominins: Wheeler's models revisited. *J. Human Evol.* 61, 169e175.

Sahnouni, M. (2006). The North African Early Stone Age and the sites at Ain Hanech, Algeria. In: Toth,

N., Schick, K. (Eds.) *The Olduwan: Case Studies into the Earliest Stone Age*, pp. 77-111. Gosport: Stone Age Institute Press.

Sandrock, O., Kullmer, O., Schrenk, F., Juwayeyi, Y.M. & Bromage, T.G. (2007). Fauna, taphonomy, and ecology of the Plio-Pleistocene Chiwondo Beds, Northern Malawi. In: Bobe, R., Alemseged, Z. & Behrensmeyer, A.K. (Eds.), *Hominin Environments in the East African Pliocene: An Assessment of the Faunal Evidence*, pp. 315-332. Berlin: Springer.

Sandgathe, D.M., Dibble, H.L., Goldberg, P., McPherron, S.P. (2011). On the role of fire in Neandertal adaptations in Western Europe: evidence from Pech de l'Azé IV and Roc de Marsal, France. PaleoAnthropology 2011, 216e242.

Shahack-Gross, R., Berna, F., Karkanas, P., Lemorini, C., Gopher, A. & Barkai, R. (2014). Evidence for the repeated use of a central hearth at Middle Pleistocene (300 ky ago) Qesem Cave, Israel. *Journal of Archaeological Science* 44: 12-21.

Shimelmitz, R., Kuhn, S.L., Jelinek, A.J., Ronen, A., Clark, A.E. & Weinstein-Evron, M. (2014). 'Fire at will': The emergence of habitual fire use 350,000 years ago. *Journal of Human Evolution* 77: 196-203.

Sponheimer, M. & Lee-Thorpe, J. (2003). Differential resource utilization by extant great apes and australopithecines: towards solving the C_4 conundrum. *Comparative Biochemistry and Physiology* 136A: 27–34.

Sponheimer, M., Lee-Thorpe, J., de Ruiter, D., Codron, D., Codron, J., Baugh, A.T. & Thackeray, F. (2005). Hominins, sedges, and termites: new carbon isotope data from the Sterkfontein valley and Kruger National Park. *Journal of Human Evolution* 48: 301-312.

Thorpe, S.K.S., Holder, R.L., Crompton, R.H., 2007. Origin of human bipedalism as an adaptation for locomotion on flexible branches. *Science* 316, 1328-1331.

Ungar, P.S. & Sponheimer, M. (2011). The diets of early hominins. *Science* 334: 190-193.

Ungar, P. S., Grine, F. E. & Teaford, M. F. (2006). Diet in early *Homo*: a review of the evidence and a new model of adaptive versatility. *Annual Review of Anthropology* 35: 209–28.

Weissner PW (2014) Embers of society: Firelight talk among the Ju/'hoansi Bushmen. *Proceedings of the National Academy of Sciences, USA* 111, 14027-14035.

Wheeler, P.E. (1984). The evolution of bipedality and loss of functional body hair in hominids. *J. Human Evol.* 13, 91-98.

Wheeler, P.E. (1991a). The thermoregulatory advantages of hominid bipedalism in open equatorial environments – the contribution of increased convective heat loss and cutaneous evaporative cooling. *J. Human Evol.* 21, 107-115.

Wheeler, P.E. (1991b). The influence of bipedalism on the energy and water budgets of early hominids. *J. Human Evol.* 21, 117-136.

Wheeler, P.E. (1992). The thermoregulatory advantages of large body size for hominids foraging in savanna environments. *J. Human Evol.* 23, 351-362.

White, M.J. (2006). Things to do in Doggerland when you're dead: surviving OIS3 at the northwestern-most fringe of Middle Palaeolithic Europe. *World Archaeology* 38: 547-575.

White, M.J. & Pettitt, P.B. (2011). The British Late Middle Palaeolithic: an interpretative synthesis of Neanderthal occupation at the northwestern edge of the Pleistocene world. *Journal of World Prehistory* 24: 25-97.

Willems, E.P. & Hill, R.A. (2009). A critical assessment of two species distribution models: a case study of the vervet monkey (*Cercopithecus aethiops*). *Journal of Biogeography* 36: 2300-2312.

Wrangham, R.W. (2010). *Catching Fire: How Cooking Made Us Human*. New York: Basic Books.

Wrangham, R.W., Jones, J.H., Laden, G., Pilbeam, D. & Conklin-Britain, N. (1999). The raw and the stolen: cooking and the ecology of human origins. *Current Anthropology* 40: 567-594.

Yetish, G., Kaplan, H., Gurven, M., Wood, B., Pontzer, H., Manger, P.R., Wilson, C., McGregor, R. & Siegel, J.M. (2015). Natural sleep and its seasonal variations in three pre-industrial societies. *Current Biology* 25: 2862-2868.

Robin I.M. Dunbar
Department of Experimental Psychology
University of Oxford
Anna Watts Building
Radcliffe Observatory Quarter
Oxford OX2 6GG
UK
robin.dunbar@psy.ox.ac.uk

Fire, the Hearth (*ocak*) and Social Life:
Examples from an Alevi Community in Anatolia

David Shankland

Gowlett's subtle and convincing presentations on the place of fire, and its significance in social life I amongst many others have found persuasive (Gowlett 2016). Stimulated by his approach, I began recently to think afresh about the role of fire in the community which I have researched in most detail. It has a different mode of production from the early societies Gowlett envisages, being a peasant village in Turkey, situated on the reverse slopes of the Black Sea Pontus mountains. Nevertheless, fire has a particular significance for the village which, as I hope to be able to show, results in the idea of a hearth being leavened in multiple ways throughout the ritual and social life of the community. It might be said, even, that the 'hearth' – in Turkish *ocak* - is the unifying, centre-point for their cosmology, that which links together their ideas of life, death, religion, ritual and social hierarchy.

The socio-ecology of the village will in part at least be very familiar to those who know rural Anatolia, or the Balkans. It occupies cleared forest land. Below it runs a substantial rift valley that marks the northern Anatolian fault and above lie the peaks and passes through which it is possible to reach Samsun and the Black Sea. Though the fields are only partially irrigated, the village has ample water for drinking, drawn from streams running down the slopes, which also serve the crops. The topography is uneven, the fields shaped by the contours of the slopes, meaning that there is little opportunity to open large level spaces for cultivation.

Susesi, the village itself, consists of about a hundred households, and is set in the middle of the village land, which stretches down to the main road below and up to the peak, where there are summer grazing pastures. Households are patrilineal and patrilocal and may contain routinely up to three generations, and exceptionally great grand-children as well. Most village households own some land, with only the very poorest bereft entirely. There are no large landowners.

The settlement is further divided into seven village quarters, or *mahalle*, which contain thirty households down to as few as six or seven. The *mahalle* acts as an important focus for traditional economic activities: field rotation take places within the *mahalle* boundary and each has a clear idea where their boundary ends, and that of the next begins. Each *mahalle* has a copse

from which they would obtain fuel, a bread oven, and a collective bath-house set into a stream. Nevertheless, there is also a strong sense of being a village: the *mahalles* join together to defend their village boundary if necessary, and also to pasture their flocks together in the summer, when the villagers combine together to go to the mountain pasture, the *yayla*, about two hours walk up the slopes above the main settlement.

Households are linked together by bilateral kinship ties, in that marriage and social links may be traced through both the father and the mother's line. However, the patrilineage or *sulale* is clearly dominant. When a woman marries, henceforth conceptionally speaking she becomes part of her husband's lineage, *sulale*. Upon marriage, it is usual to move to the husband's house, and when families divide it would be regarded as normal for two brothers to share a plot, or to build a house adjacent to one another. The *sulale* is important particularly in terms of village politics, in manoeuvring for power within the community, or when resolving or provoking a conflict. However, the *mahalle* is the largest social unit that may be regarded as being linked by a common ancestor, and usually there are several smaller, unrelated patrilineages within the *mahalle*. In practice, one of these, usually a larger one, would become dominant within the life of the *mahalle*, and other, smaller lineages become in effect clients to the dominant group.

All the villagers are joined in saying that their respective ancestors would have been transhumant, settling down from a wandering life no more than a few generations ago. One lineage may say, for example, that they came from the Black Sea coast, another from within Anatolia, and so on. What exactly is meant by 'few generations' here, is very difficult to discern; I regret very much that in the 1980s, when I first conducted research amongst the community, I was not sufficiently sensitive to archaeology to take dendrochronology samples from the oldest houses. Now, thirty years later, the previous wooden constructions have almost entirely been replaced by modern, concrete and brick houses with painted fronts, standard for the region, and the time may have passed when it is possible to do so. However, certainly living memory of being a settled community goes back more than two hundred years, and I should not be surprised if it were more than this.

Until very recent times, that is approximately until the 1990s, each household would consume largely what it had itself cultivated or ploughed. The men would plough, using oxen, reap and sow. Women would look after the household and children, cook, obtain firewood, tend the hearth, and vegetable gardens next to the home. Bread would be made in the collective ovens of each *mahalle*. Children would look after animals: milch cows, water buffalo (prized because of their milk, which makes good yoghurt), donkeys or mules, and further afield sheep or goats.

Agricultural activities other than animal husbandry generated little cash. Crops, such as wheat or maize rarely yielded any surplus that could be sold. There was no systematic marketing of other products, such as nuts or mulberries, though these flourish wild within the village territory and would be relished. Other household products could certainly be sold in a casual way, such as honey, eggs, or chicken. However, herds of sheep or goats, or tending bullocks for the market was the preferred way to invest and save. This was dependant on intense labour, the willingness of the next generation to play their part in guarding the herds day-in day-out, and impeded by stony pasture that was never quite rich or extensive enough to provide comfortable or easy grazing.

The villagers themselves are keenly aware that their soil is not as fertile as on the plains. The setting is beautiful, as the village and its neighbours overlook the valley to the river below and across to the other side. But the land, cleared from rough forest and scrub, is highly stony. Alongside the edges of fields are piles of cairns, where the stones from the fields have been removed by successive generations, but a glance reveal the stones that still continuously emerge from all but the very best fields. In a good year, with luck, a man may gain a return of some six to eight if wheat is planted, perhaps even a little more with fertilizer, but all too often it was much less than this. In a bad year, it could even hardly yield what has been sown. The return for maize is higher, but as a crop more risky, and vulnerable to damage from wild pigs coming down from the forests above.

The response of the villagers to the inability of their subsistence agriculture to yield a significant surplus has largely been to migrate: to Istanbul, to Germany, to other parts of Europe, a migration that began in the 1950s and accelerated with each passing decade. In effect this means that the village has gradually come to rely on money coming from outside through remittances, and that the greater proportion of its population lives outside. This said, the village has today in 2020 still a hundred households, or even slightly more. However, the demographic pattern has

changed: they are now very frequently made up of just a husband and his wife, with their children absent working in the cities or abroad. Often, the couple themselves are recently returned, seeking to live quietly after a life outside the village. Their children's children in turn may come back to the village to be looked after for a few years when they are very young, but the village school has closed and there is, in any case, very little opportunity for gainful employment.

Now, in 2020, no household ploughs or bakes its own bread, and it has become a necessity for survival to have some money coming in from outside, however, meagre. The better-off have a pension from Europe having returned from being workers there, or children who have done well. The fields are now largely unworked, though occasionally a man – perhaps retired early having been a civil servant or worker in the towns, and still vigorous - may set up an enterprise to use the now empty land, for example, by buying and fattening bullocks each season. Gradually though the fields are becoming overgrown.

In many respects, however fascinating in detail, the picture that I have given is one that will be familiar to those who have followed the ethnography of Anatolia and the Balkans, or even the Mediterranean lands more generally. Stirling's *Turkish Village* (1965) for example gives an account that is similar to this one in terms of the social structure of the village. The patriarchal households, the shallow patrilineages, the pattern of field holdings and ownership, are all immediately recognisable. His differs in that Sakaltutan, where he worked, is situated on a flat steppe. It also appears that the very distinctive *mahalles* which I found are much less pronounced in Sakaltutan. Further, Sakaltutan's response to modernisation differed in that it was close enough to Kayseri to become a dormitory village as possibilities for labour migration gradually become more widespread. Partly through this direct access to labour markets, it appears to have been able to develop twenty or even thirty years before *Susesi*, where I worked, and able to integrate more easily directly into the modern urban economy (Stirling 1994).

Yet, even if there are similarities, there are also great contrasts. Though both are Turkish-speaking villages, and both are Muslim Sakaltutan is Sunni, and where I worked Alevi (Shankland 1994). The differences between the dominant, Sunni majority and the Alevi minority have been the subject of a great deal of research, research which is often difficult to disentangle from the politics which inevitably becomes drawn in when discussions over faith, secularism, belief and hegemony begin to take place in the public arena. I have written about these elsewhere (Shankland 2007; also amongst many see Mélikoff (1998), Şener (1982). Here, I shall attempt instead to present the details in such a way that the ethnographic distinctiveness of the Alevi community becomes clear: that is by considering the relations between the sexes and the household, before turning to the wider social organisation and ritual life of the community.

In Alevi village life, the primary social bond may be regarded as that between husband and wife. For a man, or woman, to be fully part of the community, they need to marry, around the age of eighteen if they are a man, and a little younger if a woman. Though it is acknowledged that sometimes relationships do not work out, divorce is highly frowned upon, and very substantial social pressure is put on a husband and wife not to separate if at all possible. Divorce in the village is therefore extremely low. Monogamy is greatly preferred, and though exceptionally men might marry a second wife, this was usually for practical reasons, for example, to provide help within a household that otherwise could not cope for any reason. Upon marriage, it would be usual for a man to bring his wife to live with him and his parents, gradually as the generations pass, he would become head of the household in turn, often building a house nearby to his parents.

Within the *mahalle*, there is a dense interaction between households, a pattern of daily visiting and prestation which overlies the continual small economic interchanges necessary within a peasant household economy, whether it be to borrow small amounts of essential supplies, help with a task, or share insights into future possibilities. Here, within the *mahalle*, there is very little gender segregation: instead, most interaction takes place as a couple, men and women together. Social calls are made by the husband and wife together, and when they are received into the household of their host, equally the man and woman of the household greet them, and the couples sit together in a large, single room known simply as *ev* (house).

Within this large room is a hearth, around which the assembled company sit. From the early part of the twentieth century, it became usual to install thin sheet-iron stoves known as *soba*, into which wood is continuously fed, and on top of which food could be cooked. It is a very strong custom that food is offered to all those who come by, but if no food is taken even, tea is drunk and some small snack offered. Conversation takes place around the hearth, and before larger houses were built with separate bedrooms – around the middle of the twentieth century, the family would also sleep in that one large room, around the hearth, until rising at dawn to continue with the daily round of subsistence activity.

Social control

The Alevis place an extremely high emphasis on all within the household, and within the wider community, being able to get along with one another. This is emphasised in multiple ways within their religious culture. For example, one of their most prominent sayings is *kimse incitme*, provoke or irritate no-one. More formally, they say that their paramount rule of religious life is *eline, diline, beline sahip ol*, 'Be master of thy hands, tongue, and loins', that is, do not steal, do not tell lies or gossip, and do not commit adultery. Contrariwise, their most auspicious or sacred moments come when in a small group together, when they say that sitting in a group where all can see each other's faces enables all present to see into each other's hearts. This is known, within their religious culture, as being evocative of *muhabbet*, or divine love. They may add that, in this way, we reach a little piece of the Godhead that is present within all people, both men and women. Such a state of harmony is equally known as being pure or clean, *temiz*, and regarded as being the ideal way to exist, that is, at peace (*sulh*).

Here, it may help to offer a clarification. It goes without saying that there are, just as in any peasant society, a host of irritations and conflicts in everyday life: As I have attempted to explain, the Alevis in many respects are a typical, patriarchal peasant community. Men expect to be dominant within the home. Some men, at least, in earlier times would have no compunction in abducting another man's wife if he should be away, for example on his military service and his wife left unguarded. Adultery, if discerned, could certainly lead to violence, even death. Again, some men, if inclined that way would regard it a necessity to beat their wives if their authority should be threatened. Men, indeed, in a general sense have a strong sense of honour, a pithy if rather bawdy saying within the village is '*Namusumu yiyen sapımı yesin*'; 'Let he who abuses my honour, suck my prick'.

Nevertheless, the Alevi interpretation of religion draws very strongly on an esoteric and peaceful conception of God. They say that their understanding is that *Tanrı* (God) or *Allah*, is a God of love, rather than of jealousy or anger. Though misunderstandings and conflicts may emerge at any level of life within the community, their understanding of religion in effect is that it should be used to bring peace, to resolve existing conflicts. To make peace, *barışmak*, is in itself auspicious. To assure another of one's good will, of an absence of envy or jealousy at someone's success is essential, *hayırlı olsun!* May it be blessed (or favoured) accompanies any explanation of a piece of good fortune or major event in a person's life, such as the purchase of a new car, or house. If a neighbour or relative is sick, it is regarded as essential to visit them to wish them a speedy recovery.

The *ocak* or 'hearth' system

In seeking a wider framework to justify and support this distinctive interpretation of religious understanding, the Alevis lay stress on ordering their own affairs as much as possible, in contradistinction with the powers, and laws, of the state, which they regard as been supportive of the Sunni form of Islam, and the imposition of social control through religious mores, or the *şeriat*. This, alternative more intimate form of social control they seek to inculcate through localised holy lineages which they know as *ocak*, that is 'hearth'. The term *ocak*, in this sense, refers specifically to the patrilineage which is held to be given the right, or the privilege, to teach *Alevilik* ('Alevi-ness') by virtue of one or more of its founders or ancestors being favoured by God with the ability to perform miracles (See Shankland 2007, also Kieser 2004 and Langer, et al 2013).

It is important to stress that term *ocak* refers specifically to the institution of holy lineages. Individuals amongst the lineage who themselves become active religious leaders are referred to as *dede*, that is literally 'grandfather'. Likewise, each *ocak* has follower lineages, which are known as *talip*, or pupils. The Alevis regard these links as permanent: that is, a person, whether man or woman, by virtue of their birth will be a follower to a certain *ocak*, In turn, these links are said to be decided by Saint Hacı Bektaş, who sent forth individual *dede*s from his *tekke*, his dervish convent.

The sense of hierarchy is very clear: every lineage, whether an *ocak* lineage or not, has a lineage to which it respects as an *ocak*. Only members of an *ocak* can teach the mores of *Alevilik*, and only *dede*s can lead an Alevi religious ceremony. The links between the *ocak* and their followers are, in the region where I worked, rather localised, and a follower lineage would almost always be within a day's journey by mule from their *ocak* lineage.

In turn, there are higher ranks further afield: the patrilineal descendants of Hacı Bektaş himself are known as *efendi*s, and may be available for consultation by Alevis, or act as a higher court for problems which the villagers themselves have been unable to solve. In the village where I worked, an alternative expression for a *dede* is *rehber*, or 'guide'. In other parts of Anatolia, particularly toward the east, the duties of a *rehber* may be separated from that of *dede*, to whom he reports, and by whom he may be appointed. In *Susesi*, however, there is no such separate role.

The functions of an *ocak* 'hearth' lineage

The function of an *ocak* lineage is, then simply put, to provide individual religious leaders, or *dede*s to the follower lineages of the Alevi villages in the region where they find themselves. More specifically, they may help with advice, act as negotiators when a family seeks a spouse for their son or daughter, resolve quarrels of conflicts within the community, bless sacrifices for their followers when they wish to make a vow (*adak*), and lead collective religious ceremonies, known as *cem*.

In order to make their central place in the Alevi cosmology clearer, it may help to discuss in a little more detail the way that a *dede* resolves disputes. When a misunderstanding or quarrel emerges within the community, it may lead to immediate violence if a man is driven to lose his temper, for example if he finds someone taking his right to water for fields, or his flocks. In this case, if grave injury has resulted, it is – if feasible – best to flee for fear of retribution from the injured man's relatives. In the aftermath of the quarrel, gradually things may settle down, but the resulting grievances would usually result in a state of being on not-speaking—terms, or *kus*, between the aggrieved parties. This would mean, for example, a breaking-off of everyday communication, of civil exchanges, and of mutual visiting. In a small community, reliant on the continual exchange of small favours, this is no small matter. If between two *mahalle* or between one village and the next, it is equally not a good situation, because the resulting tension can lead to further violence, whether against body or property.

A skilled *dede*, rather than step in in the heat of the moment, waits until the stand-off has been in effect for some time, and the resulting breach, though unfortunate, at least stable. He will then either visit, or wait for an invitation to come to discuss the situation. He, as well as the relatives of the injured parties, will appeal for peace to be made. When finally they have succeeded, if the quarrel has been a serious one, it would be usual to hold a sacrifice, that is to bring together the injured parties to hold a collective ceremony which lies at the heart of Alevi religious practice, known as the *cem*. For the *cem* to be held, it is essential that all such quarrels have been resolved, and that all present are able to embrace each other in friendship.

When the *dede* is exploring the basis of quarrel and the events that led up to it, he may conduct a *görgü*, or ritual questioning. For this, the pair involved are called to the centre of the room, to an area that has been marked out by a kilim known as Ali's space: *Ali'nin meydanı*. Genuflecting, they are asked by the *dede* whether all is at peace between them. They say that this mirrors the questioning that we will all receive after our death from the Angel Gabriel as to the way that we have conducted ourselves in this life, depending on which we will be cast into heaven or into hell. To lie, in such circumstances is regarded, as being the equivalent as lying to the Angel Gabrael himself, and a grave – not to say dangerous - sin.

Though outsiders to the community are not permitted at the ceremony, it is expected that villagers will be present. Indeed, it is the public affirmation of peace between the disputants that the villagers most prefer to see, so that they can be reassured that now the quarrel is resolved. The *dede* will also seek support from the community for any solution that he has suggested, asking those present whether they are in agreement with him. Though in theory a punitive approach has said to be possible, in practice where I worked the *dede* would seek redress based on compensation for any wrong given, such restitution of firewood taken, or – if offence has been taken at the lack of a visit to a sick relative – that such a visit henceforth undertaken, and so on.

The exploration of such conflict is formalised once a year through an institution known as the village sacrifice, where everyone in the *mahalle* has to come forward, a man and a women, to go through the questioning ceremony. For this, every couple must come forward, go through Ali's space, and receive the affirmation of the community that relations within the household are cordial, and that there have been no disputes over the previous year or if there have, they are now over. This is a very powerful method of assuring that any disputes that are may have arisen are resolved, even if there is an unequal hierarchical relationship between the complainant and the accused; for example, it would be an opportunity where an old lady could ask for compensation from a careless neighbour, or where a powerful man in the village has to pay obeisance to the social niceties for the sake of the ceremony, and for peace.

If, for any reason, reconciliation cannot be made then the ceremony is abandoned, and the resulting debacle deplored, regarded by the villagers as something that it is very desirable to resolve as soon as is possible. If, on the other hand, all the questioning rituals have been successfully conducted and the whole village is 'seen' successfully, then they group together with a large village sacrifice, on which occasion a *cem* ritual is held. Traditionally, only after the village sacrifice has been completed could the winter village crops be sown. A similar stricture applies to the *dedes*, who must be seen by their *dedes* in turn before they preside over the questioning for the village as a whole.

Now, attempting to pull together and summarise these various threads: we may say that the core social relationship within Alevi society is the unit which is formed by the marriage of a man and a women. They work together in creating a household, and they share the spaces of the household, which are not not divided into men's and women's quarters, and entertaining or welcoming neighbours it undertaken equally by both men and women together, as a couple. In the central

space of the house, where traditionally the whole family would sleep, there lies a hearth around which the social life of the community would revolve, and next to which the family would rest as well as talk, debate, cook, and prepare for the tasks for the next day.

The hearth, known as *ocak*, is tended largely by women. Before modern appliances, it would be regarded as obligatory that the embers should not fail, as it was equated with the presence of life within the household. It was regarded as the height of ill luck to have to borrow an ember from a neighbour to reignite a failed fire. Likewise, before a new house was moved into, the hearth would be lit. When a hearth is not smoking, it would be said that the household itself had died.

The Alevi conception of social life lays enormous importance on peaceful harmony and co-existence within the household, and with neighbours, but if there is a situation that is not immediately resolvable, then rather than go to the state law courts or have recourse to Islamic law, they may turn to their own, domestic form of religious practice whereby the patrilineages are distinguished between those which are lay, and those which are known as 'hearth', or *ocak*.

So closely concatenated is the sense of peace and the sacred that before the main religious ceremony takes place, and before a couple can offer a sacrifice, they must undergo a public questioning from a *dede* of their particular *ocak*, so as to assure the community that they are at peace with one another, in their household, and with the wider community. Should a quarrel emerge, and indeed it is difficult to conceal a quarrel within the village, then is must be resolved or worship cannot take place.

We may ask, how does this differ from the Sunni way of doing things? The answer is profoundly. Though the Sunni household is also patriarchal and patrilineal, the social spheres of the husband and wife are distinctly separate. Rather than there being one, large room where the family as a whole congregates, women and men inhabit their spaces, and men and women divide up very readily into social groups of the same sex. This is reflected in their religious practice, where men attend the mosque and women do not, and there are distinct ceremonies for women, at which men are not present. During weddings, unlike Alevi villages, the two sexes are kept separate, with dancing taking place – if it does so –in separate houses, the musicians outside so that they cannot be seen by the women. It would not be correct to say that amongst the Alevi villages there is no gender separation – there, is, as I have emphasised a conventional division of labour and authority within the household. Further, if there are strangers present or if the gathering is very crowded, men and women may separate – which they refer to as *haremlik/selamlik* after the men's and women's quarters in the Ottoman Sultan's palace – but this is not the preferred, or default position within the Alevi household or community, where social interaction takes place with the couples from different households present together.

How does tracing the use of fire in the community help us understand this? The answer is that it does, greatly. The word house, *ev*, is used both for the overall household building, and for the main room of the house, in which the hearth lies, and around which all the social activity of the family takes place. The hearth is kept burning continuously, with the embers from the previous night kept going until the next day. Harmony, essential to keep the household together, is emphasised and reinforced through a religious institution that is known as 'hearth', which in turn is part of a wider cosmology that links the very possibility of salvation with a person, whether man or woman, acquiescing to that structure. Without the intercession of the *ocak*, the holy lineage, the ritual life of the community cannot function: there can be no *cem*, no sacrifice, and – theoretically at least – in its absence the revolution of the agricultural year, the transition from autumn to the ploughing of the winter wheat, should come to an end. One could hardly think of a more key way for the domestication of fire to become incorporated into the social life of the Alevi community.

In preparing this paper, I discussed the main contentions made by Gowlett with my friends from the village. They are entirely in accord with his thoughts. For them, fire does indeed, extend the social life of the community. In the evening, whether in the winter in the village, or in the summer in the mountain pastures, it becomes the way that the life of the day may continue after dark. More than this, however, we have seen the way that the cosmology of the village can be said almost to be constructed around the extended metaphor of the hearth, which is not only at the centre of their household, but also of their religious hierarchy.

In conclusion, we can see the way that this all comes together in the *cem* ceremony. The *cem* takes place in the largest domestic room in the house, the *ev*. Members of an *ocak* 'hearth' lineage sit next to the physical hearth itself. The sacrifice that all will share after the ceremony cooks on the hearth throughout, and forms therefore an integral part of the overall sensory perception of the ceremony: the smell of the wood, the sound of the bubbling of the pot, the changing, shifting shapes and light from the fire lie next to the *dede*. Just as the fire itself provides the essential means to cook the meat for its subsequent sharing amongst the congregation present, the *dedes* are essential to the ritual. Just as the

fire lights up the room, the role of the *dede* is said to be to enlighten his follows, *aydınlatma*.

The metaphor of the hearth, as domesticated fire, can thus be extended in multiple ways: as the physical means by which the community can survive, as life itself even – if one wishes to curse someone one might say 'May your hearth be extinguished!', *Ocağın sönsin!* but additionally, through the *ocak* 'hearth' lineage system, as the way that the Alevis themselves have domesticated the practice of religion, creating a form of religion that is ideally suited to their remote lives in a peasant village that is set off and apart from mainstream Islam.

In this article, I have not discussed explicitly the theoretical framework that led me to order the ethnography, but it will be clear to many that in the first instance it is characteristic of what may be regarded as traditional social anthropology, that is an emphasis on the relationship between culture and social order. This is perhaps best represented by Evans-Pritchard in *The Nuer* (1940) and by EP and Fortes in *African Political Systems* (1940). The Alevi villages where I worked, just as suggested in *The Nuer*, and later by Gellner (1981) with regard to the Islamic context do indeed utilize patrilineages accompanied by sacred mediators in order to achieve social control in the absence of centralised governmental administration. This has been controversial in recent decades, something that I discuss elsewhere (Shankland 2007, forthcoming), but the villages where I worked support the model unequivocally.

Introducing fire, however, into the way that the paper is structured to my surprised leads us to an earlier set of theoretical presumptions, those which structural functionalism replaced. I am referring to Malinowski, and his conception of the functional relationship between human individual needs and social institutions. I hope to return to this on a subsequent occasion, but in essence in Malinowskian terms, food would be the need that fire supplies, and tracing the way that it becomes refracted throughout the Alevis' wider social institutions enables us to describe, and appreciate their culture as a whole from many perspectives (Piddington 1957). The hearth, from this point of view, occupies an absolutely key intersection of the social and the physiological needs. How, in turn, one would turn this into a falsifiable analytical framework is not immediately clear, though it might be that it is testable simply in that presumably not all cultures place the same symbolic importance on fire. Leaving this to one side, I am clear that it provides a way that we can conceive the material culture of the community in intimate association with its ritual and social life. Perhaps with that, we should be content.

References

Andrews, P. 1999 *Ethnic Groups in the Republic of Turkey*, Wiesbaden: Dr Ludwig Reichart Verlag.

Evans-Pritchard, E. 1940 *The Nuer*, Oxford: OUP.

Fortes, M. and Evans-Pritchard, E. (eds) 1940 *African Political Systems*, Oxford: OUP.

Gellner, E. 1981 *Muslim Society*, Cambridge: CUP.

Gokalp, A. 1980 *Têtes rouges et bouches noires*, Paris: Societé d'ethnographie.

Gowlett, J. 2016. 'The discovery of fire by humans: a long and convoluted process', in Philosophical Transactions of the Royal Society, B-Biological Sciences, 371 (1696).

Kieser, H. 2004 "Alevilik as Song and Dialogue: The Village Sage Melûli Baba (1892–1989)" in David Shankland (ed.). *Archaeology, Anthropology and Heritage in the Balkans and Anatolia: The Life and Times of F. W. Hasluck, 1878–1920.* Vol. 1. Istanbul: Isis, 2004. 355–68.

Langer, R., Ağuiçenoğlu, H., Motika, R & Karolewski, J. (eds.). 2013 *Ocak und Dedelik: Institutionen religiösen Spezialistentums bei den Aleviten.* (Heidelberger Studien zur Geschichte und Kultur des modernen Vorderen Orients; 36). Frankfurt/M.: Peter Lang.

Mélikoff, I. 1998 *Hadji Bektach; un mythe et ses avatars*, Leiden: Brill.

Piddington, R. 1957 'Malinowski's theory of needs', in *Man and Culture: an evaluation of the work of Bronislaw Malinowski*, edited by Raymond Firth, London: RKP.

Şener, C. 1982 *Alevilik Olayı: Toplumsal bir Başkaldırının Kısa Tarihçesi.* Istanbul; Ant Yayınları.

Shankland, D. 2007 *The Alevis in Turkey: The Emergence of a Secular Islamic Tradition*, London Routledge.

Shankland, D. 1994 "Social Change and Culture: Responses to Modernisation in an Alevi Village in Anatolia". Chris Hann (ed.). *When History Accelerates: Essays on Rapid Social Change, Complexity and Creativitiy*, London: Athlone Press, 238–54.

Shankland, D. Forthcoming. 'Evans-Pritchard and Segmentary Lineage Systems re(re)considered', in Andre Singer (ed) *Evans-Pritchard*, London: RAI and Sean Kingston publishing.

Stirling, P. 1965 *Turkish Village*, London: Weidenfeld.

Stirling, P. 1994 "Labour Migration in Turkey: Thirty Five Years of Changes", in *Humana: Bozkurt Guvenc'e Armagan*, ed. Serpil Altuntek, Ankara: Ministry of Culture.

David Shankland
Director of the Royal Anthropological Institute

From Specialty to Specialist: A Citation Analysis of Evolutionary Anthropology, Palaeolithic Archaeology and the Work of John Gowlett 1970-2018

Anthony Sinclair

Introduction

Understanding the relationship between a scholar and their specialty is the task of the intellectual biographer. It usually involves an extensive critical reading of publications and reflecting on the impact or contribution made by these documents and their author to the research of a specialty. Of course, documents are texts that are read and then situated within a con-text supplied by the reader or biographer. Two different individuals can both read the same document and yet draw out different understandings depending on the context provided by their own biography of reading and their own understanding of a specialty that they bring to the moment of reading something new in a discipline. The success of any intellectual biography, therefore, is dependent on the expertise, knowledge and breadth of reading that can be brought to the contextualisation of a scholar's work. However, whilst the physical activity of reading and critical reflection may not have changed over the years, the identification of an effective context has become ever more difficult as a result of the growth in number of potentially relevant research outputs for any context within any speciality (see Price 1951, 1963 for an early recognition of this problem). Despite the small size of Archaeology as a discipline, the problem of document abundance is no different; the number of research outputs and, therefore, documents for contextualisation is growing exponentially (Sinclair 2016; figure 1).

There is, however, another way to see the relationship between a scholar and their specialty rather than the reading and contextualisation of a set of documents by a single biographer. The documents themselves can be used to generate a directed network (of influence) flowing 'inwards' from other published documents, researchers and collaborating colleagues, and 'outwards' through writings and contacts to other publications and scholars. Documents are joined together in a network either directly via the collaboration of authors or through the process of citation (Price 1965), or 'indirectly' or in the shared use of terms and their engagement in the same conceptual milieu. Through networks of citation we can analyse the way in which one scholar has been influenced by the work of other scholars by examining the citations made

to other publications, and we can look at the influence of one scholar within their specialty, or beyond, by examining how their own work has been cited by others. Lastly, we can look at the terms used within a set of publications to see how a specialty structures its conceptual understanding of its research interests (see Klavans and Boyack 2014). This is citation analysis.

Citation analysis is qualitatively different from intellectual biography since the process of reading and document contextualisation is distributed from a single scholar to the members of a discipline as a whole. In simplest terms, citations record the influence of one document and its authors upon another. And, even though the exact rationale and nature of citation is still actively debated (for example: Kaplan 1965; Cronin 1984; Hyland 1999; Leydesdorff 1988; Nicolaisen 2007; Davenport and Snyder 2009; Chi 2016; Boyack *et al.* 2018) there are some clear advantages to the use of this approach in the modern era. The analysis of citations, in effect, aggregates the reading of many specialists to generate a context much greater in size than the sample of publications that might be read by a single scholar whilst also reducing the impact of potentially idiosyncratic readings that might be made by any one individual. Citation analysis also facilitates the tracing of influence through a collection of publications that might number many more documents than any individual could possibly read.

Despite the development of citation data and its analysis in the 1960s (see Wouters 1999 for a succinct history), its use for understanding the humanities and social sciences has been minimal (Ardanuy 2013). Its use in archaeology has been if anything even rarer, though the few examples published illustrate its potential. These include studies of the use of citation analysis to review the use of new methods of analysis (network analysis - Brughmans 2013; microwear analysis - Dunmore *et al.* 2018), the nature of citation in the discipline (the citation of women scholars Hutson 2002; and self-citation - Hutson 2006), and for a broad historical overview (the intellectual base of archaeology 2004 to 2013 - Sinclair 2016). Missing in these examples is the use of citation analysis as a technique for examining a discipline at multiple scales, something for which citation analysis is well-suited and that is now unachievable through standard historiographic approaches due to the number of documents.

In the following study, therefore, citation analysis has been used to analyse a single discipline across three nested scales. At the broadest scale, the analysis begins with the discipline of Evolutionary Anthropology, then focuses down on the specialty of Palaeolithic archaeology and end by examining the research of a single scholar - John Gowlett. Data has been collected from a defined starting date (1970) up to the present day (2018). The 1970s are a good starting date for examining the discipline of palaeoanthropology as a self-aware, and distinguishable discipline in its own right. They begin with the publication of the first monographs on the excavations of the earliest hominins and their artefacts at Olduvai Gorge (Leakey 1971), with the appearance of the Journal of Human Evolution in 1972, the first journal specifically designed to circulate research in this field. And, whilst there has been archaeological excavation and research publications of Palaeolithic age materials since the mid 19th century (Lartet and Christy 1865-1875) one can argue that it is the combination of the external chronology of the marine oxygen isotope record along with the statistical analysis of lithic and faunal assemblages, the long-term ethnographic record of surviving hunter-gatherer populations and the first radio-carbon dates all coming together in the 1970s that gives Palaeolithic archaeology its unique character distinct from other forms of prehistoric archaeology. Finally, the first article by John Gowlett was published in 1978.

Bibliometric data and the identification and analysis of disciplines

The bibliometric data necessary for citation analysis currently exists in readily accessible through the academic citation indices first started by Eugene Garfield in the 1960s (Garfield 1955, 1964). These indices include not only the details of a publication - the author, date, title and place of publication, but now also include the reference list of any publication, key words and the abstract amongst other information. Garfield already recognised the potential for this type of information in the process of examining how disciplines and the specialties within then developed over time and called this study algorithmic historiography (Garfield and Sher 1963).

In recent years developments in computing power to process and visualise complex networked data sets has led to the development of 'science mapping' as a research technique in its own right (see Börner et al. 2003; Chen 2017 for reviews of this approach), with the potential for science maps can be made for document sets that number into the millions (see Boyack et al. 2005 for an early example). The quality of the visualisation

depends on the reliability of the bibliometric data used: specifically, whether it represents a coherent field of research. This can be problematic for a multi-disciplinary research field like Palaeoanthropology. In this study bibliometric data from the Web of Science and Scopus citation indices have been used, as well as Google Scholar. Citation indices are, by necessity, samples of currently available academic literature with none providing complete coverage of the academic literature (see Wouters 1999 for an insightful history of the first citation indices). Studies indicate that the Web of Science has better coverage of publications in the natural sciences but is known to have weaknesses in its coverage of the social sciences and particularly the arts and humanities, further exacerbated by its original focus on journals as the publication form chosen for indexing; Scopus, on the other hand, has better coverage for the social sciences, the arts and humanities as well as for publications in monograph form (Martin-Martin et al. 2018a, 2018b; Anderson and Nielsen 2018; Falagas et al. 2007).

Extracting a representative sample of bibliometric data for Palaeoanthropology with a research approach that includes everything from the physical and chemical sciences right through to the arts and social sciences, and with publications in both journal and monograph form, the use of both databases is necessary. Both the Web of Science and Scopus apply subject category codes to publication sources – journals- rather than to individual documents, following the assumption that researchers within a discipline preferentially publish in a select set of sources for that discipline (Wang and Waltman 2016). However, since neither Palaeoanthropology or Palaeolithic Archaeology has been classified as a distinct subject by the Web of Science or Scopus, the journals where its research is often published are not identifiable by a defined subject category. Moreover, research outputs in these research areas whilst often published in a common set of journals also appear in a much wider set often categorised to different research fields. Therefore, the search for bibliometric data for these two research areas, needs to identify publications in journals classified at the intersections of recognised subject fields, or via the use of terms that are exclusively used by a coherent group of researchers to identify individual documents by 'topic'. Both approaches are described in more detail below.

The networks in this study have been constructed using VOSviewer, a program designed specifically for the visualisation and analysis of networks of academic literature, created by Nees Jan van Eck and Ludo Waltman at the Centre for Science and Technology Studies at the University of Leiden (van Eck and Waltman 2010). It is available free of charge for academic use (see www.vosviewer.com). VOSviewer can construct and visualise citation-based network based on several forms of relationship: these include co-citation, bibliographic coupling or direct citation. In its visualisations, the nodes in the network can be either sources of publications (journal or monograph titles), or authors; the size of the node is indicative of the frequency of citation or use and nodes are placed closest to those with which they are most commonly associated against a measure of association strength (van Eck and Waltman 2010). The visualisations here use a co-citation relationship in which authors and sources are identified and placed together according to the number of times with which they are both included in a publication's reference list. Co-citation relationships are believed to be an effective way of understanding the intellectual base of a discipline – defined as the key publications that form the foundations of a discipline's knowledge. This is discussed in more detail in Sinclair (2016). Aside from the citation-based maps, *VOSviewer* can also generate network maps of the language of a set of publications by extracting significant terms from titles and abstracts, though at present abstracts are usually only available in the databases for publications dated after 1995. When examined online, individual maps can be magnified so that all nodes might be identified, but in the maps printed here the scale used means that many of the nodes are not specifically identified especially in the larger maps. *VOSviewer* is also able to represent the relationships ('edges') between nodes as a series of lines between nodes. This facility has been used in a number of the maps below where their presence helps identify relationships for discussion.

Further details of the process of data extraction and preparation and for mapping can be found in an earlier study examining the intellectual base of archaeology as a whole (Sinclair 2016). This study also describes the process of 'cleaning up' publication data to avoid multiple representations of the same author or source due to variations in how authors' names are recorded, and variations in publication source names caused by differing editorial policies.

Evolutionary Anthropology as a research specialty

Three maps provide a network representation for the specialty of Evolutionary Anthropology. Each map is based on bibliometric data for research outputs published in sources classified in the Web of Science as being both in the subject categories of 'Anthropology <u>AND</u> Evolutionary Biology'. This is, almost certainly, a sample of the total population of research outputs within Evolutionary Anthropology since the Web of Science does not index all journals and, for Evolutionary Anthropology, there will certainly be research outputs published in other journals than those classified according to these two subject categories. There may also be documents concerning research in

evolutionary biology beyond non-hominin species. This search generates a sample data set of bibliometric information on more than 37,000 individual documents each with links to other documents via their own cited references. Some of these other documents will already be included in the original sample set; but others will be items in journals not identified in these two subject categories and, more importantly, published in journals or books or book chapters not indexed by the Web of Science at all. As such, the maps based on the sample of bibliometric data used here probably provide a good overview of the primary networks for this research specialty. For each map, it is the relative association strength between items (nodes) that is used by *VOSviewer* to place either sources, authors or terms physically closer to each other and for determining clusters of nodes whilst ensuring that the map retains a sufficient legibility of the whole.

Figure 1 presents a co-citation network map of the 350 most frequently cited sources from more than 590 sources that have been cited at least 100 times; six discrete clusters of sources have been identified (Table 1). These sources are the primary places where Palaeoanthropologists either publish their research or go to read new research. Figure 2 presents a co-citation map of the 350 most commonly cited authors having an influence in Palaeoanthropology from more than 850 individuals cited at least 100 times with 6 discrete clusters identified (Table 2). Figure 3 presents 288 terms extracted from the titles and abstracts of the initial bibliometric document data, and once again with 6 clusters identified (Table 3).

A series of major journals provide occupy a central place for the publication of research in Palaeoanthropology: *Journal of Human Evolution, American Journal of Physical Anthropology, Yearbook of Physical Anthropology, Nature, Science,* and *Proceedings of the National Academy of Sciences of the USA.* Surrounding these journals are clusters of sources related to genetics, anatomy, primatology and animal behaviour and a cluster that is palaeontological. Finally, there are two closely overlapped clusters related to archaeology and quaternary science and another which appears to be related to the same theme but comprising journals based out of the USA or the UK, and sometimes publishing papers in languages other than English. The clustering of authors does not follow the same theme. There is a clear cluster of authors publishing research on the hominin record from east and southern Africa (*Wood, Leakey, Johanson, Tobias*) alongside a group of authors publishing research on the nature of modern human behaviour and the archaeological record (*McBrearty, Klein, Marean, Mellars*) and another cluster of researchers examining pre-modern hominins including Neanderthals (*Hublin, Wolpoff, Smith, Stringer*). The three remaining clusters contain authors publishing research on aspects of

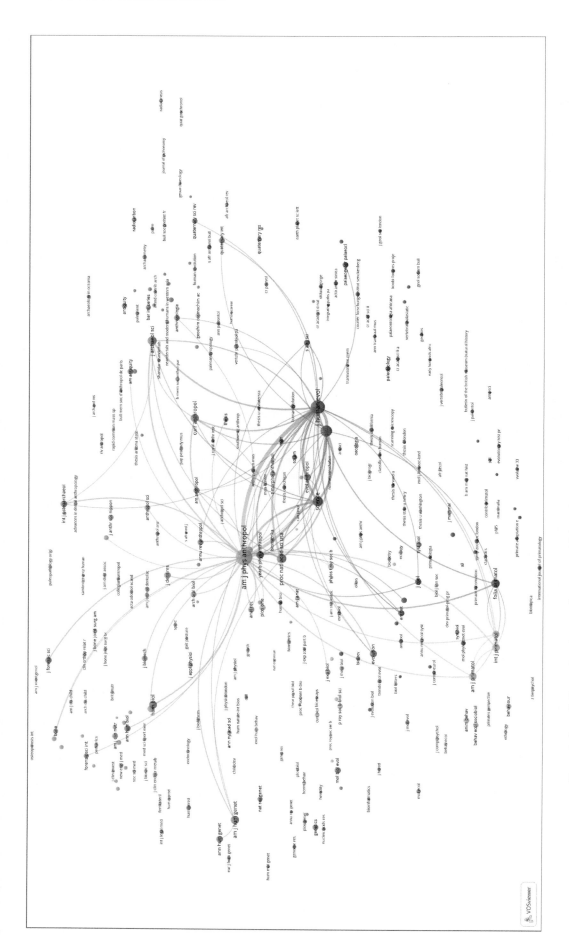

Figure 1. A co-citation network map of sources publishing documents in the discipline of Evolutionary Anthropology 1970-2018.

Table 1. Clusters of sources identified in the co-citation network map of sources publishing documents in the discipline of Evolutionary Anthropology 1970-2018.

Cluster No.	No. of Nodes	Example Nodes
1	86	Journal of Human Evolution, Current Anthropology, Journal of Archaeological Science, Quaternary International, Palaeogeography, Palaeoecology & Palaeoenvironment, Evolutionary Anthropology, Nature
2	73	American Journal of Physical Anthropology, Journal of Zoology, American Journal of Anatomy, Anthropological Sciences, Journal of Dental Research, Science
3	49	International Journal of Osteoarchaeology, Journal of Biomechanics, American Journal of Clinical Nutrition, Journal of Experimental Biology, Anatomical Record
4	45	International Journal of Primatology, American Journal of Primatology, Primates, Philosophical Transactions of the Royal Society B: Biological Sciences, Animal Behaviour, PLOS One
5	45	American Journal of Human Genetics, Annual Review of Human Genetics, Forensic Science International, Molecular Evolution, Trends in Genetics, Proceedings of the National Academy of Sciences of the United States of America

Table 2. Clusters of authors cited more than 100 times identified in the co-citation network map of authors of documents in the discipline of Evolutionary Anthropology 1970-2018. (For authors cited more than 100 times)

Cluster No.	No. of Nodes	Example Nodes
1	74	Wood, White, Blumenschine, Klein, Hublin, Binford, Marean, Leakey LSB, Leakey MD, Mellars, McBrearty
2	68	Fleagle, Hylander, Walker, Wrangham, McGrew, Pickford, Darwin, Jolly, Goodall, Pilbeam
3	60	Ruff, McHenry, Day, Leakey REF, Lovejoy, Washburn, Zihlman
4	55	Wolpoff, Brace, Arsuaga, Dean, Aiello, Corrucini, Relethford
5	45	Stringer, Trinkaus, Morwood, Rightmire, Hrdlicka, Boule, Lahr

Table 3.Clusters of terms used more than 100 times identified in the association network map of terms from titles and abstracts of documents in the discipline of Evolutionary Anthropology 1970-2018. (For terms used more than 100 times)

Cluster No.	No. of Nodes	Example Nodes
1	68	Hominoid, locomotion, bipedalism, gorilla sp, fossil, anatomy, a. africanus, a. afarensis
2	65	Child, weight, infant, health, mortality, death, race, demography, nutritional status
3	55	Assemblage, cave, Palaeolithic, sediment, carnivore, burial, occupation
4	48	Person, migration, haplogroup, marker, mitochondrial DNA, ethnic group, gene, cluster, marker
5	44	Disease, lesion, pathology, trauma, warfare, violence, geometric morphometrics

anatomy (*Dean, Conroy, Macho, Mays and Relethford*) and two largely overlapping groups considering broader evolutionary approaches to human evolution (*Andrews, Crompton, McHenry, Zilhman*) and another more focussed on primates (*Boesch, Fleagle, Goodall, McGrew and Whiten*). The language of Palaeoanthropology comprises clusters of terms centred on population, morphology, species, primate, growth and evolution. The terms or concepts of these themes largely overlap with smaller clusters of specialised conceptual vocabulary such as the species names of primates and hominins, terms related to disease, death and burial and a quite distinct set of vocabulary related to genetics. Underlying these

clusters of terms is a more dispersed set of terms related to the physical recovers of the fossil remains and their position in ancient landscapes.

Palaeolithic research as a specialty

Documents published in the subfield of Palaeolithic archaeology were located by performing a topic search in the Web of Science. Specifically, a search was made for documents containing either 'Palaeolith* AND/OR Paleolith*' in their titles or abstracts. The combination of two different spellings was used to help collect bibliometric data for documents written either in

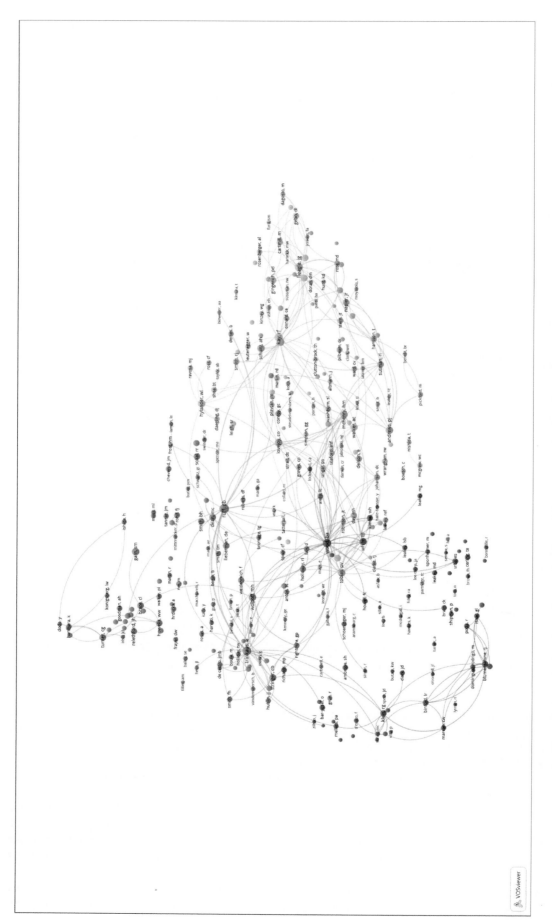

Figure 2. A co-citation network map of authors of documents in the discipline of Evolutionary Anthropology 1970-2018.

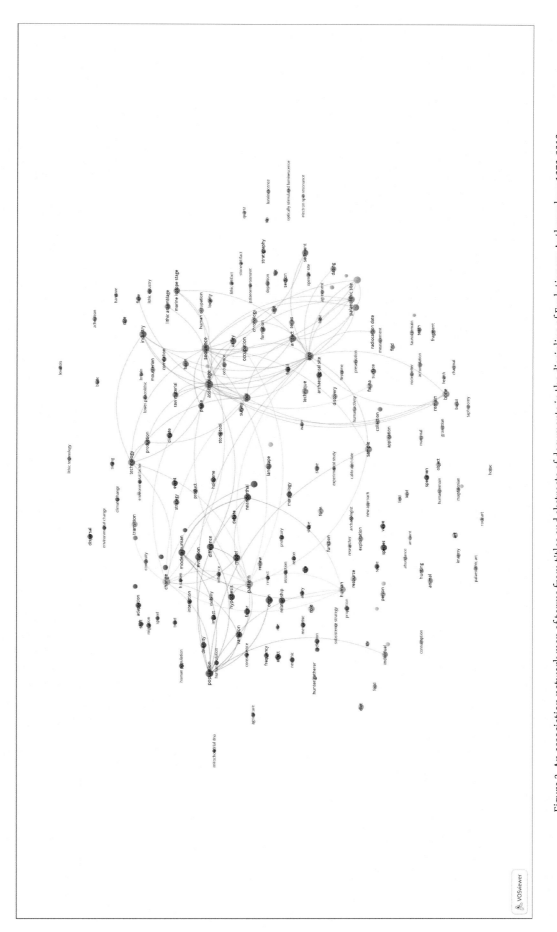

Figure 3. An association network map of terms from titles and abstracts of documents in the discipline of Evolutionary Anthropology 1970-2018.

English and/or a range of European languages. This search gathered bibliometric data for approximately 6500 research outputs published between 1970 and 2018. The known coverage of the Web of Science means that this data set is likely to be missing information on monographs as well as articles published in smaller journals or not in English, but a good number of these will be picked up through the further citations made in these documents. Using this dataset, network maps of sources, authors and terms were constructed and are presented here with each map restricted to the most commonly used terms or cited authors for the sake of legibility.

The network map of sources (Figure 4) for Palaeolithic research necessarily overlaps with that of Evolutionary Anthropology. Six clusters of sources are identified (Table 4) with two distinctive clusters for quaternary science and dating, and for physical anthropology, genetics and animal behaviour. The remaining four clusters overlap considerably making it difficult to determine how the specific combinations of sources have been clustered at first. Closer inspection, however, suggests that clustering reflects the region of study (Northwestern Europe versus France and Spain) and possibly topic (emergence of modern

humans and modern human behaviour). In the map of terms (Figure 5), six distinct clusters are identified (Table 5). One clear cluster of concepts relates to the interpretation of lithic technology and lithic assemblages (*assemblage, artefact, industry, tool, technology, industry*), with two further clusters related to the chronological interpretation of sites (*sediment, section, layer, occupation, age, sample*) and how we interpret their contents (*cave, bone, remain, resource, fauna, taphonomy*). A further cluster relates to theme of species evolution and movement (*population, transition, modern human, Neanderthal, dispersal, origin, Aurignacian*) and another to hominin social formations (*individual, hunter gather, role, difference, size, diet*). The final and smallest cluster relates to the discovery and interpretation of symbolic materials (*rock art, Palaeolithic art, pigment, imagery, technique*). In contrast to the maps for Palaeoanthropology, the overlapping nature of the maps of terms and sources clearly identifies Palaeolithic archaeology as a conceptually coherent and largely unified area of research generating papers that might be offered to a broad range of sources unless focused on genetics or the hard science aspects of quaternary research. By contrast the network co-citation map of authors seems more clearly defined (Figure 6, Table 6) with seven clusters

Table 4. Clusters of sources identified in the co-citation network map of sources publishing documents in the specialty of Palaeolithic Archaeology 1970-2018. (For sources cited more than 50 times)

Cluster No.	No. of Nodes	Example Nodes
1	65	Journal of Human Evolution, Current Anthropology, South African Archaeological Bulletin, Palaeorient, Palaeoanthropology
2	60	American Journal of Physical Anthropology, American Journal of Human Genetics, Human Biology, American Journal of Clinical Nutrition, Nature, Science, PLOS One
3	45	Quaternary International, Quaternary Science Reviews, Quaternary Research, Palaeogeography, Palaeoecology & Palaeoenvironment, Journal of Archaeological Science, Archaeometry, Geoarchaeology
4	42	Radiocarbon, American Antiquity, Cambridge Archaeological Journal, American Anthropologist, World Archaeology
5	42	Journal of Quaternary Science, Proceedings of the Prehistoric Society, ERAUL, Lithics
6	41	Anthropologie, Bulletin de la Societie Préhistorique Française, Journal of Archaeological Method and Theory, Gallia Préhistoire

Table 5. Clusters of authors identified in the co-citation network map of authors of documents in the specialty of Palaeolithic Archaeology 1970-2018. (For authors cited more than 50 times)

Cluster No.	No. of nodes	Example Nodes
1	94	Binford, Stiner, Kuhn, Zilhao, Straus, Gamble, Villa, Dibble, Breuil, Leroi-Gourhan, Clottes
2	64	Mellars, Bordes, Svoboda, Hahn, Klima, Hublin, Hedges
3	50	Bridgeland, Ashton, White, Roebroeks, Moncel, Boeda, Wymer, Roe, de Lumley, McNabb
4	46	D'Errico, Trinkaus, Ruff, Holliday, McBrearty, Henshilwood, Klein
5	46	Bar-Yosef, Stringer, Shea, Petraglia, Goren-Inbar, Dennell, Petraglia, Gowlett

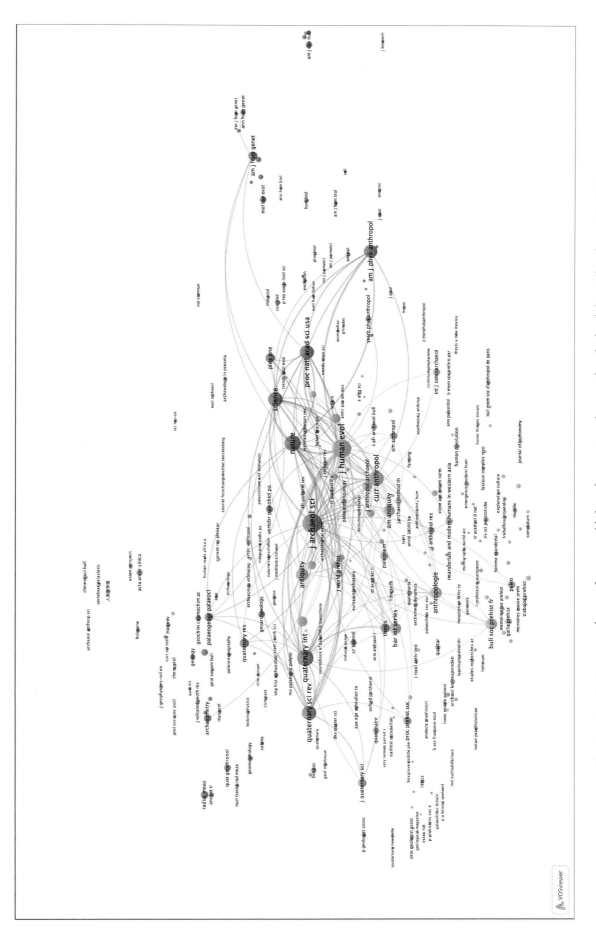

Figure 4. A co-citation network map of sources publishing documents in the specialty of Palaeolithic Archaeology 1970-2018.

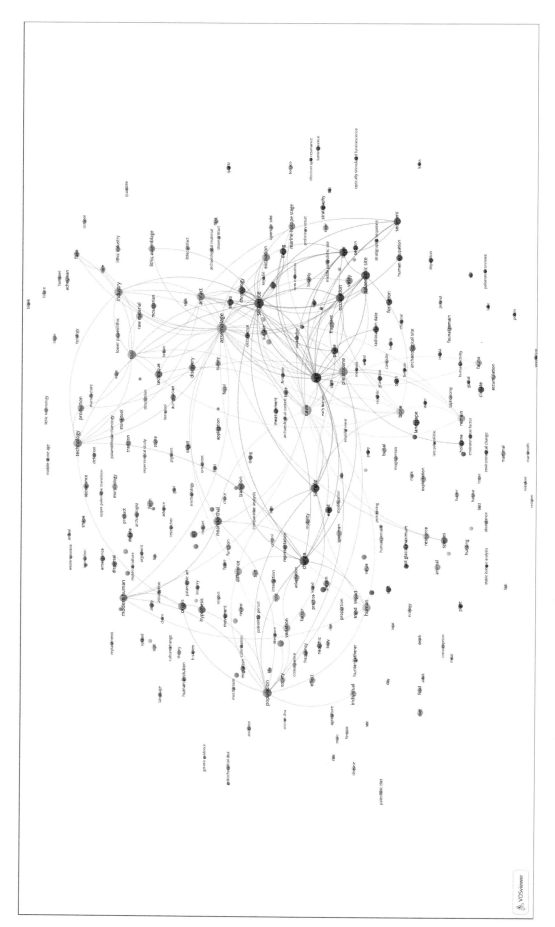

Figure 5. A co-citation network map of authors of documents in the specialty of Palaeolithic Archaeology 1970–2018.

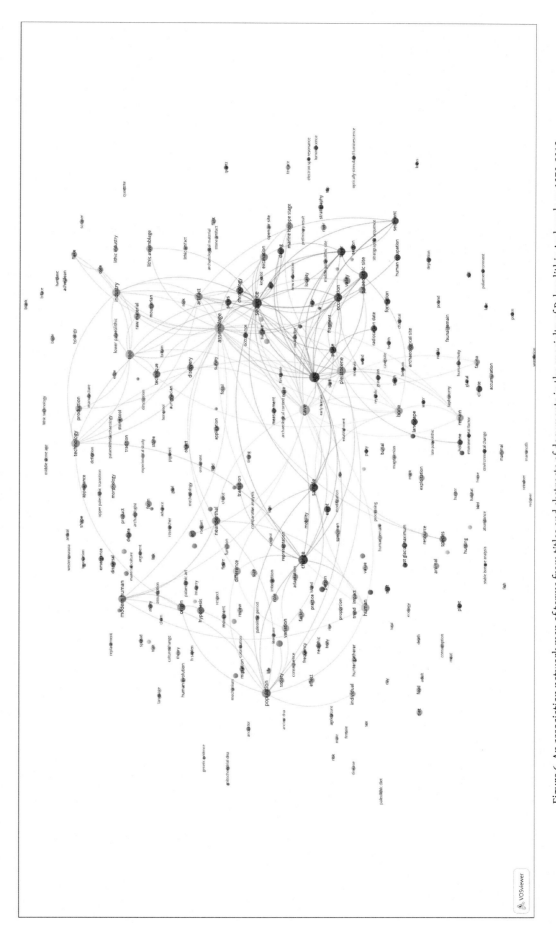

Figure 6. An association network map of terms from titles and abstracts of documents in the specialty of Palaeolithic Archaeology 1970-2018.

Table 6. Clusters of terms identified in the association network map of terms from titles and abstracts of documents in the specialty of Palaeolithic Archaeology 1970-2018. (For terms used more than 50 times)

Cluster No.	No. of nodes	Example Nodes
1	69	Sequence, age, layer, dating, sediment, Pleistocene, occupation, chronology, sample, stratigraphy
2	65	Bone, burial, carnivore, resource, fauna, remain, trend, diet, food, role
3	41	Acheulean, technology, tool, artefact, industry, core, flake, stone tool, raw material, function
4	39	Modern human, origin, evolution, human, change, Neanderthal, Aurignacian, Gravettian, dispersal
5	32	Cave, art, form, Palaeolithic art, hypothesis, prehistory, painting, rock art

of coherent author groupings where the placement of individual authors is determined by the preponderance of citations to an author's publications by their peers. One cluster includes authors writing about hominin morphology and the transition between species (*Stringer, Ruff, Trinkaus, Wolpoff, Lieberman, Hublin*). Another cluster relates to authors on rock art research specifically (*Bednarik, Breuil, Leroi-Gourhan, Clottes, Conkey* and *White*) and the interaction between material culture and human social relations more generally (*Shennan, Gamble, Hodder, Mithen, Wobst*). A third cluster is about the earlier Palaeolithic of northern Europe (*Roebroeks, Ashton, Bridgeland, McNabb*), with a fourth relating to the interpretation of faunal assemblages (*Binford, Stiner, Speth, Grayson*). The remaining three clusters gather authors publishing work on early hominin technologies and the Acheulean in particular (*Goren-Inbar, Wynn, Potts, and Shea*) another concerned with the interpretation of lithic technology more generally (*Dibble, Kuhn, Bordes, Odell* and *Bleed*) and, finally, two clusters of authors working on the dating and impact of modern humans in Western Europe (*Higham, Zilhao, Teyssander*) and in Central and Eastern Europe (*Conard, Derevianko, Klima, Mania, Valoch*).

The work and influence of John Gowlett

Bibliometric data can also be used to drill down into the work of a single researcher, developing what White has termed an 'ego-centred citation analysis' (White 2000). In this approach White identifies 4 distinct '*ego-alter*' relationships split into two broad types; the *citation identity* of an author comprising (1) the author's collaborators in the form of co-authors, and (2) the other scholars and their publications that an author cites and the *citation image* of an author comprising (3) the network of other scholars who directly cite the author, and (4) the network of scholars who are co-cited along with the author in other publications. The distinction between the *citation identity* that is created by an author's own agency and the *citation image* that is cumulatively created by the agency of the readers of the author's work makes it possible to see whether the documents of an author are read and contextualised by readers in the same way that they have been by their author. Since research develops over time, change in an author's networks of co-authors, and their citation identity may also change across time. Despite the potential for using bibliometric data for this form of

Table 7. Forms of Ego-Alter relationship, their relevant document sets and the possible nature of variation through time (modified after White 2000, 2001).

	Form of '*Ego-Alter*' Relationship	Document set used for bibliometric data source	Possible Variation over Time
Ego's Citation Image	Ego and his/her research collaborators	Ego published documents – author and co-authors	Long-term versus short-term project collaborators (*Ego as a developing collaborator over time*)
	The set of researchers cited by Ego	Ego published documents – authors of cited documents in reference lists	Potential change through time as ego's research interests develop / change (*Ego as a developing researcher and citer over time*)
Ego's Citation Identity	The set of researchers who cite publications by Ego	Ego-citing documents – authors	Changes in the citation form / context over Ego's knowledge claims over time (*Life-history of Ego's individual knowledge claims*)
	The total set of authors cited in publications which cite publications by Ego	Ego-citing documents – all authors of documents in reference lists	Variation in Ego's knowledge claims as contributions to the understanding of knowledge in a particular research topic / area (*Life-history of Ego's individual knowledge claims in context*)

analysis, there are few if any published examples of this approach used outside of short pieces on the work of bibliometric researchers themselves (Bar-Ilan 2006, Leydesdorff 2010). Here we look at the work of John Gowlett.

Extracting a full set of bibliometric data for a single author should be easily acheivable. Unfortunately, the potential effect of the sampling process used in the citation indices has its greatest impact at an individual level. In the case of John Gowlett, this is evident in the fact that both Scopus and the Web of Science contain bibliometric data on 75 documents, whereas Google Scholar, which uses an automated trawling process to identify and collect basic bibliometric data across a wide range of source types including websites, presents details on 195 documents extracted using the 'Publish or Perish' software created by Anne-Wil Harzing (available at www.harzing.com/resources/publish-or-perish). Once duplicates, simple book reviews and commentaries to articles of the type published in Current Anthropology are removed, 139 documents remain as individual research publications (Appendix A). These include research articles, book chapters, encyclopaedia entries, a single-author book, several edited books and four edited thematic sections to World Archaeology (Table 8). A number of the articles and book chapters are clearly the published products of research originally presented at conferences, but since it is not possible to identify all such pieces, these documents are not specifically identified as conference proceedings papers except for two abstracts for papers given to the annual meeting of the Palaeoanthropology Society. A distinct set of papers, written mostly in the 1980s, are related to John Gowlett's appointment to the radiocarbon dating unit in the Research Laboratory of

Table 8. The format of publications for research outputs by John Gowlett (1978-2018) used in this study. (Individual document details are set out in Appendix 1).

Publication format	No. of examples
Journal articles	58
Book chapters	59
Conference Abstract	2
Encyclopaedia entries	10
Book (single author)	1
Edited Book	4
Edited Journals	4
Thesis	1
Note / Invited commentaries	3
Extended Book Reviews	2
TOTAL	144

Archaeology at the University of Oxford at the time of the development and use of the United Kingdom's first accelerator mass spectrometry radiocarbon dating facility from 1983. Another set of papers derives from 'Lucy to Language', one of the British Academy's Centenary Research Projects, co-directed by John Gowlett, Clive Gamble and Robin Dunbar between 2003 and 2010. These papers however align themselves in subject matter with Gowlett's long-term research interests discussed below.

The Citation Identity of John Gowlett

The Google Scholar-derived document set can be used to map John Gowlett's co-authors / research collaborators (Figure 7). This network map includes 73 individuals alongside John Gowlett himself, with chronological changes in co-authorship networks from 1980 to 2020. Five distinct groups can be seen. The two oldest are associated with John Gowlett's work for the Oxford radiocarbon dating (Hedges, Gillespie) and with early fieldwork in Africa (Harris). Another distinct group relates to research and publication with a group of research students of data from the site of Beeches Pit in East Anglia (Bell, Brant, Chambers, Hallos) around 2000. Two final groups relate to 'Lucy to Language' (Dunbar, Gamble) and a second phase of fieldwork in Kenya (Brink, Herries, Hoare, Rucina) since 2010.

Unfortunately, Google Scholar cannot collect bibliometric data on abstracts, keywords or references cited, so the bibliometric data mapped for the authors cited and concepts used by Gowlett must be based on the smaller Scopus data set. The network map of authors co-cited in John Gowlett's own papers presents the 100 authors most cited (Figure 8). They fall into two primary groups, with a set of 19 authors related to John's work at the Oxford Radiocarbon Dating Unit and both clustered and separated to one side, and a larger set of the remaining 81 authors in three clusters to the other side. Of these three clusters, one appears to be related to fire and the colonisation of northern Europe (Wrangham, Brain, Roebroeks, Gamble), another relates to cognition, symbolism and communication (Dunbar, Wynn, Aiello and Goren-Inbar) and a final group related to lithic analysis, the Acheulean and Africa (Isaac, Clark, Leakey, Roe). Citation numbers alone suggest that the major influences have been Dunbar, JD Clark, Gamble, Wynn, Isaac, Wrangham, Aiello, Goren-Inbar and Binford.

Using the titles and abstracts of the Scopus data set, VOSviewer identifies 64 items divided into 4 major clusters (Figure 9). One cluster includes a series of terms related to research on fire, fire use and fire control, including the sites of Chesowanja and Beeches Pit where the identification of fire has been so significant. This cluster also includes terms related to the cooking

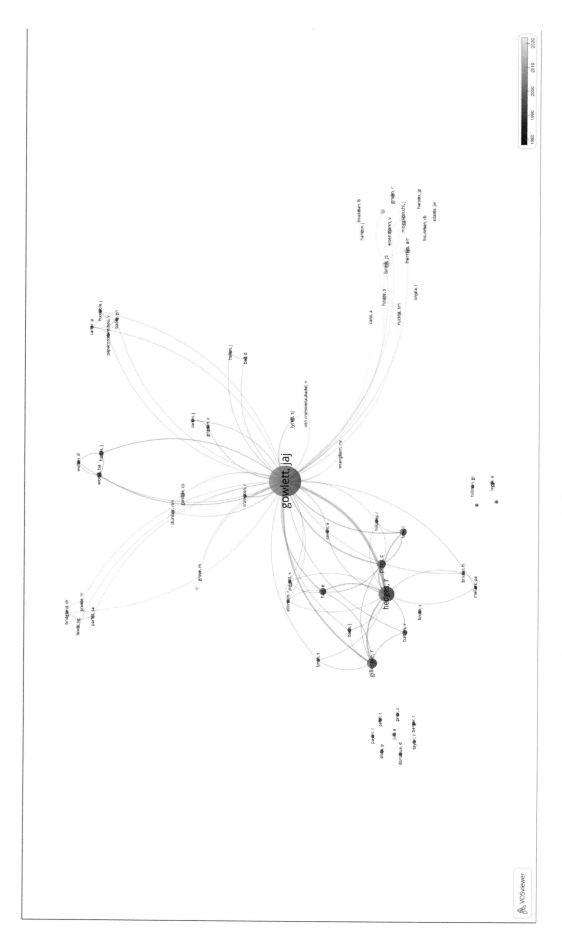

Figure 7. An association map of authors writing documents with John Gowlett.

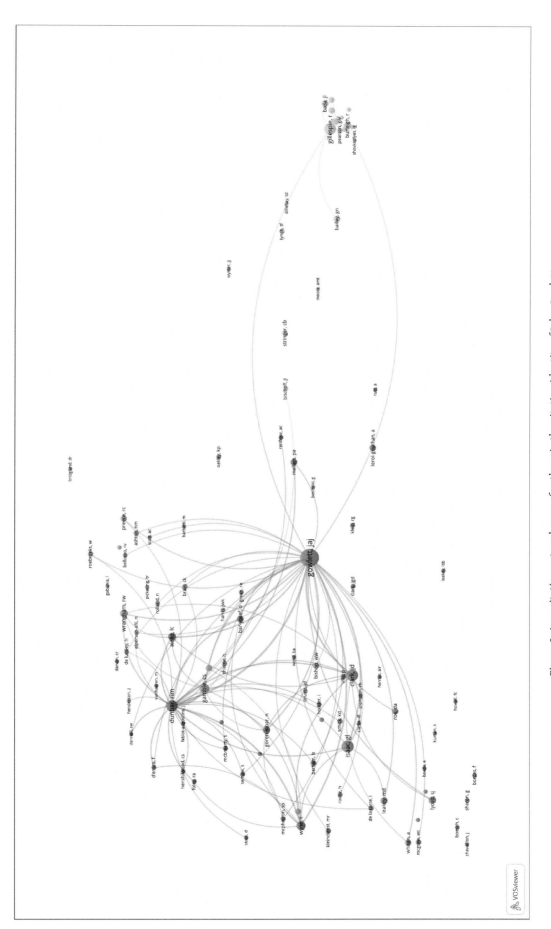

Figure 8. A co-citation network map of authors in the citation identity of John Gowlett

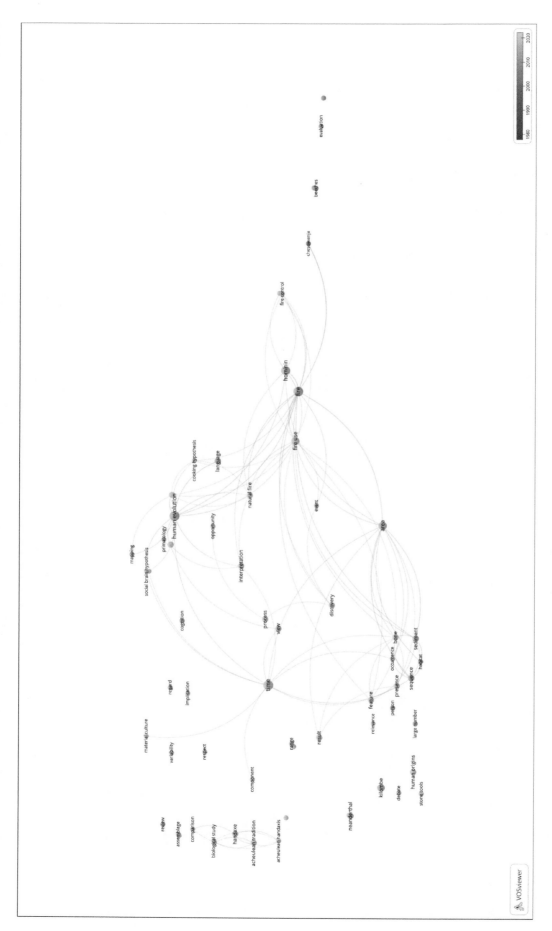

Figure 9. An association network map of terms used in abstracts and titles of documents published by John Gowlett

hypothesis and the role of diet in the *social brain hypothesis*. Another cluster relates to John's research on the *Acheulean tradition*, the *handaxe* and *material culture*. *Process* and *language* are also in this set of terms. A third set of terms relates to John Gowlett's research interest in chronology with terms specifically related to the integrity of dating samples: *area, discovery, event, result, range, sequence* and *sediment*. A final set of terms is more general, and probably derives from the broader pieces written about human origins. This set includes *Neanderthal, debate, stone tools, view* and *human origins*.

The Citation Image of John Gowlett

For examining John Gowlett's Citation Image, Scopus provides its users with information on the number of times a document has been cited and then access to the bibliometric data of these citing documents themselves. The collection of data for these citing documents - to generate the network maps for White's type 3 and 4 ego-alter relationships - has been restricted to those documents that have been cited ten times or more. The assumption made is that frequent citation of a document is likely to indicate that the document has a consistent influence within a particular area of knowledge; for documents where citations are fewer, an idiosyncratic reading and citation will skew the broader pattern of relationships. Thus, data was collected for 1609 citing documents making citations to 53 papers in John Gowlett's oeuvre.

Whilst the name Gowlett does not appear in the network map for authors' in Palaeoanthropology (Figure 3), he is present in the map of most-cited authors for Palaeolithic archaeology (Figure 6) placed close to a cluster of authors conducting research on Acheulean lithic assemblages (*Goren-Inbar, Sharon, Petraglia, Wynn*), and another cluster of researchers looking at the earlier Palaeolithic record of northern Europe (*Roebroeks, Ashton, Roe*), and a third cluster producing research on the broader theory of interpretation of lithic technology (*Boeda, Dibble*). Gowlett, himself, is clustered amongst the Acheulean researchers indicating that during the time period of this analysis (1970-2018) it is Gowlett's Acheulean research that is most consistently cited. The network map of the citation image of authors (those co-cited along with John Gowlett) provides greater nuance still (Figure 10). It contains 250 names, with Gowlett in the centre. These authors are clustered into 6 groupings. One group contains a series of names related to discussions of dating (*Hedges, Grun, Schwarz, Valladas, Gillespie, Stuiver, Renfrew*), a second group contains authors associated with research about the European Palaeolithic (*Straus, Zilhao, Gamble, Kozlowski, Bailey*). A third group includes authors associated with the earlier Palaeolithic record of Europe (*Ashton, White, Carbonell, Goren-Inbar, Roebroeks*). There are two

clusters of authors associated with the interpretation of lithic technology. One is very African focused (*Isaac, de la Torre, Leakey, Schick, Toth, Roche*) the other is more about technology and manufacture in general (*Shott, Lycett, Shea, Dibble, McPherron*). The final cluster, in which Gowlett is himself clustered, seems to be more general in character and concerned with the nature of change in human evolution (*Clark, Klein, Binford, Bordes, Wrangham, Wadley, Mellars*).

Conceptually, *VOSviewer* identifies 250 terms from the titles and abstracts of the papers citing documents by Gowlett clustered into 5 groups (Figure 11). The first can be associated with dating issues (*radiocarbon date, carbon, charcoal, archaeometry datelist, bone, period, age*). The second is the general conceptual terminology of excavation and analysis of Palaeolithic sites (*environment, landscape, occupation, excavation, fossil, specimen, sediment*). The three final clusters are more directly related to the research topics in John Gowlett's own papers. These include a cluster examining hominin forms of communication and sociality (*language, material culture, nature, mind, social brain*), another looking at fire and food (*evolution, fire, hearth, cooking, homo erectus, cooking, cognition*) and the final and largest cluster examining lithic technology and its form (*Acheulean, handaxe, assemblage, morphology, manufacture, reduction, variability, skill*).

Discussion

The network maps presented here visually demonstrate that research in evolutionary anthropology is extraordinarily diverse spanning fields such as biology and anatomy, primates, animal behaviour, environment, cognition, and more. More significant than the range, however, is the clear network of connections between research published in each of these distinct fields and the publication of research in the primary journals (*Journal of Human Evolution, American Journal of Physical Anthropology*). Evolutionary Anthropology is not an artificially constructed area of research. At each of the levels explored here (discipline, specialty and specialist) this same combination of diverse research areas comes together. Whilst a specialty such as Palaeolithic Archaeology may be defined by its recovery and analysis of artefactual remains, and a specialist may focus on the understanding of a smaller subset of such remains, it is clear that each cannot conduct research without reference to the wider range of fields, and consider the impact of their interpretations back to this range.

A second clear disciplinary quality present in these maps is the reduced pattern of obsolescence of sources. For example, this network map of most cited authors for Palaeolithic archaeology contains the names of many, deceased scholars including *Isaac, Garrod, Leroi-Gourhan*,

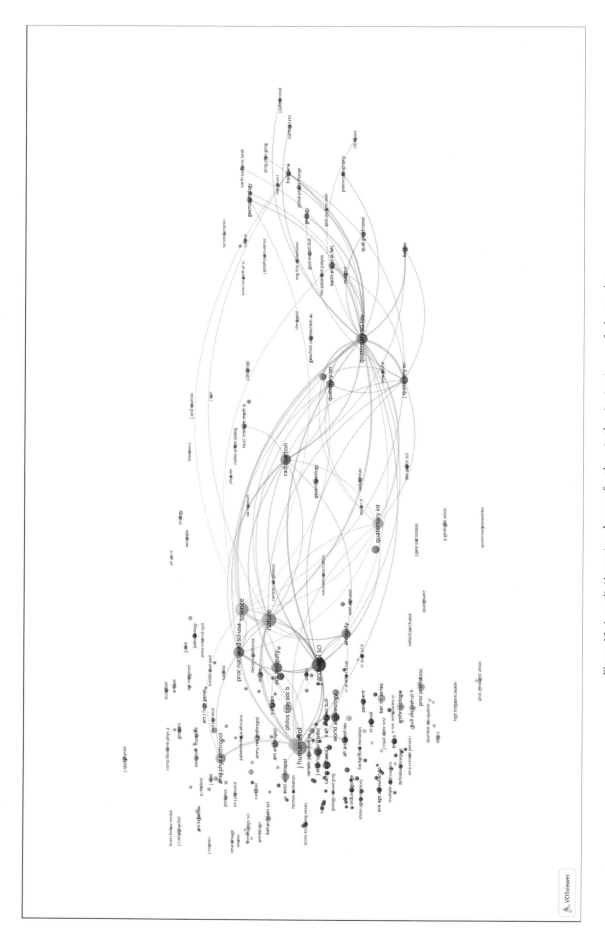

Figure 10. A co-citation network map of authors in the citation image of John Gowlett

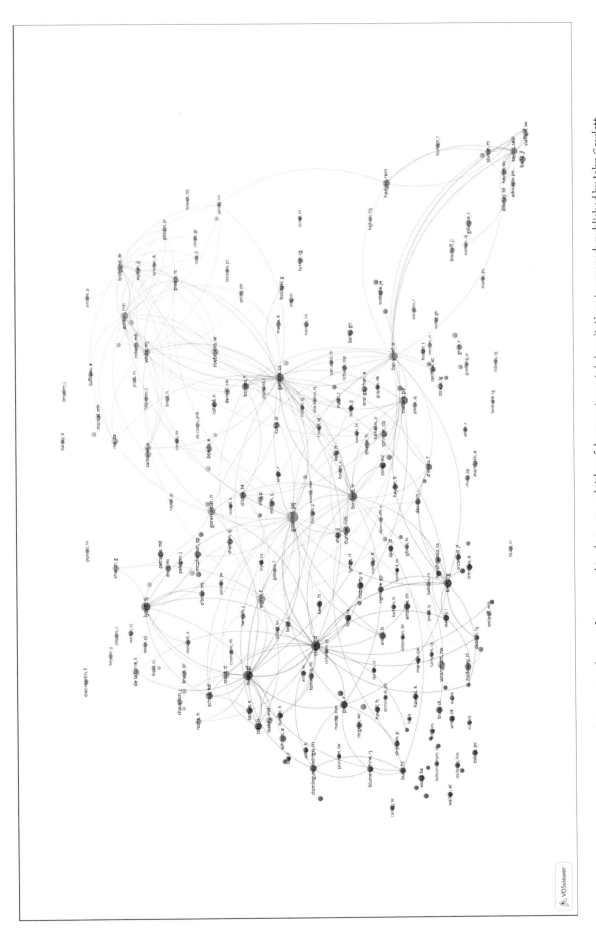

Figure 11. An association network map of terms used in abstracts and titles of documents containing citations to research published by John Gowlett

Leakey (Louis and *Mary), Wymer, Movius* and *Breuil.* Even though deceased, the published research of these authors still actively contributes to the contemporary intellectual base of Palaeolithic research. The research of these scholars has retained a long 'use-life' in apparent contrast to that observed through citation in the natural and biological sciences. Research outputs in Evolutionary Anthropology whilst seemingly similar in content and form to those in the natural and biological sciences have a long 'use life' just like research literature in the social sciences, arts and humanities (Cole 1983), where classic pieces if research are often subsumed into the common knowledge of disciplinary practitioners. In Garfield's terms, these documents do not become obliterated (in terms of later citation) through the incorporation of their contents into the commonly accepted knowledge of practitioners that no longer requires citation (Garfield 1975). If the required working knowledge of practitioners in the discipline requires detailed familiarity with such older documents, there are clear implications for the teaching of Evolutionary Anthropology in the context of the quasi exponential growth in number of research outputs. As a first step, however, a more detailed study is needed to explore the exact nature of the citation of these older documents to see how they are used in current knowledge claims examining how their life histories as knowledge claims (Cozzens 1985) has changed and whether the detail of their contents is still cited or whether their citation acts as a conceptual symbol of a broader approach (Small 1978).

Finally, whilst White (2000) suggested that a researcher may move from one separate research problem to another, the contrary impression seems to arise from the research interests of John Gowlett as visualised in the network of terms used. The most 'central' terms in John Gowlett's research lexicon are *time* and *process*. But these should perhaps be seen in their broadest sense as evidenced by an interest in the Palaeolithic as a period characterised by multiple scales of time and action from the macro to the micro ('High-definition archaeology'). His research is also characterised by a series of consistent research problems - technology and fire. However, as can be seen from the list of papers themselves (Appendix A), each is explored through different perspectives over time. In the case of fire, the earlier papers are concerned with the methodology and epistemology of demonstrating human use of fire at great time depths, whilst recent papers theoretically explore the social and cognitive impacts of fire as a technology of environmental transformation. In the case of technology, the earlier papers explore technology as a cognitive process, and later papers develop methodologies for demonstrating the expression of cognitive processes in archaeological artefacts.

Whilst philosophers and historians of scientific research might envisage discrete problems that are investigated and solved, the examination of Palaeoanthropology from discipline to sub-field to scholar suggests that research in the evolution of early hominins does not proceed in that way. The complex interrelationships theorised between the basic material objects and the processes of social and cognitive development result in a continuous process of hermeneutic re-engagement with, and possibly refinement, of the relationship between interpretation and data rather than a definitive resolution.

References

Andersen, J.P., and Nielsen, M.W., 2018. Google Scholar and Web of Science: examining gender differences in citation coverage across five scientific disciplines. *Journal of Informetrics* 12: 950-959.

Ardanuy, J., 2013. Sixty years of citation analysis studies in the humanities (1961-2010). *Journal of the American Society for Information Science and Technology* 64: 1751-1755.

Bar-Ilan, J., 2006. An ego-centric citation analysis of the works of Michael O. Rabin based on multiple citation indexes. *Information Processing and Management* 42: 1553-1566

Börner, K., Chen, C. and Boyack, K.W., 2003. Visualizing knowledge domains. *Annual Review of Information Science and Technology* 37, 179-255.

Boyack, K.W., van Eck, N.J., Colavizza, G., and Waltman, L., 2018. Characterizing in-text citations in scientific articles: a large-scale analysis. *Journal of Informetrics* 12: 59-73.

Boyack, K.W., Klavans, R. and Börner, K., 2005. Mapping the backbone of science. *Scientometrics* 64: 351-374.

Brughmans, T., 2013. Networks of networks: a citation network analysis of the adoption, use and adaptation of formal network techniques in archaeology. *Literary and Linguistic Computing* 28: 538-562.

Chen, C., 2017. Science mapping: a systematic review of the literature. *Journal of Data and Information Science* 2: 1-40.

Chi, P-S., 2016. Differing disciplinary citation concentration patterns of book and journal literature. *Journal of Informetrics* 10: 814-829

Cole, S., 1983. The hierarchy of the sciences? *American Journal of Sociology* 89: 111-139.

Cozzens, S., 1985. Life history of a knowledge claim: the opiate receptor case. *Science Communication* 9: 511-529.

Cronin, B., 1984. *The citation process: the role and significance of citations in scientific communication.* London: Taylor Graham.

Davenport, E., and Snyder, H., 2007. Who cites women? Whom do women cite? An exploration of gender and scholarly citation in sociology. *Journal of Documentation* 51: 404-410.

Dunmore, C.J., Pateman, B., and Key, A.J.M., 2018. A citation network analysis of lithic microwear research. *Journal of Archaeological Science.* 91: 33-42.

Falagas, M.E., Pitsouni, E.I., Malietzis, G.A. and Pappas, G., 2007. Comparison of PubMed, Scopus, Web of Science, and Google Scholar: strengths and weaknesses. *The FASEB Journal* 22: 338-342.

Garfield, E., 1955. Citation indexes for science: a new dimension in documentation through Association of Ideas. *Science* 122: 108-111.

Garfield, E., 1964. Science citation index-a new dimension in indexing. *Science,* 144: 649-654.

Garfield, E., 1975. The 'Obliteration Phenomenon' in science and the advantage of being obliterated. *Current Contents* 51-52: 5-7

Garfield, E., and Sher, I. H., 1963. New factors in the evaluation of scientific literature through citation indexing. *American Documentation,* 14: 195-201.

Gilbert, G.N., 1977. Referencing as persuasion, *Social Studies of Science* 7: 113-122.

Hutson, S., 2002. Gendered citation practices in American Antiquity and other archaeology journals. *American Antiquity* 67: 331-342.

Hutson, S., 2006. Self-citation in archaeology: age, gender, prestige and the self. *Journal of Archaeological Method and Theory* 13: 1-18.

Hyland, K., 1999. Academic attribution: citation and the construction of disciplinary knowledge, *Applied Linguistics* 20, 341-367.

Kaplan, N., 1965. 'The norms of citation behaviour: prolegomena to the footnote. *American Documentation* 16, 179-184.

Klavans, R., and Boyack, K.W., 2014. Mapping altruism. *Journal of Informetrics* 8: 431-447.

Lartet, E., and Christy, H., 1865-1875. *Reliquae Aquitainae; being contributions to the archaeology and palaeontology of Périgord and the adjoining provinces of southern France.* London, Williams and Northgate.

Leakey, M., 1971. *Olduvai Gorge excavations in beds I and II.* Cambridge: Cambridge University Press.

Leydesdorff, L., 1998. Theories of citation? *Scientometrics* 43, 5-25.

Leydesdorff, L., 2010. Eugene Garfield and algorithmic historiography: co-words, co-authors, and journal names. *Annals of Library and Information Science* 57: 248-260.

Martin-Martin, A., Orduna-Melea, E., Thelwall, M., and López-Cózar, E.D., 2018a. Google Scholar, Web of Science, and Scopus: a systematic comparison of citations in 252 subject categories. *Journal of Informetrics* 12: 1160-1177.

Martin-Martin, A., Orduna-Melea, E., and López-Cózar, E.D., 2018a. Coverage of highly cited documents in Google Scholar, Web of Science and Scopus: a multidisciplinary comparison. *Scientometrics* 116: 2175-2188.

Nicolaisen, J., 2007. Citation analysis. *Annual Review of Information Science and Technology* 41, 609-641.

Price, D.J. de Solla, 1951. Quantitative measures of the development of science. Archives. *Internationales d'Histoire des Sciences* 4, 85-93.

Price, D.J. de Solla, 1963. *Little science, big science.* New York: Columbia University Press.

Price, D.J. de Solla, 1965. Networks of scientific papers. *Science* 149, 510-515.

Sinclair, A., 2016. The intellectual base of archaeological research 2004-2013: a visualisation and analysis of its disciplinary links, networks of authors and conceptual language. *Internet Archaeology* 42. https://doi.org/10.11141/ia.42.8

Small, H., 1978. Cited documents as concept symbols. *Social Studies of Science* 8: 327-340

Van Eck, N.J. and Waltman, L., 2010. Software survey: VOSviewer, a computer program for bibliometric mapping. *Scientometrics* 84, 523-538.

Wang, Q., and Waltman, L., 2016. Large-scale analysis of the accuracy of journal classification systems of Web of Science and Scopus. *Journal of Informetrics* 10: 347-364.

Anthony Sinclair

Archaeology, Classics and Egyptology, University of Liverpool

Appendix A. Publications by John Gowlett 1978-2018

1. JAJ Gowlett, 1978. Kilombe—an Acheulian site complex in Kenya. In ed. WW Bishop (ed.) *Geological Background to Fossil Man.* Edinburgh, Scottish Academic Press Pp 337–60

2. JAJ Gowlett, 1979. Complexities of cultural evidence in the Lower and Middle Pleistocene. *Nature* 278: 14-17

3. JWK Harris & JAJ Gowlett, 1979. Kenya. A Preliminary Report on Chesowanja. *Nyame Akuma* 14: 22-27

4. JAJ Gowlett, 1979. *A contribution to studies of the Acheulean in East Africa with especial reference to Kilombe and Kariandus.* PhD thesis, University of Cambridge.

5. JAJ Gowlett, 1980. Acheulean sites in the central Rift Valley, Kenya. In RE Leakey and BA Ogot (eds.) *Proceedings of the 8th PanAfrican Congress of Prehistory and Quaternary Studies.* Nairobi TILLIMAP Pp 213-217.

6. JWK Harris, JAJ Gowlett, 1980. Evidence of early stone industries at Chesowanja, Kenya. In RE Leakey and BA Ogot (eds.) *Proceedings of the 8th Panafrican Congress of Prehistory and Quaternary Studies.* Nairobi TILLIMAP Pp 208-212

7. JAJ Gowlett, 1981. Earliest technology. In WG Mook & HT Waterbolk (eds.) *Proceedings of the First International Symposium 14 C and Archaeology.* Groningen. Council of Europe

8. JAJ Gowlett, 1981. Contributions to British and European prehistory: the scope and problems of 14C accelerator dating. In WG Mook & HT Waterbolk (eds.) *Proceedings of the First International Symposium 14 C and Archaeology.* Groningen. Council of Europe

9. JWK Harris, JAJ Gowlett, D Walton, & BA Wood, 1981. Palaeoanthropological studies at Chesowanja. In *Las Industrias Mas Antiguas.* UISPP 10

10. JAJ Gowlett, JWK Harris, D Walton, & BA Wood. 1981. Early archaeological sites, hominid remains and traces of fire from Chesowanja, Kenya. *Nature* 294:125-129

11. JAJ Gowlett, 1981. Confusing script*Nature*291: 104

12. JAJ Gowlett, 1982. Procedure and Form in a Lower Palaeolithic Industry: Stoneworking at Kilombe, Kenya. In D Cahen (ed.) *Studia Praehistorica Belgica. Tailler. Pour quoi Faire ? Prehistoire et Technologie* II Pp 101-110

13. JAJ Gowlett, JWK Harris, & BA Wood, 1982. Early hominids and fire at Chesowanja, Kenya (reply). *Nature* 296: 870

14. JAJ Gowlett, 1982. Updating the Old Stone Age. *Nature* 298: 204

15. JWK Harris, JAJ Gowlett, R Blumenschine, & JE Maiers, 1983. Chesowanja—a summary of the early Pleistocene Archaeology. In In RE Leakey and BA Ogot (eds.) *Proceedings of the 9th PanAfrican Congress of Prehistory and Quaternary Studies.* Nairobi TILLIMAP

16. R Gillespie & JAJ Gowlett, 1983. Archaeological sampling for the new generation of radiocarbon techniques. *Oxford Journal of Archaeology* 2: 379-382

17. JAJ Gowlett, 1984. *Ascent to Civilization: The Archaeology of Early Man.* London, Collins.

18. JAJ Gowlett, 1984. Mental abilities of early man: a look at some hard evidence. In R Foley (ed.) *Hominid Evolution and Ecology.* London, Academic Press Pp 167-192

19. JL Bada, R Gillespie, JAJ Gowlett, & REM Hedges, 1984. Accelerator mass spectrometry radiocarbon ages of amino acid extracts from Californian palaeoindian skeletons. *Nature* 312: 442-444

20. R Gillespie, JAJ Gowlett, ET Hall, & REM Hedges, 1984. Radiocarbon measurement by accelerator mass spectrometry: an early selection of dates. *Archaeometry* 26: 15-20

21. R Gillespie, JAJ Gowlett, & REM Hedges, 1984. Recent developments in archaeological dating using an accelerator. *Nuclear Instruments and Methods in Physics Research* B 5: 308-311

22. REM Hedges & JAJ Gowlett, 1984. Radiocarbon Dating: accelerating carbon dating. *Nature* 308: 403-404

23. JAJ Gowlett, 1985. Kilombe (Kenya). *Nyame Akuma* 26: 1-22

24. R Gillespie, JAJ Gowlett, ET Hall, REM Hedges & C Perry, 1985. Radiocarbon dates from the Oxford AMS system: Archaeometry datelist 2. *Archaeometry* 27: 237-246

25. TF Lynch, R Gillespie & JAJ Gowlett, 1985. Chronology of Guitarrero Cave, Peru. *Science* 229: 864-867

26. RE Taylor, LA Payen, CA Prior, PJ Slota Jnr, R Gillespie, JAJ Gowlett, REM Hedges, AJT Jull, TH Zabel, DJ Donohue & R Berger, 1985. Major revisions in the Pleistocene age assignments for North American human skeletons by C-14 acceleratyor mass spectrometry: none older than 11,000 C-14 years BP. *American Antiquirty* 50: 136-140

27. JAJ Gowlett, 1986. Radiocarbon accelerator dating of the Upper Palaeolithic in North-West Europe: a provisional view. In SN Colcutt (ed.)

The Palaeolithic of Britain and its Nearest Neighbours: recent trends. Sheffield, Sheffield University Press Pp 98-102

28. JAJ Gowlett, 1986. Culture and conceptualisation: the Oldowan-Acheulian gradient. In GN Bailey & P Callow (eds.) *Stone Age Prehistory: studies in memory of Charles McBurney.* Cambridge, Cambridge University Press Pp 243-260

29. JAJ Gowlett, R Gillespie, ET Hall, & REM Hedges, 1986. Accelerator radiocarbon dating of ancient human remains from Lindow Moss. In IM Stead, J Bourke, & D Brothwell (eds.) *Lindow Man, The body in the Bog.* London, British Museum Publications Pp 22-24

30. JAJ Gowlett & REM Hedges (eds.), 1986. *Archaeological Results from Accelerator Dating.* Oxford, Oxford University Committee for Archaeology Monograph Series 11

31. JAJ Gowlett, 1986. Problems in dating the early human settlement of the Americas. In JAJ Gowlett & REM Hedges (eds.), 1986. *Archaeological Results from Accelerator Dating.* Oxford, Oxford University Committee for Archaeology Monograph Series 11 Pp51-59

32. JAJ Gowlett & REM Hedges, 1986. Lessons of context and contamination in dating the Upper Palaeolithic. In JAJ Gowlett & REM Hedges (eds.), 1986. *Archaeological Results from Accelerator Dating.* Oxford, Oxford University Committee for Archaeology Monograph Series 11 Pp63-72

33. GN Bailey, CS Gamble, HP Higgs, C Roubet, DP Webley, JAJ Gowlett, DA Sturdy & C Turner, 1986. Dating results from Palaeolithic sites and palaeoenvironments in Epirus (North-west Greece). In JAJ Gowlett & REM Hedges (eds.), 1986. *Archaeological Results from Accelerator Dating.* Oxford, Oxford University Committee for Archaeology Monograph Series 11 Pp99-108

34. R Gillespie & JAJ Gowlett, 1986. The terminology of time. In JAJ Gowlett & REM Hedges (eds.), 1986. *Archaeological Results from Accelerator Dating.* Oxford, Oxford University Committee for Archaeology Monograph Series 11 Pp157-162

35. RJ Batten, CR Bronk, R Gillespie & JAJ Gowlett, 1986. A review of the operation of the Oxford Radiocarbon Accelerator Unit. In JAJ Gowlett & REM Hedges (eds.), 1986. *Archaeological Results from Accelerator Dating.* Oxford, Oxford University Committee for Archaeology Monograph Series 11

36. RM Jacobi, JAJ Gowlett, REM Hedges & R Gillespie, 1986. Accelerator Mass Spectrometry Dating of Upper Palaeolithic Finds, with the Poulton Elk as an Example. In DA Roe (ed.) *Studies in the Upper Palaeolithic of Britain.* Oxford, British Archaeological Reports S296 Pp: 121-128

37. JAJ Gowlett, ET Hall & REM Hedges, 1986. The date of the West Kennet long barrow. *Antiquity* 60: 143-144

38. JAJ Gowlett, ET Hall, REM Hedges & C Perry, 1986. Radiocarbon dates from the Oxford AMS system: Archaeometry datelist 3. *Archaeometry* 28: 116-125

39. JAJ Gowlett, REM Hedges, IA Law & C Perry, 1986. Radiocarbon dates from the Oxford AMS system: Archaeometry datelist 4. *Archaeometry* 254: 100-107

40. REM Hedges & JAJ Gowlett, 1986. Radiocarbon dating by accelerator mass spectrometry. *Scientific American* 254: 100-107

41. RJ Batten, R Gillespie, JAJ Gowlett & REM Hedges, 1986. The AMS dating of separate fractions in archaeology. *Radiocarbon* 28: 698-701

42. Moore, A, Gowlett JAJ, Hedges REM Hillman G, Legge A & P Rowley-Conwy, 1986. Radiocarbon (AMS) Dates for the Epipalaeolithic Settlement at Abu Hureyra, Syria. *Radiocarbon* 28: 1068-1076

43. JAJ Gowlett & REM Hedges, 1987. Radiocarbon dating by Accelerator Mass Spectrometry-Applications to Archaeology in the Near East. In O Aurenche & J Evin (eds.) *Chronologies in the Near East: relative chronologies and absolute chronology 16,000 - 4,000 BP.* Oxford, British Archaeological Reports, International Series 379 Pp 121-144

44. A Saville, JAJ Gowlett & REM Hedges, 1987. Radiocarbon dates from the chambered tomb at Hazleton (Glos.): a chronology for Neolithic collective burial. *Antiquity* 61: 108-119

45. JAJ Gowlett, 1987. The coming of modern man. *Antiquity* 61: 213-219

46. JAJ Gowlett, 1987. New dates for the Acheulean age. *Nature* 329: 200

47. JAJ Gowlett, 1987. The archaeology of radiocarbon accelerator dating. *Journal of World Prehistory* 1: 127-170

48. JAJ Gowlett, REM Hedges, IA Law & C Perry, 1987. Radiocarbon dates from the Oxford AMS system: Archaeometry datelist 5. *Archaeometry* 29: 125-155

49. PA Mellars, HM Bricker, JAJ Gowlett & REM Hedges, 1987. Radiocarbon accelerator dating of French Upper Palaeolithic sites. *Current Anthropology* 29: 128-132

50. JAJ Gowlett, 1988. Human adaptation and long-term climatic change in Northeast Africa: An archaeological perspective. In D Johnson & DM Anderson (eds.) *The Ecology of Survival. Case*

Studies from Northeast African History. Boulder CO, Westview Pp 27-45

51. JAJ Gowlett (ed.), 1988. New Directions in Palaeolithic Archaeology. *World Archaeology* 19(3)

52. JAJ Gowlett, 1988. A case of Developed Oldowan in the Acheulean? *World Archaeology* 19: 13-26

53. JAJ Gowlett, 1988. Culture and Conceptualisation: the Oldowan-Acheulian Gradient. In I Tattersall, E Delson & J Van Couvering (eds.), *Encyclopedia of Human Evolution and Prehistory.* London, St James Press.

54. JAJ Gowlett, 1989. Introduction. In B. Isaac (ed.) *The Archaeology of Human Origins: Papers by Glynn Isaac.* Cambridge, Cambridge University Press Pp 1-10

55. C Grigson, JAJ Gowlett & J Zarins, 1989. The camel in Arabia—a direct radiocarbon date, calibrated to about 7000 BC. *Journal of Archaeological Science* 16: 355-362

56. JAJ Gowlett, REM Hedges & IA Law, 1989. Radiocarbon accelerator (AMS) dating of Lindow Man. *Antiquity* 63: 71-79

57. JAJ Gowlett, 1990. Archaeological studies of human origins and early prehistory in Africa. In P.T. Robertshaw (ed.) *A history of African Archaeology.* London, Heinemann Pp 13-30

58. JAJ Gowlett, 1990. Indiana Jones: crusading for archaeology? *Antiquity* 64: 157

59. JAJ Gowlett, 1990. Technology, Skill, and the Psychosocial Sector in the Long Term of Human Evolution. *Archaeological Review from Cambridge* 9(1): 82-103

60. JAJ Gowlett, 1991. Kilombe—Review of an Acheulian site complex. In JD Clark (ed.) *Cultural beginnings: approaches to understanding early hominid life ways in the African savannah.* Bonn, R. Habelt Pp 129-136

61. JAJ Gowlett (ed.), 1991. Chronologies. *World Archaeology* 23(2)

62. JAJ Gowlett, 1992. Tools—the Palaeolithic record. In RD Martin (ed.) *The Cambridge Encyclopaedia of Human Evolution.* Cambridge, Cambridge University Press Pp 350-360.

63. J Huxtable, JAJ Gowlett, GN Bailey, PL Carter & V. Papaconstantinou, 1992. Thermoluminescence dates and a new analysis of the early Mousterian from Asprochaliko. *Current Anthropology* 33: 109-114

64. JAJ Gowlett, 1993. Chimpanzees deserve more than crumbs of the palaeoanthropological cake. *Cambridge Archaeological Journal* 3: 297-300

65. JAJ Gowlett, 1993. Le site Acheuleen de Kilombe: stratigraphie, geochronology, habitat et industrie lithique. *L'Anthropologie* 97: 69-84

66. RH Crompton & JAJ Gowlett, 1993. Allometry and multidimensional form in Acheulean bifaces from Kilombe, Kenya. *Journal of Human Evolution* 25: 175-199

67. JAJ Gowlett (ed.), 1994. Communication and Language. *World Archaeology* 26(2)

68. JAJ Gowlett & RH Crompton, 1994. Kariandusi: Acheulean morphology and the question of allometry. *African Archaeological Review* 13: 3-42

69. JAJ Gowlett, 1995. Psychological Worlds within and without: Human-Environment Relations in Early Parts of the Palaeolithic. In H. Ulrich (ed.) *Man and Environment in the Palaeolithic.* Liege, ERAUL, 62: 29-42

70. JAJ Gowlett, 1995. A Matter of Form: Instruction Sets and the Shaping of Early Technology. *Lithic* 16: 2-16

71. JAJ Gowlett, 1996. The frameworks of early hominid social systems: how many useful parameters of archaeological evidence can we isolate. In J Steele & S Shennan (eds.) *The archaeology of human ancestry: power, sex and tradition.* London, Routledge Pp 135-183

72. JAJ Gowlett1996Rule systems in the artefacts of Homo erectus and early Homo sapiens: constrained or chosen. In PA Mellars & KR Gibson (eds.) *Modelling the early human mind.* Cambridge, McDonald Institute Pp 191-215

73. JAJ Gowlett, REM Hedges & RA Housley, 1997. Klithi: the AMS radiocarbon dating programme for the site and its environs. In GN Bailey (ed.) *Klithi: Palaeolithic Settlement and Quaternary Landscapes in Northwest Greece Volume 1.* Cambridge, McDonald Institute Pp 27-39

74. JAJ Gowlett & P Carter, 1997. The basal Mousterian of Asprochaliko rockshelter, Louros Valley". In GN Bailey (ed.) *Klithi: Palaeolithic Settlement and Quaternary Landscapes in Northwest Greece Volume 2* Cambridge, McDonald Institute Pp 27-40

75. A Sinclair, EA Slater & JAJ Gowlett (eds.), 1997. *Archaeological Sciences 1995.* Oxford, Oxbow Monographs 64.

76. M Farid Khan & JAJ Gowlett, 1997. Age-depth relationships in the radiocarbon dates from Sanghao Cave, Pakistan. In A Sinclair, EA Slater & JAJ Gowlett (eds.), 1997. *Archaeological Sciences 1995.* Oxford, Oxbow Monographs Pp 182-187

77. RH Crompton & JAJ Gowlett, 1997. The Acheulean and the Sahara: allometric comparisons between

North and East African sites. In A Sinclair, EA Slater & JAJ Gowlett (eds.), 1997. *Archaeological Sciences 1995*. Oxford, Oxbow Monographs Pp 400-405

78. SA Andresen, DA Bell, J Hallos, TRJ Pumphrey & JAJ Gowlett, 1997. Approaches to the analysis of evidence from the Acheulean site of Beeches Pit, Suffolk, England. In A Sinclair, EA Slater & JAJ Gowlett (eds.), 1997. *Archaeological Sciences 1995*. Oxford, Oxbow Monographs Pp 389-394

79. T Brown, AG Latham & JAJ Gowlett, 1997. Uranium-series dating of fossil Nile Oyster from a Palaeolithic site, Mweya, Uganda. In A Sinclair, EA Slater & JAJ Gowlett (eds.), 1997. *Archaeological Sciences 1995*. Oxford, Oxbow Monographs Pp 174-181

80. JAJ Gowlett (ed.), 1997. High Definition Archaeology. *World Archaeology* 29()2)

81. JAJ Gowlett, 1997. High Definition Archaeology: ideas and evaluation. *World Archaeology* 29(2): 151-171

82. JAJ Gowlett, 1997. Why the muddle in the middle matters: the language of comparative and direct in human evolution. In CM Barton & GA Clark (eds.) *Rediscovering Darwin: evolutionary theory in archaeological explanation.* Archaeological Papers of the American Anthropological Association 7 (1): 49-65

83. JAJ Gowlett, 1998. Unity and diversity in the early stone age. In N Ashton, F Healey & P Pettitt (eds.) *Stone Age Archaeology: Essays in honour of John Wymer.* Oxford, Oxbow Books Pp 59-66

84. JAJ Gowlett, JC Chambers & J Hallos, 1998. Beeches Pit: First views of the archaeology of a Middle Pleistocene site in Suffolk, UK, in European context. *Anthropologie* (Brno) 36: 91-97

85. JAJ Gowlett, 1999. Lower and Middle Pleistocene archaeology of the Baringo Basin. In P Andrews & P Banham (eds.) *Late Cenozoic environments and hominid evolution: a tribute to Bill Bishop.* Geological Society London Pp 123-141

86. RC Chiverrell, PJ Davey & JAJ Gowlett, 1999. Radiocarbon dates for the Isle of Man. In PJ Davey (ed.) *Recent archaeological research on the Isle of Man.* Oxford, British Archaeological reports S278

87. JAJ Gowlett, 1999. Paleoclimate and evolution, with emphasis on human origins. *Journal of Quaternary Science* 14: 99-100

88. JAJ Gowlett, 1999. The Lower and Middle Palaeolithic, transition problems and hominid species: Greece in broader perspective. In GN Bailey, E Adam, E Panagopoulou, C Perles, & K Zachos (eds.), *The Palaeolithic Archaeology of Greece and Adjacent Areas.* British School at Athens Studies Pp 43-58

89. JAJ Gowlett, 1999. The Work and Influence of Charles McBurney. In T Murray (ed.) *Encyclopaedia of Archaeology II.* Oxford, ABC-Clio Pp 713-726

90. JAJ Gowlett & J Hallos, 2000. Beeches Pit: overview of the archaeology. In SG Lewis, CA Whiteman, & RC Preece (eds.) *The Quaternary of Norfolk and Suffolk: Field Guide.* Quaternary Research Association Pp 197-206

91. JAJ Gowlett, DA Bell & J Hallos, 2000. Beeches Pit: archaeology of a Middle Pleistocene site in East Anglia, UK, 1996-1999 seasons. Abstracts for the Palaeoanthropology Society Meeting, Philadelphia, USA, 4-5 April 2000. *Journal of Human Evolution* 38(3) p. A13

92. JAJ Gowlett, RH Crompton & Li Yu, 2001. Allometric comparisons between Acheulean and Sangoan large cutting tools at Kalambo Falls. In JD Clark (ed.) *Kalambo Falls Prehistoric Site Volume III.* Cambridge, Cambridge University Press Pp 612-619

93. JAJ Gowlett, 2001. Archaeology: Out in the Cold - News and Views. *Nature* 413: 33-34

94. JAJ Gowlett, 2002. Apes, hominids and technology. In CS Harcourt & BR Sherwood (eds.) *New Perspectives in Primate Evolution and Behaviour.* Linnean Society Westbury Academic and Scientific Publishing Pp 147-171

95. JAJ Gowlett, 2003. The AMS radiocarbon dates: an analysis and interpretation. In P Parr (ed.) *Excavations at Arjoune, Syria.* Oxford, British Archaeological Reports International Series 1134

96. JAJ Gowlett, 2003. What actually was the Stone Age diet? *Journal of Nutritional & Environmental Medicine* 13(3): 143-147

97. J Hallos, JAJ Gowlett, V Brant & S Hounsell, 2004. Missing Links: Refitting studies at Beeches Pit as an approach to understanding tool production in the Middle Pleistocene. Abstracts for the Paleoanthropology Society Meeting, , Montreal, Canada 29–31 March 2004. PaleoAnthropology PAS 2004 Abstracts, p. A51.

98. JAJ Gowlett, 2005. Seeking the Palaeolithic individual in East Africa and Europe during the lower-middle Pleistocene. In CS Gamble & M Porr (eds.) *The hominid individual in context: archaeological investigations of Lower and Middle Palaeolithic Landscapes, Locales and Artefacts.* London, Routledge Pp 50-67

99. JAJ Gowlett, J Hallos, S Hounsell, V Brant & N Debenham, 2005. Beeches Pit: archaeology,

assemblage dynamics and early fire history of a Middle Pleistocene site in East Anglia, UK. *Eurasian Prehistory* 3(2): 3-28

100. JAJ Gowlett, 2006. The Elements of Design Form in Acheulian Bifaces. In N Goren-Inbar & G Sharon (eds.) *Axe Age: Acheulian Tool-Making from Quarry to Discard.* London, Routledge Pp 203-222

101. JAJ Gowlett, 2006. Archaeological dating. In J Bintliff (ed.) *A Companion to Archaeology.* Oxford, Blackwell Pp 197-205

102. JAJ Gowlett, 2006. Chronology and the Human Narrative. In J Bintliff (ed.) *A Companion to Archaeology.* Oxford, Blackwell Pp 206-234

103. JAJ Gowlett, 2006. The Early Settlement of Northern Europe: Fire History in the Context of Climate Change and the Social Brain. In H de Lumley (ed.) *Climats, Cultures et sociétés aux temps préhistoriques, de l'apparition des Hominidés jusqu'au Néolithique.* C.R. Palevol 5(1-2): 299-310

104. RC Preece, JAJ Gowlett, SA Parfitt, D Bridgeland & D.R. Lewis, 2006. Humans in the Hoxnian: habitat, context and fire use at Beeches Pit, West Stow, Suffolk, UK. *Journal of Quaternary Science* 21: 485-496

105. JAJ Gowlett, 2008. Deep roots of kin: developing the evolutionary perspective from prehistory. In NJ Allen, H Callan, R Dunbar & W James (eds.) *Early Human Kinship: From Sex to Social Reproduction.* London, Blackwell Pp 41-57

106. JAJ Gowlett & R Dunbar, 2008. A brief overview of human evolution. In NJ Allen, H Callan, R Dunbar & W James (eds.) *Early Human Kinship: From Sex to Social Reproduction.* London, Blackwell Pp 21-24

107. SJ Lycett & JAJ Gowlett, 2008. On questions surrounding the Acheulean tradition. *World Archaeology* 40: 295-315

108. JAJ Gowlett, 2009. The longest transition or multiple revolutions? Curves or steps in the record of human origins. In M Camps & PR Chauhan (eds.) *Sourcebook of Palaeolithic Transitions: Methods, Theories and Interpretations.* Heidelberg, Springer Pp 65-78

109. JAJ Gowlett, 2009. Boucher de Perthes: pioneer of Palaeolithic Prehistory. In R Hosfield, F Wenban-Smith & M Pope eds.) *Great Prehistorians: 150 years of Palaeolithic Research 1859-2009.* Lithic 30: 13-24"

110. JAJ Gowlett, 2009. Artefacts of apes, humans, and others: towards comparative assessment and analysis. *Journal of Human Evolution* 57: 401-410

111. JAJ Gowlett, 2010. The future of lithic analysis in Palaeolithic Archaeology: a view from the Old World. In SJ Lycett & PR Chauhan (eds.) *New Perspectives on Old Stones: Analytical Approaches to Palaeolithic Technologies.* Heidelberg, Springer Pp 295-309

112. RIM Dunbar, CS Gamble & JAJ Gowlett, 2010. *Social Brain, Distributed Mind.* London, British Academy

113. SJ Lycett, N von Cramon-Taubadel & JAJ Gowlett, 2010. A comparative 3D geometric morphometric analysis of Victoria West cores: implications for the origins of Levallois technology. *Journal of Archaeological Science* 37: 1110-1117

114. JAJ Gowlett, 2011. The empire of the Acheulean strikes back. In J Sept & D Pilbeam (eds.) *Casting the net wide: Studies in honor of Glynn Isaac and his approach to human origins research.* Cambridge Mass, Peabody Museum Harvard University Pp 93-114

115. JAJ Gowlett, 2011. The Vitale Sense of Proportion. *PaleoAnthropology (Special Issue: Innovation and the Evolution of Human Behavior)* Pp 174-187

116. JAJ Gowlett, 2012. Shared intention in early artefacts: an exploration of deep structure and implications for communication and language. In SC Reynolds & A Gallagher (eds.) *African Genesis: perspectives on hominin evolution in Africa.* Cambridge, Cambridge University Press Pp 506-530

117. JS Brink, AIR Herries, J Moggi-Cecchi & JAJ Gowlett, 2012. First hominine remains from a ~1.0 million year old bone bed at Cornelia-Uitzoek, Free State Province, South Africa". *Journal of Human Evolution* 63: 527-535

118. JAJ Gowlett, 2013. Elongation as a factor in artefacts of humans and other animals: an Acheulean example in comparative context. *Philosophical Transactions of the Royal Society B - Biological Sciences* 368: 20130114

119. JAJ Gowlett & RW Wrangham, 2013. Earliest fire in Africa: towards the convergence of archaeological evidence and the cooking hypothesis. *Azania: Archaeological Research in Africa* 48: 5-30

120. RIM Dunbar, C Gamble, JAJ Gowlett (eds.), 2014. *Lucy to language: the benchmark papers.* Oxford, Oxford University Press

121. CS Gamble, JAJ Gowlett & R Dunbar, 2014. Thinking big: the archaeology of the social brain. In RIM Dunbar, C Gamble, JAJ Gowlett (eds.) *Lucy to language: the benchmark papers.* Oxford, Oxford University Press

122. RIM Dunbar & JAJ Gowlett, 2014. Fireside chat: the impact of fire on hominin socioecology. In RIM Dunbar, C Gamble, JAJ Gowlett (eds.) *Lucy to language: the benchmark papers.* Oxford, Oxford University Press

123. RIM Dunbar, J Lehmann, AH Korstjens & JAJ Gowlett, 2014. The road to modern humans: time budgets, fission-fusion sociality, kinship and the division of labour in hominin evolution. In RIM Dunbar, C Gamble, JAJ Gowlett (eds.) *Lucy to language: the benchmark papers*. Oxford, Oxford University Press

124. JAJ Gowlett, 2014. Human Evolution: Use of Fire. In C Smith (ed.) *The encyclopaedia of global archaeology*. New York, Springer.

125. S White, JAJ Gowlett & M Grove, 2014. The place of the Neanderthals in hominin phylogeny. *Journal of Anthropological Archaeology* 35: 32-50

126. JAJ Gowlett, 2015. Terra Amata: a view of the assemblages in the wider Acheulian domain. In H de Lumley (ed.) *Terra Amata, Nice, Alpes-Maritimes, France. Volume 4, Part 1, Les industries acheuléennes: étude de l'outillage, planches de dessins et de photographies de l'industrie lithique.* Paris CNRS Editions Pp 793-794

127. JAJ Gowlett & JS Brink, 2015. At the heart of the African Acheulean: the physical, social and cognitive landscapes of Kilombe. In F Coward, R Hosfield, & F Wenban-Smith (eds.) *Settlement, Society and Cognition In Human Evolution.* Cambridge, Cambridge University Press Pp 75-93

128. JAJ Gowlett, 2015. Les origins de l'utilisation du feu par les hommes: hypotheses actuelles et indices les plus anciens. In H de Lumley (ed.) *Sur le chemin de l'hunanité. Via humanitatis: les grandes étapes de l'évolution morphologique et culturelle de l'Homme: émergence de l'être humain.* Paris: Académie Pontificale des Sciences / CNRS Pp 171-197

129. JAJ Gowlett, 2015. Variability in an early hominin percussive tradition: the Acheulean versus cultural variation in modern chimpanzee artefacts. *Philosophical Transactions of the Royal Society B - Biological Sciences* 370-20140358

130. JAJ Gowlett, 2016. The discovery of fire by humans: a long and convoluted process. *Philosophical Transactions of the Royal Society B - Biological Sciences* 371: 20150164

131. JAJ Gowlett, JR Brink, AIR Herries, S Hoare & S Rucina, 2017. The small and short of it: mini-bifaces and points from Kilombe, Kenya, and their place in the Acheulean. In D Wojtczak, M Al Naijar, R Jagher, H Elsuede & M Otte (eds.) *Vocation préhistoire: hommage à Jean-Marie Le Tensorer.* Liege, ERAUL 148: 121-132

132. JAJ Gowlett, JS Brink & SM Hoare, 2017. A major event in the Middle Pleistocene? In M Pope, J McNabb & CS Ganble (eds.) *Crossing the Human Threshold: dynamic transformation and persistent places during the Middle Pleistocene.* Oxford, Routledge Pp 252-266

133. JAJ Gowlett, JS Brink, A Caris & S Hoare, 2017. Evidence of burning from bushfires in southern and east Africa and its relevance to hominin evolution. *Current Anthropology* 58: S206-216

134. JAJ Gowlett, 2018. Dating, Archaeological. In H Callan (ed.) *The International Encyclopaedia of Anthropology.* Wiley Online Library

135. JAJ Gowlett, 2018. Kinship (Early Human), the Archaeological Evidence for. In H Callan (ed.) *The International Encyclopaedia of Anthropology.* Wiley Online Library

136. JAJ Gowlett, 2018. Hugo Oliveira. In H Callan (ed.) *The International Encyclopaedia of Anthropology.* Wiley Online Library

137. JAJ Gowlett, 2018. Fire, Early Human Use of. In H Callan (ed.) *The International Encyclopaedia of Anthropology.* Wiley Online Library

138. JAJ Gowlett, 2018. Archaeological Approaches in Anthropology. In H Callan (ed.) *The International Encyclopaedia of Anthropology.* Wiley Online Library

139. T. Wynn & JAJ Gowlett, 2018. The Handaxe Reconsidered. *Evolutionary Anthropology* 27: 21-29